O NEILL GROUND WATER ENGINEERING
7 SOUTH MAIN STREET
NAAS, CO. KILDARE
TEL: 353-45-895668
FAX: 353-45-881705
Email: info@groundwatereng.ie

NITRATES IN GROUNDWATER

SELECTED PAPERS ON HYDROGEOLOGY

5

INTERNATIONAL ASSOCIATION OF HYDROGEOLOGISTS

Nitrates in groundwater

Selected papers from the European meeting of the International Association of Hydrogeologists, Wisla, Poland, 4–7 June 2002

Edited by

Lidia Razowska-Jaworek
Polish Geological Institute

Andrzej Sadurski
Polish Geological Institute

A.A. BALKEMA PUBLISHERS   LEIDEN / LONDON / NEW YORK / PHILADELPHIA / SINGAPORE

Library of Congress Cataloging-in-Publication Data

Applied for

UNESCO and the International Association of Hydrogeologists (IAH) coordinated the preparation of this book as an IAH contribution to the 5th phase of the UNESCO International Hydrogeological Programme Project 3.4 'Groundwater pollution due to urban development'.

Design of the cover picture: Gabor Lorinczy
Cover design: Studio Jan de Boer
Typesetting: Charon Tec Pvt. Ltd, Chennai, India
Printed in Great Britain

Published by: A.A. Balkema Publishers, Leiden, The Netherlands, a member of Taylor & Francis Group plc.
www.balkema.nl / www.tandf.co.uk

ISBN 90 5809 664 5

Contents

Foreword

This book provides an overview of a wide spectrum of current studies of nitrates in groundwater. Although a number of papers dealing with nitrate pollution of groundwater have appeared in different conference proceedings for the last 30 years, this is the first book devoted entirely to the problem of nitrate risk and behaviour in groundwater. This volume includes 27 papers selected from those presented during the Euromeeting of the International Association of Hydrogeologists (IAH) 'Nitrate in groundwater in Europe'. The meeting, organized by the Polish Geological Institute and the Polish Chapter of the IAH was held in the health resort Wisła in Poland in 2002. In spite of the title of the meeting, the problems presented and discussed in Wisła covered all the aspects of nitrate pollution of groundwaters. Therefore, the scope of this book extends beyond the European continent. Only the section with articles describing the case studies is limited to groundwater in Europe. The remaining sections treat the main chemical and physical processes, which affect nitrate in groundwater as well as describe nitrate distribution in various aquifers. These are the essential problem areas of groundwater resource protection.

The papers have been grouped in five sections:

Section 1. *Origin of nitrates in groundwater* – contains chapters dealing with diffuse and point sources of nitrates in groundwaters. The authors describe:
- Time evolution of nitrates in waters
- Denitrification and redox processes
- Depth profiles of nitrate content

Section 2. *Experimental investigations of nitrates in groundwater* – comprises chapters describing field and laboratory experimental investigations of nitrate behaviour in groundwaters. These investigations were conducted in:
- Saturated and unsaturated zones
- Renaturalized and managed peatlands
- Changing flow conditions
- Porous and karstic aquifers
- Different agricultural soil use

Section 3. *Modelling of nitrate transport and chemistry* – all the chapters in this section present the results of nitrate studies based on the transport and chemical modelling. Different modelling software tools were applied, such as:
- DAISY, MIKE-SHE, MIKE11 (Hubrechts *et al.*)
- MODFLOW, MT3D (Mioduszewski *et al.*)
- SWAP, ANIMO, FLOTRANS (Mioduszewski *et al.*)
- WAVE (Mioduszewski *et al.*)
- GLEAMS (Pintar *et al.*)
- WEKU (Kunkel *et al.*)

Section 4. *Nitrate pollution of groundwater – national and regional studies–* presents case studies from different countries in Europe showing the spatial distribution of nitrates in groundwater. The topics presented by the authors are mainly related to:

- Nitrate pollution of shallow groundwaters
- Nitrate pollution of deeper aquifers
- Groundwater age dating
- Aquifers vulnerability for nitrate pollution

Section 5. *Monitoring and protection of groundwater from nitrate pollution –* these chapters concentrate on:

- Regional monitoring of nitrates in groundwaters
- Strategies of protection of groundwater from nitrate pollution in the agricultural areas
- Nitrate Sensitive Areas
- Legislative works

Most of the chapters deal with the pollution of the Quaternary aquifer, but nitrate in shallow and deeper zones of the Tertiary, Jurassic, Cretaceous and Triassic aquifers have also been introduced. The area of the studies described in this book covers western, central and southern parts of Europe. The authors come from different countries:

- western Europe (UK, The Netherlands, Denmark, Belgium, France)
- central Europe (Germany, Poland)
- southern (Spain, Italy, Greece, Slovenia)
- eastern (Bulgaria, Romania)

Data presented in the chapters confirm that nitrate pollution in groundwaters has been observed in all countries in Europe. Deterioration of groundwater quality is also associated with an increase in the abstraction of groundwater. The basic geochemical concepts, modelling and mass transport in aquifers are often discussed in these articles but there remains the lack of a universal theory of behaviour of nitrate compounds in the soil. At this boundary between biotic and abiotic environment the chemical composition of water is strongly modified before reaching the aquifer. A large collection of problems is still ahead for researchers to clarify and apply in practice.

We do hope that the implementation of the Nitrate Directive of the EU will stimulate a better understanding of geochemistry of groundwater and presentation of different approaches to the nitrate hazard in many countries of Europe. This was the aim of our meeting in Wisla and also the purpose of the papers published in this book.

Dr. Lidia Razowska-Jaworek
Prof. Dr. Andrzej Sadurski

SCIENTIFIC COMMITTEE OF THE EUROMEETING

Prof. Wojciech Ciężkowski	University of Wrocław, Poland
Prof. Józef Górski	University of Poznań, Poland
Prof. Bohdan Kozerski	Technical University of Gdańsk, Poland
Prof. Aleksandra Macioszczyk	University of Warsaw, Poland
Prof. Andrzej Różkowski	University of Silesia, Sosnowiec, Poland
Prof. Andrzej Sapek	Institute for Land Reclamation, Falenty, Poland
Prof. Jadwiga Szczepańska	University of Mining and Metallurgy, Cracow, Poland
Prof. Stanisław Witczak	University of Mining and Metallurgy, Cracow, Poland

LOCAL ORGANIZING COMMITTEE OF THE EUROMEETING

Andrzej Sadurski	Polish Geological Institute, Warsaw, Poland
Lidia Razowska-Jaworek	Polish Geological Institute, Sosnowiec, Poland
Igor Brodziński	Polish Geological Institute, Sosnowiec, Poland
Martyna Guzik	Polish Geological Institute, Sosnowiec, Poland
Andrzej Pacholewski	Polish Geological Institute, Sosnowiec, Poland
Jadwiga Wagner	Polish Geological Institute, Sosnowiec, Poland
Albin Zdanowski	Polish Geological Institute, Sosnowiec, Poland
Janusz Jureczka	Polish Geological Institute, Sosnowiec, Poland

Technical edition in the Polish Geological Institute:
Lidia Razowska-Jaworek, Igor Brodziński

Introductory remarks

It is a real pleasure for me to provide the introductory remarks to the scientific output of this conference for two reasons:

- first, as IAH Vice-President for Western Europe, I should congratulate our Polish Chapter in having the vision and efficiency to organize an event that has attracted as much interest from 'my side of the continent' as from their own
- second, in the early 1970s, I was one of the first in Europe to recognize the significance − and research the processes − of leaching of nitrate from agricultural soils to groundwater, and have retained an interest in the subject ever since.

Whilst the topic has thus been with us for a number of decades, with the advent of the EU Water Policy Framework Directive, and supplementary Groundwater Quality Directive, it is one of renewed interest and concern at European environmental policy level.

HISTORICAL CONTEXT − 'TOO MUCH OF A GOOD THING'

Most, but certainly not all, nitrate leaching to groundwater has been an insidious consequence of the enormous changes in European agriculture over the past 50 years, especially the (largely successful) efforts to increase grain, milk and meat production. Although fertilizer nitrogen use has been decreasing in recent years, overall more than a 20-fold increase was applied to achieve a 3-fold increase in food production, with the balance being incorporated in soils from where it was either lost to the atmosphere following denitrification or leached to water bodies as nitrate.

Given the essentially diffuse character of the agricultural activity generating the subsurface contaminant load, pollution control has presented (and continues to present) a complex regulatory challenge in legal, technical, economic and social terms.

THE SCIENTIFIC PAY-OFF

Important broader advances in hydroscientific concepts have been achieved incidentally over the past decades as a result of the pursuit of research on diffuse agricultural pollution of groundwater including recognition of:

- the intimate, but concealed, relationship between agricultural land-use and groundwater quality
- the significance of the vadose zone and potentially very slow average rates of downward transport of soluble contaminants (like nitrate), even in the fractured porous strata, introducing the possibility of large time-lags in the impact of contamination events on deeper aquifers
- the importance of subsurface microbial activity and the *in-situ* denitrification capacity of some hydrogeological environments, which where active could result in less pressure on the agricultural sector to reduce nitrate leaching losses.

ADVANCES OF PRESENT PUBLICATION

In addition to providing the opportunity for the coordinated presentation of groundwater nitrate and related data from a number of areas on which little was previously available, this conference and its associated publication includes papers on a wide variety of highly pertinent themes. The publication includes work from some 13 EU (and designate EU) countries, and covers not only the underpinning knowledge base but also operational issues such as monitoring and prediction of groundwater quality trends.

I have distilled out a number of important scientific conclusions and messages for EU water policy:

- while certain agricultural cultivation practices are the predominant source of increasing nitrate in groundwater urban wastewater discharge (by one route or another) and effluents from livestock rearing can be also highly significant locally, but scientific methods are available to diagnose the situation under most field conditions
- it is mistaken to focus narrowly on nitrate, there being widespread evidence of related increasing trends in one or more of the following groundwater pollutants (depending on circumstances) – sulphate, ammoniacal nitrogen, chloride, microbiological contaminants (including Cryptosopridium), and certain trace elements
- most groundwater systems show a major time-lag in their response to changes in nitrate input – which arises from a combination of soil processes, vadose zone transport, saturated zone dispersion and dilution, and in certain cases 'subsurface chemical fronts' – and in consequence many systems are in a chemically-dynamic state taking various decades to adjust to external changes, which makes the establishment of water-quality trends in groundwater bodies a challenging task
- it is of much more practical importance to monitor the *average quality of contemporary recharge* than the *average quality of groundwater in storage* but excessive dependence on deep public water-supply boreholes often frustrates this ideal, and the construction of purpose-drilled sampling piezometers located by a sound conceptual model of the groundwater flow regime and chemical controls is an urgent priority
- significant advances in the quantification of natural nitrate reduction *in-situ* within aquifers continue to be made, and it is now clear that this phenomena occurs under anoxic subsurface conditions associated variously with peaty organic matter and ferrous minerals in shallow formations and sulphide minerals (notably pyrite) in deeper formations, whose spatial distribution may be in clearly-defined 'redox fronts' or in 'localized pockets' within a generally aerobic matrix with not all of the solid-phase reducing mineral species having access to nitrate in the flowing groundwater
- a great complexity in detail of vadose zone processes is apparent which is related to the interaction of various factors including lateral heterogeneity in the soil zone, 'preferential flow' in a fairly wide variety of geological media, locally anoxic conditions, deep-rooted vegetation and its effect on the position of the 'zero flux plane', and water-table and capillary zone fluctuations
- there have been major advances in the numerical modelling of groundwater flow and pollutant transport in the various components of the soil water–groundwater system, and in general these have outstripped the availability of field data on which to apply them.

AN EYE TO THE FUTURE

It is likely that agriculture will experience an accelerated rate of technological innovation in the coming decades, with the introduction of 'new chips, new genes and new molecules'. This will result both in significant opportunities, but also significant threats, where diffuse pollution of groundwater is concerned. The most significant changes are likely to be:

- the increased availability of low-cost sensors (backed by computers) which could enable micro-management of soil moisture and agrochemicals at farm and field level, with the opportunity of much more control over soil leaching losses
- the potential widespread introduction of genetically-modified crops, which could be far more efficient in terms of nutrient uptake: however concern has to be expressed because these may require much higher applications of specific pesticides and have a potential effect on soil bacteria and soil contaminant attenuation.

Another contrasting trend is likely to be significant increases in organic crop production (without use of inorganic fertilizer applications), but for the present this is confined to relatively limited land areas and its ability to reduce soil nitrate leaching losses is still open to question.

At the hydrogeological research level, it is hoped that uncertainty in the quantification of attenuation capacity will be significantly reduced and that this capacity will be used intelligently in the interests of protecting groundwater quality. At the environmental management level, it is hoped that selective risk-based land-use controls will become more acceptable to the farming community and that special taxes on fertilizers and pesticides could be raised to provide funds for compensation of those affected.

Prof. Dr. Stephen Foster
IAH Vice-President — Western Europe

Origin of nitrates in groundwater

CHAPTER 1

Agricultural activities as a source of nitrates in groundwater

A. Sapek

Institute for Land Reclamation and Grassland Farming at Falenty, 05-090 Raszyn, Poland

ABSTRACT: Agriculture is the main source of groundwater pollution by nitrate. This is a result of the rising use of synthetic nitrogen fertilisers and the increasing consumption of animal protein by the population in developed countries. Nitrate is not a component of rock minerals or soil particles; it is a product of nitrogen turnover in soil with short live time. Once nitrate appears in soil, it is quickly absorbed by growing plants or bacteria to form new organic substances. The nitrate concentration in cropped soils varies greatly during the hydrological year and is at its highest in late autumn, when the main volume of water is percolating to the saturated zone. The observed and calculated data show that, in lowland Poland, the annual leaching rarely exceeds $25\,kg\,ha^{-1}a^{-1}$ NO_3-N. Such leaching, and higher, is likely to occur on sandy soils, prevailing in Poland. The best way to mitigate the groundwater pollution by nitrate from agricultural sources is through the implementation of Best Agricultural Management Practices and the effective education of farmers and rural area population.

INTRODUCTION

Agriculture activities in developed countries are generally accepted as a main source of nitrate in groundwater. Nitrate leaching is a natural process, but the leached load depends on the accelerated nitrogen cycle in soil resulting from agricultural operation and/or land use changes. The soil processes that control the nitrogen turnover in soil and the resulting nitrate concentration in it are numerous, but the most important are: nitrogen uptake by growing plants, nitrogen inputs with fertiliser and manure, inorganic nitrogen immobilization in soil organic matter, mineralization of soil organic matter, nitrification and denitrification. The temporary nitrate concentration in soil is an entire result of the above activities, but could be quickly changed if one process becomes more effective at any moment.

The nitrate leaching from soil profile is not the sole function of its concentration. A second hydrological process is necessary — water percolation down the soil profile. The last event occurs from a few to some scores of days a year, and mostly during the late autumn and winter months. Therefore, the activities aimed to mitigate the nitrate leaching from agricultural soils should be oriented so that the nitrate concentration in soil is the least, when the possibility of percolation events is the greatest.

The second major source of water pollution is animal husbandry, where animal waste is dispersed around the farmstead and its vicinity. This subject is discussed in Chapter 4 of this book (B. Sapek).

The land use changes are another important source of nitrate leaching, since in some ecosystems a huge accumulation of soil organic matter rich in nitrogen may be observed. This pool of organic matter is a result of special conditions, such as abundance of water, the kind of plants growing and the soil management systems. When these conditions are changed, the oxidation of a part of organic matter may follow and comprised organic nitrogen compounds would be mineralized to ammonia and then to nitrate.

The aim of this paper is to highlight the agriculture activities on cropped soils as a driving force in the nitrate leaching to groundwater.

Nitrogen in environment

The nitrogen flux in the natural environment has its beginning in the biological fixation of atmospheric N_2 by specialized bacteria, or directly in atmosphere during lightning, where the molecular nitrogen is oxidized to form NO_x. Human activities supported some new sources of fixed nitrogen such as increased area of nitrogen-fixing legume crops, industrial synthesis of ammonia and combustion of fuels from natural resources, where ammonia and nitrogen oxides are the abundant by-products. Human activity is nowadays a greater source of fixed nitrogen than natural processes (Table 1). The biological fixation by agricultural legume crops is twice as much as by natural vegetation. The electrical discharge in the atmosphere produces the same quantity of fixed nitrogen as do the industrial processes or combustion of fuel. Nevertheless, the main sources of anthropogenic fixed nitrogen is the synthesis and use of fertilisers, which is equal to all other sources, if we do not consider the oceans. The greatest amount of fixed nitrogen is deposited in soil organic matter with a world pool of about 30 billion tons N. This pool is equal to about 4 tons N per hectare of average arable soil.

The nitrogen balance in agriculture comprises nitrogen bounded by soil organic matter, mineral nitrogen in soil — nitrate and ammonium, nitrogen inputs from external sources — biological fixation, fertilisers, atmospheric deposition etc., and nitrogen outputs with sold farm products or losses to atmosphere or water bodies. The nitrogen pool in soil does not change much from year to year, but the content of its mineral forms changes during the year. Nitrogen inputs and outputs depend generally on the farming goals and practices. Entire nitrogen from the external inputs into the farming system is the beginning of nitrogen flux in the human food chain that could be divided on the crop and

Table 1. Global sources of nitrogen fixation (Pacyna and Graedel 1995).

Sources	Million tons N per year
Biological fixation	
• cultivated soils	89
• natural terrestrial ecosystems	49
• oceans	20–120
Industrial	
• mineral fertilisers	93
• other industrial products	21
• combustion processes	21
Other	
• atmospheric lightning	20

Table 2. Hypothetical nitrogen flux in the human food chain in Poland.

Subsequent steps in the human food chain	N (kg ha^{-1})	N (kg capita^{-1})
Inputs into agriculture with fertilisers, imported fodder, atmospheric precipitation and by biological fixation	72	36
Harvested with crops	34	17
Sold from farm with plant and animal products	15	7.5
Recycled from food processing works to agriculture	4	2
Sold to market	11	5.5
Purchased by consumers	10	5
Consumed	8	4
Throw away to garbage	2	1

Table 3. Emissions of nitrogen gaseous compounds from different sources in Poland in 1996 (Sapek 2001).

Emitted compound	Emission sources	
	Industry, transportation, energy production (load − thousand tons N annually)	Agriculture (load − thousand tons N annually)
Ammonia (NH_3)	6	323
Nitrogen oxides (NO_x)	340	20
Nitrous oxide (N_2O)	12	45
Total emissions	358	388

animal production, food processing, food consumption and waste management. Each of these steps is leaky and some losses of nitrogen into the environment occur. Thus, barely about 14% of nitrogen inputs to the farming system in Poland are purchased by the population in Poland, and only about 12% is consumed. At least 62 kg ha^{-1} N annually is dispersing from agriculture in Poland into the environment (Table 2).

Whereas, the nitrogen inputs to agriculture in Poland could be easily estimated on the basis of official statistical data, the inventory of nitrogen losses deals only with emissions of gaseous nitrogen compounds into the atmosphere (Table 3) and nitrogen inflow into the Baltic Sea (Table 4). Bogacka (1999) has made an inventory of nitrogen load to surface water based on the monitoring data in geographical information system (GIS). It was estimated that the total nitrogen load carried into surface waters in Poland was 209,384 tons N in 1993; 51.4% of this load originated from non-point source (107,700 tons N), including 66,343 tons N from agriculture. The load carried into groundwater in Poland was not estimated.

Nitrate turnover in nature

Nitrate is not a component of rock minerals or soil particles. If we do not consider the rare geological deposits of sodium or potassium nitrate, they occur in nature almost entirely in water bodies and soil solution, always in soluble form. In both systems, the presence and concentration of nitrate is a result of the nitrogen mass flow complex and

Table 4. Nitrogen load to Baltic Sea from Poland territory (GUS, 2000).

Year	Water inflow (km^3)	N-NO$_3$ (thousand tons N)	Total nitrogen (thousand tons N)
1990	38	43	104
1991	40	49	123
1992	39	95	166
1993	43	97	171
1994	57	157	254
1995	54	122	207
1996	53	108	242
1997	58	94	208
1998	64	139	229
1999	70	146	245
2000	62	122	194
2001		118	191

Table 5. Nitrogen cycle from organic matter to organic matter through mineral nitrogen forms.

	Organically bound nitrogen (soil organic matter, plant residues, living organisms)	
Mineralization process	$\Downarrow$ NH$_4$	Absorption by growing plants
Nitrification	$\Downarrow$ NO$_3$	and/or by micro-organisms
Denitrification	$\Downarrow$ N$_2$	
Biological N fixation	$\Downarrow$ Organically bound nitrogen in growing plants or micro-organisms	
Plant ripening or harvesting	$\Downarrow$ Nitrogen in plant residue	
Decay of plant residues	$\Downarrow$ Nitrogen in soil organic matter	

cyclic processes, mostly of a biochemical nature. The main processes are organic matter mineralization and renewed nitrogen biological fixation to organic compounds. The released inorganic nitrogen forms, ammonium and nitrate, are transition forms with a short lifetime in soil or water bodies (Table 5).

This nitrogen flux is linked with some reverse processes and losses of inorganic nitrogen to other ecosystems. Ammonium released from organic matter could be absorbed by growing plants or micro-organisms to form some new organic substances or could also be emitted as ammonia into the atmosphere. A part of released ammonium is oxidised to nitrate. This process, named nitrification, is responsible for producing nitrate in soil. Nitrate in soil is absorbed quickly by living organisms (plants and bacteria) to form biomass and/or is reduced to molecular nitrogen (N$_2$) due to process of denitrification, if only a reductive regime in soil appeared. Thus, the quantity of nitrate in soil is a result of above processes; if mineralization process prevails − some nitrate accumulation in soil will occur; if the absorption and/or denitrification processes prevail − this accumulation is rather reduced.

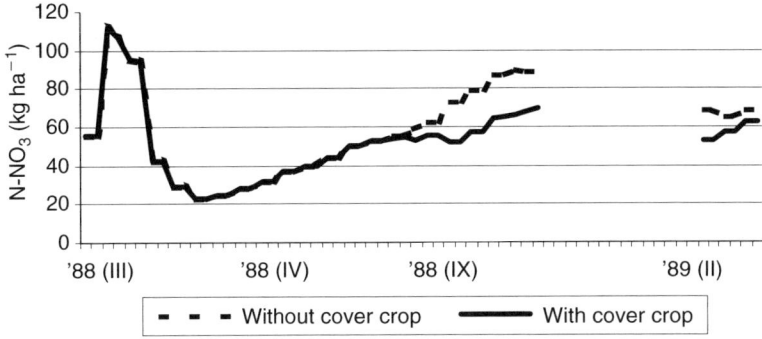

Figure 1. Nitrate content in 1 m soil layer and under spring barley crop followed by cover crop (Falenty 1988).

The fate of nitrogen from manure and mineral fertilisers added to soil is similar. Organic nitrogen contained in manure undergoes mineralization much faster than the nitrogen compounds in soil organic matter. The produced ammonia is volatilized, absorbed or nitrified. Ammonia and nitrate applied in the soil with mineral fertilisers are promptly involved in the processes of absorption, nitrification and denitrification described above.

Nitrate is not a conservative compound in soils and water bodies; its lifetime in soil is rather short. Thus, higher concentrations of nitrate can be observed only in soils without vegetation or if the biological activity there is low. That may occur after the growing season or if the soil is too dry. Quite opposite conditions are created in soils that are too wet, where the denitrification process prevails and only traces of nitrate could be found.

RESULTS

Nitrate content in soil

The nitrate content in soil may be specified as nitrate ion NO_3^{-1} or nitrate nitrogen NO_3-N. Both modes can be presented as units $mg\,kg^{-1}$ or $mg\,dm^{-3}$ of soil. The last unit is more useful as we may use it to calculate the nitrate content in an area of soil, for instance $kg\,ha^{-1}$ NO_3-N in defined soil layer. The most common nitrate concentrations in cropped soils ranged between 2 and $20\,mg\,kg^{-1}$ NO_3-N or recalculated on content in 100 cm soil top layer between 30 and $300\,kg\,ha^{-1}$ NO_3-N. Typical changes of nitrate content in arable soil during vegetation season are presented on an example of barley crop in Falenty (Figure 1).

The actual nitrate content in cultivated soil depends mostly on the intensity of crop grown. That could be clearly presented on the example of this content in 1 m layer of topsoil under spring barley. The nitrate content was low just after the wintertime. The application of $120\,kg\,ha^{-1}$ N in the form of ammonium nitrate at the end of March increased this content particularly in the surface of the soil. The intense nitrate absorption by growing plants reduced this content down to about $30\,kg\,ha^{-1}$ NO_3-N before harvesting time in the middle of July. Next, this content steadily increased, mostly in deeper layers, up to about $100\,kg\,ha^{-1}$ NO_3-N at the end of the autumn.

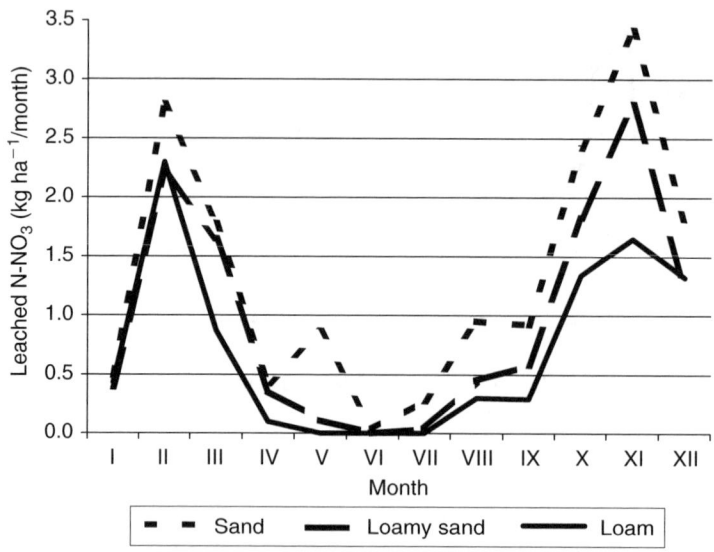

Figure 2. Simulated nitrate leaching from typical crop rotation and fertiliser rates according to standard recommendation – mean values from 20 years (CREAMS model was used).

The above example demonstrates that there appear to be two maxims in nitrate content in cultivated soils: first, just after fertilizer application, and second, after crop maturity. Nitrate from applied fertilizers accumulates in topsoil, which is far from groundwater and the probability of its leaching is low. The main risk of nitrate leaching happens after harvesting time, when there are no growing crops and nitrate also accumulates in deeper layers of soil. Therefore, much attention should be paid to keep the nitrate content in soil as low as possible at that time. The actions involved in the EU Nitrate Directive predict two proposals: first, the ban of applying manure after 1 October to prevent the increase of mineral nitrogen content in soil, and second the obligation to keep most arable soils under green cover.

The impact of cover crops on decreasing nitrate content in soil was demonstrated in the above described experiment with barley crop, where just after harvest in middle of August **the field was sown with white mustard**. The nitrate content in the 1 m soil layer below the cover crop was decreased by about $30 \, kg \, ha^{-1}$ NO_3-N at the end of year (Figure 2).

Nitrate leaching from soil

From the agricultural point of view, the investigations connected with nitrate losses to groundwater are mostly concerned with the question of its leaching below the root zone. The quantity of nitrate percolated below the root zone is supposed to be transferred in total to groundwater and is lost to agricultural production. The root system of arable crops reaches on average the depth of up to 1 m, and of grassland sward up to 60 cm. Nevertheless, the bulk of roots in arable soils are present in tillage top layer, which commonly is not deeper than 30 cm. That is the layer where the intense nitrogen turnover

Table 6. Velocity of nitrate transport down the soil profile in relation to water field capacity and accumulated drainage (Deutsche Bodenkundliche Gesellschaft 1992).

Field capacity (mm dm^{-3})	Yearly accumulated drainage (mm)					
	50	100	150	200	300	400
	← Mean velocity of nitrate transport (m/year) →					
10	0.50	1.00	1.50	2.00	3.00	4.00
15	0.33	0.67	1.00	1.33	2.00	2.67
20	0.25	0.50	0.75	1.00	1.50	2.00
25	0.20	0.40	0.60	0.80	1.20	1.60
30	0.17	0.33	0.50	0.67	1.00	1.33
40	0.13	0.25	0.38	0.50	0.75	1.00
50	0.10	0.20	0.30	0.40	0.60	0.80

occurs and the complete pool of nitrate is produced or introduced with fertilizer or precipitation. That is also the layer from where the nitrate starts to move to the deeper soil layers and to groundwater. There are three processes responsible for nitrate moving: i) percolation, ii) water diffusion, and iii) nitrate diffusion in soil solution. Generally, only the percolation is blamed for nitrate leaching. That is true in deep soil profile, but the diffusing processes play an important role in nitrate movement from topsoil to deeper layers of root zone, and down to the level of groundwater. The velocity of nitrate movement down the soil profile depends on soil porosity and the volume of water percolated (Table 6). In Poland, the most common annual percolation of water below root zone ranges from 100 to 200 mm. Thus, the annual transfer of nitrate deeper than 1 m can be observed only on porous sandy soil, but in most cropped soil this transfer could be expected less than $1 \, m \, a^{-1}$.

The transport and fate of nitrate in the unsaturated zone below the root zone is not taken into consideration in the presented paper.

The quantity of nitrate leached below root zone is a result of two factors: i) actual nitrate content in this zone and ii) volume of percolated water. The highest nitrate content in soil occurs in late autumn and winter months. At this very time, the probability of water percolation is the greatest, as the water balance in soil is then positive, due to the small evapotranspiration rate. The nitrate leaching from the previously described spring barley field (Figure 1) was simulated by means of CREAMS **C**hemicals, **R**unoff, and **E**rosion from **A**gricultural **M**anagement **S**ystems (1980) model. The deep percolation appeared only in 4 months during the year of observations. The highest percolation and nitrate leaching occurred in February, just after the snow had thawed. The highest precipitation and minor nitrate leaching in August shows that usually the nitrate leaching during summer months is insignificant. The accumulated nitrate leaching exceeded $15 \, kg \, ha^{-1} \, a^{-1}$ NO_3-N and cover crops resulted in decreasing this leaching at $5 \, kg \, ha^{-1} \, a^{-1}$ NO_3-N (Table 7). The decreasing of leaching by $5 \, kg \, ha^{-1} \, a^{-1}$ NO_3-N may be assumed as positive effect, as in many cases this is sufficient to decrease the nitrate concentrations in groundwater below threshold value according to EU Nitrate Directive (Table 8). The simulated nitrate losses due to the denitrification process were similar to the amount of leached nitrate. Such a relationship was observed in many experiments.

Table 7. Simulated nitrate leaching and denitrification losses on a spring barley field at Falenty (1988) — model CREAMS (1980) was used.

	Aug. '88	Nov. '88	Dec. '88	Feb. '89	Total
Precipitation (mm)	117.6	39.9	65.0	106.2	329.8
Percolated (mm)	5.4	23.4	39.9	72.4	141.1
Nitrate leached (kg ha^{-1} N-NO$_3$) — without cover crop	0.4	2.7	4.5	7.6	15.2
Nitrate leached (kg ha^{-1} N-NO$_3$) — with cover crop	0.4	1.8	3.0	4.7	9.9
Denitrification (kg ha^{-1} N-NO$_3$) — with cover crop	1.1	2.8	3.2	5.1	12.2
Denitrification (kg ha^{-1} N-NO$_3$) — without cover crop	1.1	1.6	2.1	3.1	7.9

Table 8. Nitrate concentration in groundwater in relation to nitrogen leached and accumulated drainage (Deutsche Bodenkundliche Gesellschaft 1992).

Nitrate leached (kg ha^{-1} a^{-1})	Accumulated drainage (mm/year)		
	100 mm (mg dm^{-3} NO$_3$-N)	200 mm (mg dm^{-3} NO$_3$-N)	400 mm (mg dm^{-3} NO$_3$-N)
10	10*	5	2.5
20	20	10*	5.0
50	50	25	12.5*
100	100	50	25
200	200	100	50

* The highest permitted nitrate concentration according to EU Nitrate Directive is 11.3 mg dm^{-3} NO$_3$-N (=50 mg dm^{-3} NO$_3$).

The use of models to estimate the nitrate leaching is very helpful to trace the trends of changes due to different management systems, but not to obtain some real data. Nevertheless, it also rather difficult to obtain experimental data those are close to natural. The last is particularly true for bigger areas. Most leaching studies do comprise lysimeters or are located on small plots. The catchment's studies are dealing with nitrate losses to surface waters.

CREAMS (1980) model was used to compare the simulated and experimental nitrate leaching on a long-term experiment. CREAMS is a field scale model comprising two submodels: hydrology and nutrient. Submodel Hydrology simulates water balance in soil and one set of results comprises deep percolation below root zone. Submodel Nutrients aims to describe the nitrate leaching and simulates nitrogen turnover in soil; the results comprise nitrate content in soil, nitrate leaching and denitrification losses after each rain event. A field experiment with different nitrogen management on grassland was established in 1987 at the Experimental Station in Falenty (Sapek and Sapek 1993). Observations were on 60 m^2 experimental plots equipped with a lysimetric system covering 9 m^2 made up of a series of gutters buried below the root zone of soil. The effect of following three rates of nitrogen fertilization was compared: 120, 240 and 360 kg ha^{-1} a^{-1} N was applied in the form of ammonium nitrate. The presented mean values covered data from 1988–1995 (Table 9). The simulated data were greater by 30–60% than measured, but the trends were maintained. The denitrification losses were greater than nitrate leaching and such a relationship is typical on grassland.

Table 9. Simulated and experimental nitrate leaching and simulated denitrification losses from grassland soils fertilised with different rates of nitrogen. Mean values from 1988 to 1995.

Nitrogen losses (kg ha^{-1} a^{-1} N-NO$_3$)	Fertiliser rates (kg ha^{-1} a^{-1} N) in form of NH$_4$NO$_3$		
	120	240	360
Simulated leaching	10.3	18.7	25.3
Measured leaching	6.4	13.2	19.1
Simulated denitrification losses	16.2	24.4	32.4

Table 10. Simulated nitrate leaching from typical crop rotation versus soil kind (Sapek and Sapek 1995), kg ha^{-1} a^{-1} NO$_3$-N.

Soil kind	Sand	Loamy sand	Loam
N applied with fertilisers and manure			
Mean values over 20 years	16.2	11.6	8.6
Maximum	25.4	18.2	14.6
Minimum	6.9	4.9	2.5

CREAMS model was also used to simulate the nitrate leaching from arable soils on Poland lowland from typical crop rotation comprising about 70% of cereals. The meteorological data from the observation station located in Chrząstowo, Kisielice and Kórnik from 1965 to 1985 were used. The applied nitrogen fertiliser rates were used according to the actual Polish fertilization recommendation system. Three kinds of soil were considered – sand, loamy sand and loam. Nitrate leaching occurred mainly during winter and autumn months (Figure 2). The greatest nitrate leaching was estimated for sandy soils, usually with a greater proportion of potatoes in crop rotation. Fewer nitrates were leached from heavy loamy soils (Table 10). The estimated mean leaching from sandy soils may create a real risk to groundwater quality increasing nitrate content above the highest permitted concentration according to EU Nitrate Directive, if we consider the values proposed by German Soil Science Society (1992) (Table 8). The nitrate leaching from heavy soils in Poland may result in a lower threat to groundwater quality. This may be the reason for the common view in Poland that the nitrate leaching to groundwater from agricultural land is small and does not cause any greater risk on greater areas. Nevertheless, sandy soils prevail in Polish agriculture.

Critical groundwater pollution by nitrate was observed after land use changes, particularly due to the ploughing of old permanent grassland followed by accelerated soil organic matter mineralization and the accumulated release of nitrogen in it. Strebel, Böttcher and Renger (1984) observed that after the turnover of old grassland to arable land, about 5,000 to 6,000 kg ha^{-1} N was realized from the organic pool in the soil during the first two years. According to their calculation, about 80% of realized nitrogen was washed out as nitrate into groundwater. Lloyd (1992) described similar losses in the United Kingdom. The changing of old grassland to arable land meant up to 400 kg ha^{-1} year^{-1} on NO$_3$-N could be realized, as was reported. Whitemore, Bradbury and Johnson (1992) observed that the losses of nitrate from grassland soil during the 25 years after ploughing showed an exponential decline. Half of the decline took place in the first 5 years and 90% in 18 years. Almost 4 t N/ha from soil organic matter were lost from the top 0–25 cm layer.

They also found that water drained from a field of ploughed grass contained up to $450\,mg\,dm^{-3}$ NO_3-N in the first season after ploughing.

Each tillage operation on arable or grassland soil accelerates the nitrogen mineralization and nitrate leaching. Thus, many best management practices recommend limiting the tillage to a minimum. A new system of cropping maize and some cereals recommends non-till technologies in which soils are not ploughed before sowing and after harvest; weed control is limited by the use of chemicals.

Mitigation of nitrate leaching

The improving of nitrogen management in agricultural production is the main approach to reduce the nitrate leaching. The strategy is to increase the nitrogen efficiency in agricultural production as a way to reduce nitrate losses. This covers the entire agricultural production – cropping systems as well as livestock husbandry. The increase in efficiency can be achieved through rationalization of each step of production by implementation of the Best Management Practices (BMP). They take into account among others: crop rotation, rates and time of fertiliser and manure application, manure spreading technologies, livestock feeding, animal waste storage and handling, grazing systems etc. Usually, the BMP do not limit the yield or farmers' benefit, and in many cases increase both.

Most of BMP are not obligatory for farmers and they are implemented through agricultural advisory services and in the course of different educational activities. Nevertheless, some of them were embraced in a set of regulations:
• farms are obliged to hold sealed tanks for animal waste with adequate storage capacity,
• ban manure application during some seasons of year,
• ban its application during unfavourable meteorological conditions,
• designation of protection areas next to surface water bodies etc.

The regulations do not result directly from BMP, but they are supposed to be helpful in mitigating nitrate leaching. One set of regulations limits the livestock density on farms; another assigns the areas of green cover on arable fields during autumn and winter. There is always an administrative pressure to increase the regulation system, but experience has shown that much more can be improved through the education of farmers and rural area population.

SUMMARY

Nitrate is not a component of rock minerals or soil particles; it is a product of nitrogen turnover in soil. Once nitrate appears in soil it is quickly absorbed by growing plants or bacteria to form new organic substances. Part of the nitrate pool in soils is used as an energy donor and is decomposed to molecular nitrogen due to the process of denitrification. Nevertheless, a defined pool of nitrate is still present in soil and this pool could be leached below the root zone, if favourable conditions appear. The nitrate concentration varies greatly during the hydrological year and is at its highest in late autumn, when the main volume of water is percolating to the saturated zone.

The most frequent nitrate content in cropped soils ranged between 30 and $200\,kg\,ha^{-1}$ NO_3-N in 100 cm layer of topsoil, but only a part of this pool could be leached by favourable conditions. The observed and calculated data show that in the Polish lowland, the annual leaching can reach as much as $25 \, kg \, ha^{-1} a^{-1} \, NO_3$-N. Such high leaching may occur especially on the sandy soils prevailing in Poland. Mean precipitation in Poland lowland seldom exceeds 600 mm annually, thus the water percolation does not often surpass 150 mm and the nitrate concentrations in leachates may be high, which creates some risk of groundwater pollution.

The best way to mitigate the groundwater pollution by nitrate from agricultural sources is the implementation of Best Agricultural Management Practices and the effective education of farmers and rural area population.

REFERENCES

Bogacka T., 1999. Strategy and action mitigating the load of nitrogen and phosphorus aiming to surface waters from non-point sources (orig.title, Strategia i działania ograniczające ładunki azotu i fosforu odprowadzane do wód powierzchniowych ze źródeł obszarowych.). In: Podstawy naukowe strategii ochrony wód w Polsce w świetle przystapienią do Unii Europejskiej, Ed. Gromiec M., IAWQ, Warszawa. 221–337.

CREAMS, *A field scale model for Chemicals, Runoff, and Erosion from Agricultural Management Systems.* Knisel, W.G. (ed.): 1980. US Department of Agriculture, Science and Education Administration, Conservative Research Report 26.

Deutsche Bodenkundliche Gesellschaft 1992: *Strategien zur Reduzierung standort- und nutzungs-bedingter Belastungen des Grundwassers mit Nitrat*, Wilhelmstraße 19, 2900 Oldenburg: pp. 1–42.

Lloyd, A. 1992: Nitrate leaching under arable land ploughed out from grass. *The Fertilizer Society.* Proceedings No. 330. 1–32.

Pacyna, J.M. and Graedel, T.E. 1995: Atmospheric emissions inventories: Status and prospects. *Annual Review of Energy and the Environment 20*: 265–300.

Sapek, A. and Sapek, B. 1993: Assumed non-point water pollution based on the nitrogen budget in Polish agriculture. *Water Sci. Technology 28*, No. 3–5: 483–488.

Sapek, A. and Sapek, B. 1995: *Application of nitrogen balance and CREAMS model to describe and foresee the nitrogen losses in Polish agriculture.* GI-Fachausschuß 4.6 "Informatic im Umweltschutz" Band 7. Proceeding of 9th International Symposium on Computer Science for Environmental Protection CSEP'95 *Space and Time in Environmental Information Systems. Part I.* Metropolis-Verlag, Marburg, pp. 219–226.

Sapek A. 2001. The evaluation of environmental risk resulting from nitrous oxide emissions from the agriculture (orig. title, Ocena ryzyka środowiskowego wynikającego z emisji podtlenku azotu z rolnictwa). In: Obieg pierwiastków w przyrodzie. B. Gworek, A. Mocek Eds. Instytut Ochrony Środowiska, Warszawa. V. 1, 286–294.

Sapek, A., Sapek, B. and Pietrzak, S. 2002: Cycle and budget of nitrogen in Polish agriculture (orig. title, Obieg i bilans azotu w rolnictwie polskim). *Nawozy i Nawożenie*, **4** (1(10)): 100–121.

Statistical Year Book, 2001. Environment 2001. Information and statistical papers. Central Statistical Office, Warszawa.

Strebel, O., Böttcher, J. and Renger, M. 1984: Einfluss von Boden und Bodennutzung auf die Stoffanlieferung an das Grundwasser. Proceedings of the International Symposium *Recent Investigations in the Zone of Aeration*, Munich, West-Germany, October 1984.

Udluft, P., Merkel, B. and Prösl, K.H. 1984: *Einfluss von Boden und Bodennutzung auf die Stoffanlieferung an das Grundwasser.* Department for Hydrology and Hydrochemistry, Technical University of Munich, FRG. 2: 663–669.

Whitemore, A.P., Bradbury, N.J. and Johnson, P.A. 1992: Potential contribution of ploughed grassland to nitrate leaching. *Agriculture, Ecosystems and Environment 39*: 221–233.

CHAPTER 2

Nitrate time-evolution in the waters of the Quaternary aquifer of Vitoria-Gasteiz (Basque Country, Northern Spain) – influence of wetlands

C. García-Linares[1], M. Martínez[1], I. Antigüedad[1] and J.M. Sánchez-Pérez[2]

[1]*Hydrogeology Group, University of the Basque Country, 48940 Bizkaia, Basque Country, Spain*
[2]*Laboratoire d'Ecologie des Hydrosystèmes (LEH, FRE CNRS-UPS 2630), Université Paul Sabatier, 31055 Toulouse, France*

ABSTRACT: The Quaternary aquifer of Vitoria-Gasteiz has been recently declared by the Basque Government as a Vulnerable Zone according to the 91/676/CEE European Law related to the farming origin nitrate pollution in groundwater. This is the only aquifer in the Basque Country under this legal consideration.

Because of the change from dry to irrigated farming and the need to enlarge cultivable land, a series of land transformations, with an important increase in the drainage network, took place 40 years ago. These transformations have led to a notable loss of resources in the aquifer, evaluated to 40%, and an intense contamination by nitrogen compounds. This contamination decreased and seems to have stabilized during the last years as a result of some changes in the use of agricultural fertilizers and in the origin of waters used in the irrigation. Nevertheless, nitrate presence in groundwaters continues to be higher ($60-80\,mg\,l^{-1}$) than reasonable. As a result of the agriculture intensification both the forests and the wetlands disappeared along with their biogeochemical functionality.

Nitrate concentration in surface water remains high ($50-70\,mg\,l^{-1}$) during winter–spring time, but at the end of summer the nitrate content decreases strongly (less than $10\,mg\,l^{-1}$) due to the nitrate assimilation by plants that entirely cover the river course.

On the other hand, wetlands have been recently restored in the peri-urban area of Vitoria-Gasteiz city closing main ditches and conducting to elevation in the local piezometric level. Restoration allowed its biogeochemical function recovery, reducing nitrate from groundwaters. Conditions near wetland are conductive to denitrification: organic matter-rich soil and clay presence allowing a local semiconfined flow.

INTRODUCTION

In the last decades, contamination of surface and groundwater in areas with large portions of agricultural land has reached very high levels. In a great number of cases, water resources have been affected in such a way that they are not suitable for further use without previous treatment.

Impact of farming activities on groundwater quality in European alluvial valleys is well known (Gustafson 1983; Andersen and Kristiansen 1984; Bernhard, Carbiener,

Cloots *et al.* 1992; Böhlke 2002). In these areas, agricultural activities combined with water re-circulation have caused an increase in concentrations of nitrogen compounds and pesticides in the groundwater (Ritter, Chirnside and Scarborough 1990; Arrate, Sánchez-Pérez, Antigüedad *et al.* 1997).

The Quaternary aquifer of Vitoria-Gasteiz (Figure 1) represents a well-documented example of water resource losses and degradation of groundwater quality due to various land transformations. A change in agricultural practices combined with the diversion of principal rivers traversing the aquifer and an increment of the drainage network of the aquifer has driven to a rapid intense contamination by nitrogen compounds and to a reduction of existing resources (Arrate 1994; Arrate *et al.* 1997).

Several actions have been carried out recently to diminish this problem. Thus, in 1999 the East Sector of the Quaternary aquifer was designated as a Nitrate Vulnerable Zone according to the 91/676/CEE European Law and a Code of Good Practices was approved. On the other hand, wetlands close to Vitoria-Gasteiz were restored, so they have recovered its biogeochemical functionality (García-Linares, Martínez, Sánchez-Pérez *et al.* 2003).

STUDY SITE: QUATERNARY AQUIFER

Geology of the aquifer

The Quaternary aquifer of Vitoria-Gasteiz is formed by fluvial and alluvial deposits. The area extends approximately 90 km^2 with a thickness of approximately 5 m. These deposits form a permeable aquifer with intergranular porosity of free character although local phenomena of semi-confinement may exist. The storage coefficient and average transmissivity are respectively 0.2 and 40–150 m^2/day. The formation presents two sectors clearly individualized: The West Sector extends over approximately 40 km^2 and has a thickness that ranges from 1 m to 4 m. It is basically formed by sandy–muddy materials. Borders and substratum are constituted, in the northeast area, by a high permeable carbonated formation known as the Apodaka Karst, and in the southern area, by a marly series of medium permeability. Both formations are hydraulically connected to the Quaternary aquifer (Arrate *et al.* 1997). The East Sector (the area considered in this study) has an extension of approximately 50 km^2 with a thickness ranging from 2 m to 10 m. It is formed by heterometric gravels with an argillaceous sandy matrix. Borders and substratum are virtually impermeable marls. This Sector is shown in Figure 1, along with the Vulnerable Zone polygon and the network implemented to control surface and groundwater quality.

Groundwaters in this sector flow into the Alegria River; finally, all the waters leave the Sector through the A4 point, where a narrowing in the Quaternary deposits is observed.

Functioning of the aquifer until the 1950s

Recharge mainly proceeded from precipitation over the Quaternary deposits and, in minor quantities, from infiltration of surface runoff from the borders of the aquifer, and from the fluvial network, in flood periods, during which rivers are influent.

The outlets of the system flowed towards the fluvial network. This drainage was partially favoured by the existence of several trenches 0.5 m deep, historically built with

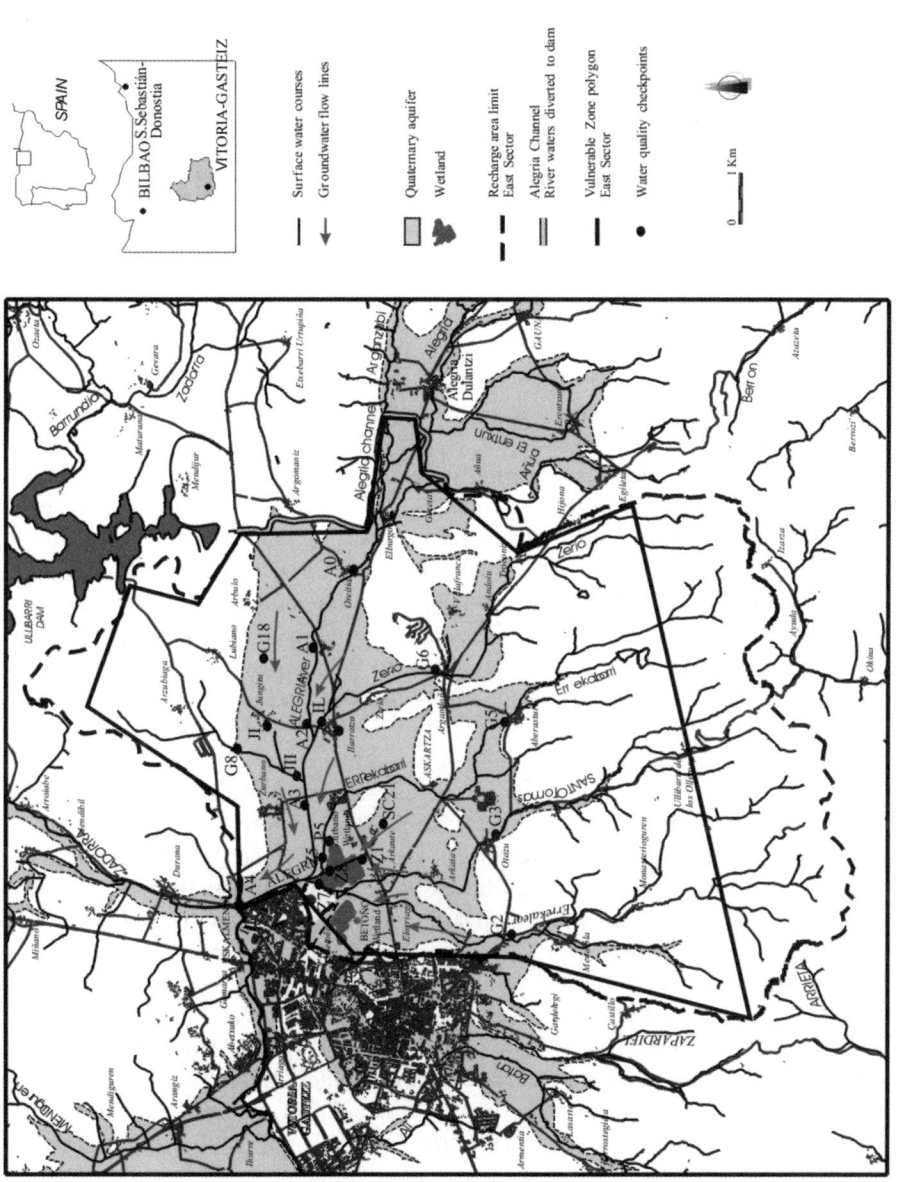

Figure 1. East Sector of the Quaternary aquifer of Vitoria-Gasteiz and Vulnerable Zone.

the purpose of avoiding flooding in agricultural areas. However, various wetland zones were still found in discharge areas of the aquifer unit. At the time, there were small-sized pumping stations built to meet domestic demands of small communities situated on the aquifer. The water table was found at a depth ranging from 0 m to 1.5 m. Total reserves, estimated at the beginning of dry seasons, have been evaluated to be 25 hm^3.

Concerning land use, the major part of the aquifer area was dedicated to dry farming, fundamentally cereal. In the area under study it is interesting to highlight the existence of extended zones of meadowland and broadleaf forest, relics of pre-existing autochthonous vegetation.

Land transformations on the aquifer after the 1950s

The existence of changes in agrarian practices and water-resource management in the basin of the Alegria River, required several modifications that changed appreciably the dynamics of the Quaternary aquifer.

To satisfy demands of a larger cultivable area, practically the entire existing autochthonous forest on the aquifer was progressively felled. Thus, in 1954 the forest area represented 747 ha, in 1968, 395 ha and in 1982, only 67 ha. The current situation is more or less similar to that in 1982.

At the onset of the 1960s, the last existing wetland (Zurbano wetland in Figure 1) was drained by means of deepening and modification of several watercourses. In order to avoid possible floods, deepening and widening of the rest of water courses in the aquifer were carried out as well, adapting them to the geometry of agricultural terrain. Trenches were, on several occasions, more than 2.5 m deep. The drainage density of the analysed area went from 5.3 km^{-1} in 1954 to 7.7 km^{-1} in 1982, a similar situation to the present one.

Moreover, a crop change from traditional cereal to the present potato and sugar beet (Figure 3) meant a greater demand of water during the summer months. Such a need was satisfied with the aquifer water in most of the East Sector. As a result, nitrate contents increased very rapidly due to the recirculation of groundwaters in a system almost closed. However, in the West Sector, groundwater resources were not sufficient to meet demands so it was necessary to take water from nearby small rivers. Moreover, the crop change entailed the use of large quantities of fertilizers and pesticides on the aquifer surface.

The area occupied by irrigated farming successively went from less than 1% (1954) to 11.5% (1968) and to 67.5% (1982) of the area of study. This increase has evolved to the detriment of dry farming areas that have passed from 78.9% in 1954 to only 11.4% in 1982, a similar situation to the present one.

On the other hand, in order to increase water supplies to Vitoria-Gasteiz city at the beginning of the 1970s, complete diversion of the Alegria River and two of its tributaries was carried out at the entrance of the East Sector of the aquifer (Figure 1) sending supplies towards the Zadorra dam system. The average annual volume that was diverted was of the order of 18 hm^3. Also, in 1999 the Santo Tomás and Errekaleor rivers (located at the West Sector near the wetlands) were diverted and now they flow into the Alegria River.

The consequences of all these changes were the fall of the water table, by $1-2.5$ m, the disappearance of wetlands, with the resulting loss of its biogeochemical functionality, and an increase in the nitrate contents in groundwaters (frequently more than $150-200$ mg l^{-1}). Figure 2 represents the evolution in nitrate contents from 1990 up to

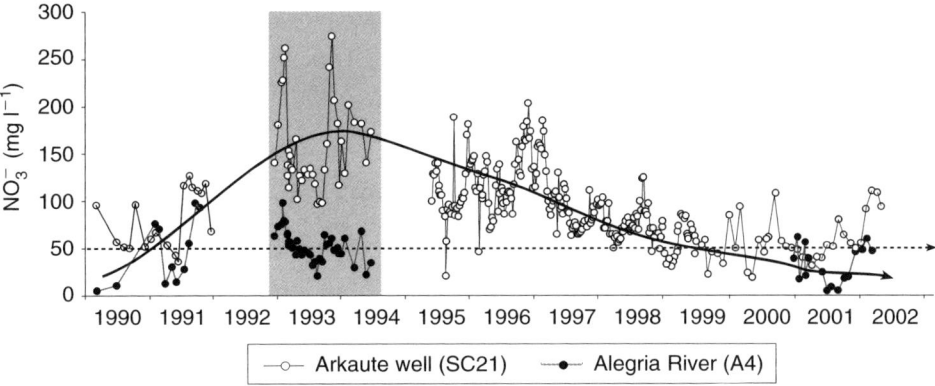

Figure 2. Time-evolution of nitrate contents (mg l^{-1}) in well SC21 and at the outlet of the Alegria River (A4).

date in the well SC21 (Arkaute well), representative of the East Sector and part of the Groundwaters Control Basic Network (Basque Government), and in the outlet of the Alegria River (A4). The location of both points is shown in Figure 2. The highest contents appear in the 1993–1994 period. In recent years the nitrate contents are on the decrease due to a possible rationalization in the use of fertilizer (this is not easy to prove because of lack of data) and the change of the origin of water used for irrigation that now comes from artificial pools (surface water stored outside of the Quaternary aquifer) avoiding the recirculation of waters. Figure 2 shows clearly that water in the outlet (A4) has fewer nitrates than groundwaters; this fact will be explained in the next section.

CURRENT SITUATION: DATA AND DISCUSSION

Surface and groundwaters were sampled systematically (at least every month) in 26 points (most of them shown in Figure 1) that have been assembled according to their origin and geographical setting in six groups: waters of the main river (Alegria) and its tributaries (making a difference between the East and the Central Zone), groundwater in cultivated areas and ditches and groundwater in the wetland. Table 1 shows the results of the analyses from samples taken over a period of time between January 2001 and March 2002.

Several principal components analyses (PCA) have been made in order to evidence differences in the chemistry characteristics of the surface and groundwaters as a way to improve the knowledge of the aquifer dynamic:

PCA 1 – the Quaternary as a whole
A principal components analysis was made on the basis of analytical data taken from samples of surface as well as groundwaters (366 analyses in all) taken over a period of time between January 2001 and March 2002. The factorial plane I–II (Figure 3) reflects the most important information, with factor I (45.8% of the variance) characterized by bicarbonates, sulphates, calcium and magnesium, and factor II (22.3%) characterized by nitrates. It is clearly appreciated how the waters with a higher nitrate content correspond to

Table 1. Mean contents ($mg\,l^{-1}$) of the main chemical elements in the different zones of the East Sector.

	Alegría River[1]	Groundwater in cultivated areas[2]	Surface water West Zone[3]	Surface water Central Zone[4]	Wetland groundwater[5]	Wetland ditches water[6]
HCO_3	318.5	318.8	297.8	307.9	608.3	395.6
SO_4	57.8	69.6	45.1	54.4	190.7	66.8
NO_3	39.0	63.6	32.1	50.4	8.7	3.3
Cl	37.1	44.8	26.6	34.6	44.7	27.1
Ca	123.1	143.1	104.3	121.9	193.1	122.1
Na	12.7	13.8	13.2	12.0	54.6	15.2
Mg	8.2	7.2	6.8	7.3	31.2	10.0

[1] A0, A1, A2, A3 and A4; [2] Ilarratza spring, well SC21, Z1 and Z2; [3] G2, G3, G5, Z5 and Z7; [4] G6, G8, G18, IL, JI and JII; [5] P5; [6] Z3, Z4 and Z8 (the outlet). See Figure 1 for location points.

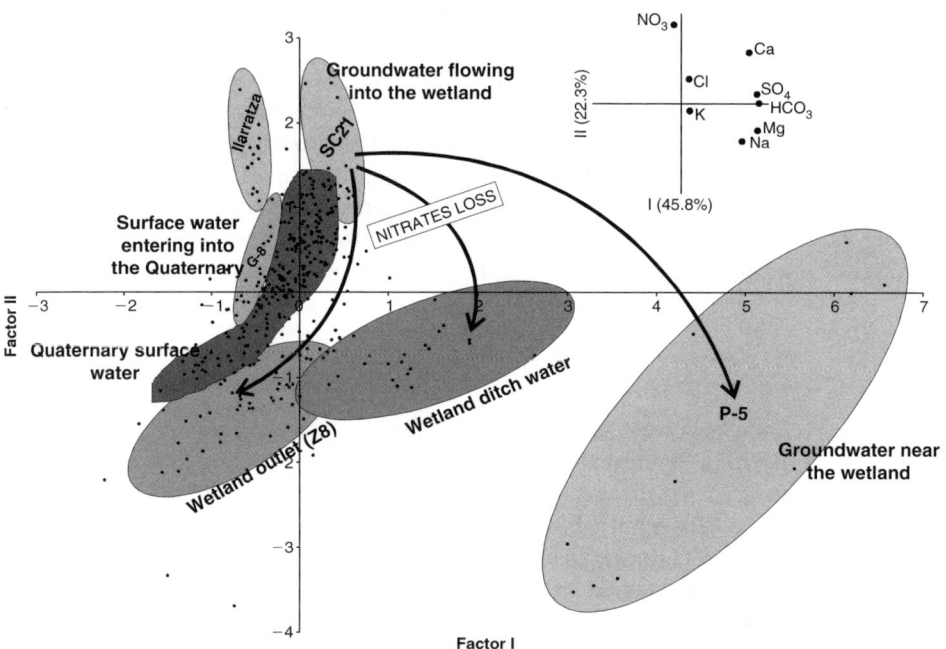

Figure 3. PCA 1 – all the waters of the Quaternary.

those samples taken from the Ilarratza spring and from well SC21, both representative of the groundwaters of the Quaternary aquifer in cultivated areas (see Table 1); in particular, well SC21 represents the groundwaters that enter the Zurbano wetland. In the wetland area there are nitrate losses, a fact that is reflected in plane I–II: the points of both groundwaters (from piezometer P5 and drainage ditches Z3 and Z4) and the waters at the outlet of the wetland (Z8) are clearly nitrate-impoverished. In spite of this loss in nitrates in the wetland area, the waters present very different mineralizations, which imply different

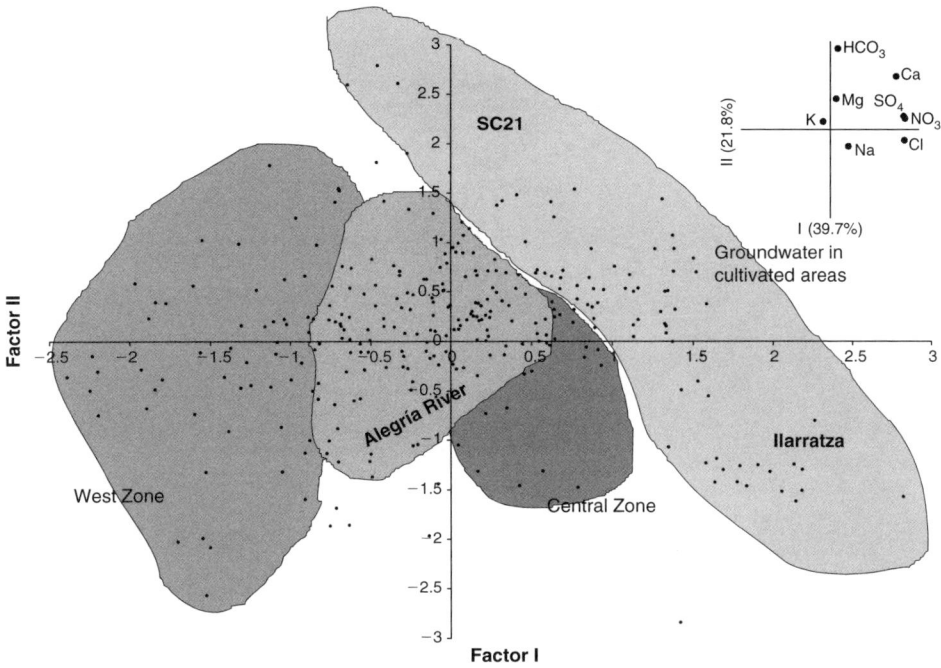

Figure 4. PCA 2 – waters of the Quaternary without waters of the wetland area.

processes involved in this loss. Thus, at point P5, the higher mineralization of the waters (see Table 1) is accounted for by the existence of a reducing environment (clay silty soil and abundance of organic matter) during a good part of the year. The waters at the outlet (Z8), for their part, are a combination of different subterranean inputs. The rest of the points, corresponding to surface waters within the Quarternary aquifer and surface waters coming from outside it – the latter with a lower mineralization – are grouped in an intermediate position on plane I–II as regards nitrate content. So as to have a clearer view of the chemical characteristics of the surface and groundwaters of the Quaternary aquifer and of their spatial and temporal distribution, another principal components analysis (PCA 2) was carried out, in which the points having a low nitrate content, that is, the points in the wetland area, have been dispensed with.

PCA 2 – the Quaternary without the wetland area
Once more, the most important information is reflected in factorial plane I–II (Figure 4), in which factor I (39.7% of the variance) is characterized by chlorides, sulphates, nitrates and calcium, and factor II (21.8%) by bicarbonates. On one hand, all the groundwaters (well SC21 and Ilarratza spring) fall into a group that is differentiated from the rest by the presence of a higher mineralization in nitrates, above all, but in sulphates and chlorides as well. The surface waters show less mineralization by comparison. A division may be established according to location on the plane: the waters that are less mineralized in terms of the characterizing elements of factor I are those coming from the rivers of the West Zone (Table 1) of the Quaternary aquifer (the Santo Tomás, Errekaleor and Errekabarri rivers), whereas the more mineralized waters are those from the Central Zone (the Zerio and Jungitu rivers); the waters from the principal river, the Alegria, occupy an

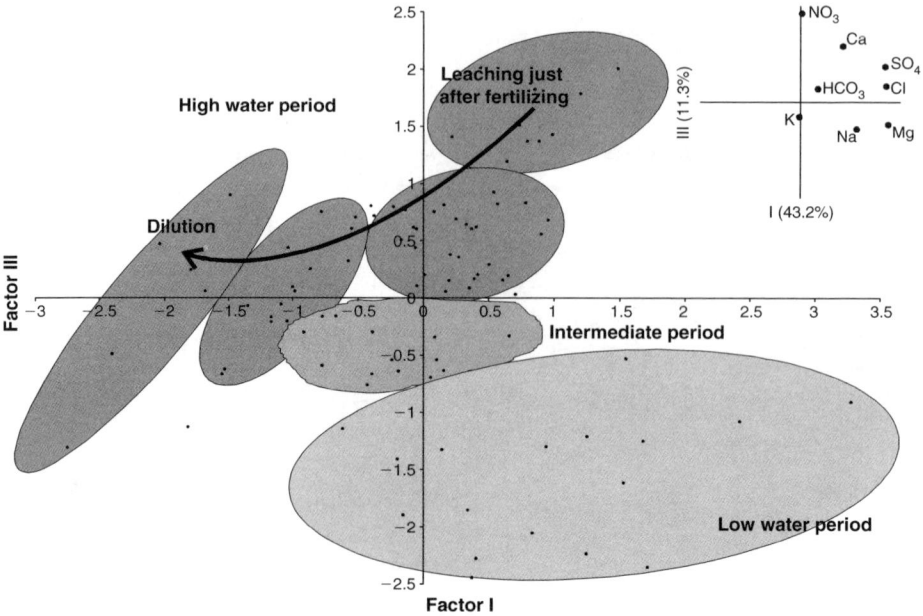

Figure 5. PCA 3 – waters of the Alegria River.

intermediate position, and a new PCA was carried out in order to better observe their evolution. The higher mineralization of the waters of the Central Zone is explained by the fact that these are the rivers into which the groundwaters of the aquifer flow, whereas in the West Zone, the rivers are disconnected from the aquifer throughout a great part of their course. This factorial plane is a good representation of data shown in Table 1.

PCA 3 – waters along the Alegria River
This analysis covers the water course of the Alegria River. It only deals with factorial plane I–III (Figure 5), since this depicts the data that best represents the temporal distribution of the waters in terms of compounds that are fundamentally agricultural in origin. This plane represents 54.5% of the variance expressed, with factor I (43.2%) being represented by chlorides, sulphates and magnesium, and factor III by nitrates. Generally, throughout the course of the river, the waters richest in nitrates appear during the seasons of heavy rain that arise right after the main application of the fertilizers that occurs between February and March. The rapid infiltration and leaching of nitrogenated compounds is notable then. Subsequent rains, nonetheless, give rise to a generalized decrease in mineralization that includes nitrates; this dilution is even more significant during periods of heavy rain that occur a long time after the fertilizing, or before this takes place.

   The lower part of the graph depicts the waters corresponding to the period of low water (July–October), characterized by a relatively significant overall mineralization, albeit with a lower content in nitrates. The decrease in nitrates, which does not affect the other elements of agricultural origin (sulphates and chlorides, above all), is explained by the notable loss of nitrates that occurs due to its assimilation by the vegetation that proliferates abundantly on the riverbed. The variations along the Alegria River, from its entry into the Vulnerable Zone (A0) up to its point of exit (A4), is better observed in Figure 6.

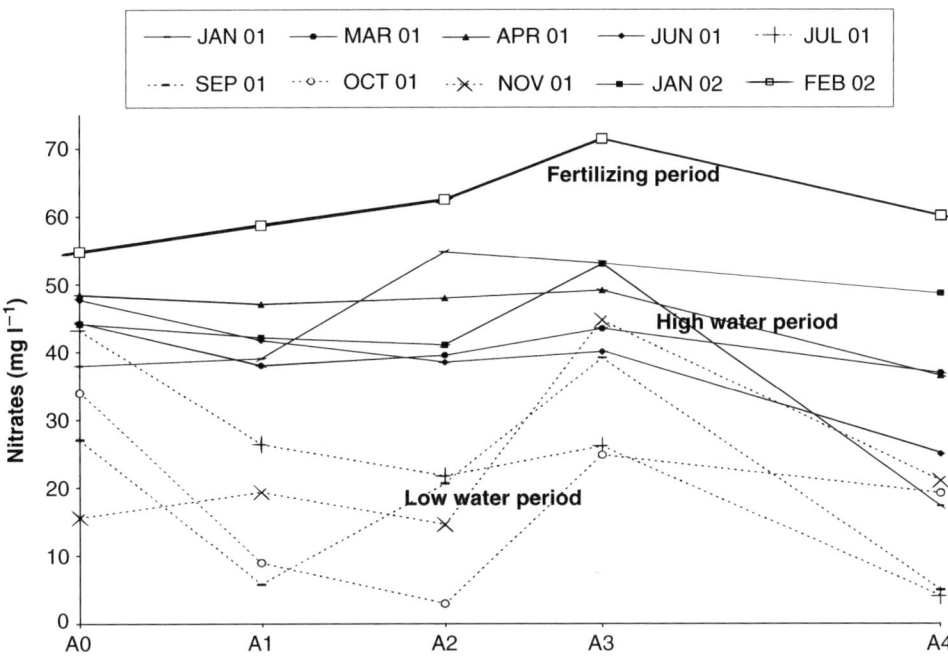

Figure 6. Nitrate content evolution in waters along the Alegria River. See Figure 1 for points location.

Figure 6 represents the spatial and temporal evolution of nitrate content along the Alegria River, the principal watercourse of the Quaternary aquifer and makes a complementary interpretation of the data shown in ACP 3 (Figure 5). The highest values of nitrate concentration correspond to the fertilizing seasons (principally in February), in which values higher than $55 \, mg \, l^{-1}$ at all points of the river are reached. During the months of winter–spring, the concentration in nitrates remains quite constant, at around $35-50 \, mg \, l^{-1}$. The lowest contents are reached during the low water season (Figure 2), above all at points A1 and A2, due to the assimilation of the nitrates by the plants, which, at those periods of low waters, totally occupy the riverbed; this decrease does not affect the rest of the predominant components in solution as Figure 5 explained. It must be pointed out that during the year 2001, the period of low waters extended into the months of October–November.

As regards spatial variation, the increase frequently observed in nitrate content between A2 and A3 and the subsequent decrease between A3 and A4 is relevant. Between points A2 and A3, the Alegria River receives waters from the Zerio River and the Arbulo-Jungitu drainage (Figure 1); nonetheless, these inputs do not account for the increase in nitrates observed, since the concentrations of these waters are either similar or lower (Central Zone in Table 1). Thus, between A2 and A3, it is possible that a diffuse discharge into the river from the waters of the aquifer is taking place – a hypothesis supported by the existence of a threshold in the impermeable marls of the substrate in this zone (Arrate 1994) that might be provoking the ascent and discharge of groundwaters. The decrease in nitrate contents observed between A3 and A4 (the outlet of the Quaternary system) may be due to several reasons, depending on the hydrological situation: during high water periods, nitrate decrease is due to an effect of dilution owing to the mixture of the waters of the Alegria River with those of the Santo Tomás, Errekaleor and Errekabarri Rivers and the waters coming from the discharge of the Zurbano wetland (West Zone), all of them with lower nitrate contents than the waters of the Alegria River (Figures 3 and 4). During low water periods, nonetheless, the decrease is due to the assimilation of nitrates by vegetation, which practically covers the bed of the Alegria River, sufficiently wide at this stretch and thus subject to a slower flow rate.

CONCLUSIONS

Nitrate concentration in the groundwaters of the Vulnerable Zone of the Vitoria-Gasteiz Quaternary aquifer reached the highest values at the beginning of the 1990s, when nitrate contents higher than $200 \, mg \, l^{-1}$ were measured in many wells. Afterwards, nitrate concentration decreased and in recent years seems to remain stable, between $50-70 \, mg \, l^{-1}$. Apart from a possible rationalization in the fertilizer's use, the change in the origin of waters used for irrigation must be exposed in order to explain the observed decrease; in fact, waters are taken now from artificial pools outside of the Quaternary aquifer and not from the aquifer as in the previous situation.

As regards the current situation nitrate content is higher in groundwaters than in surface waters. After the recent wetland restoration nitrates loss is evident near the wetland area, although the processes involved are not well known in the detail; nevertheless, there is a reducing environment favoured by the presence of clay silty soils and abundance of organic matter. Discharge from the wetland helps to decrease nitrate contents in the Alegria River. Another nitrate loss process is evident just along the Alegria River in low water periods, when the vegetation that proliferates on the riverbed assimilates nitrate compounds.

The nitrate content of the surface waters of the West and the Central Zone is different according to the relation between these waters and the groundwaters. In fact, the Central Zone river presents a higher nitrate concentration because groundwaters are flowing into them; thresholds in the impervious marls of the substrate seem to control local groundwater pattern favouring discharge into the surface waters. Time and spatial evolution of nitrate contents along the Alegria River is the result of all the above-mentioned processes.

ACKNOWLEDGEMENTS

This work was funded by the HID 99-0333 Project (CICYT, Spanish Government). Authors thank contribution from I. Arrate (EVE, Basque Government) and A. Alonso (Council of Vitoria-Gasteiz).

REFERENCES

Andersen, L.J. and Kristiansen, R. 1984: Nitrate in groundwater and surface water related by land use in the Karup Basin, Denmark. *Environmental Geology 5*: 207–212.
Arrate, I. 1994: *Estudio hidrogeológico del acuífero cuaternario de Vitoria-Gasteiz* (Araba, País Vasco). Tesis Doctoral. University País Vasco.

Arrate, I., Sánchez-Pérez, J.M., Antigüedad, I., Vallecillo, M.A., Iribar, V. and Ruiz, M. 1997: Groundwater pollution in Quaternary aquifer of Vitoria-Gasteiz (Basque Country, Spain). *Environmental Geology 30*: 257–265.

Bernhard, C., Carbiener, R., Cloots, A.R., Froehlicher, R., Schenk, Ch. and Zilliox, L. 1992: Nitrate pollution of groundwater in the Alsatian plain (France). A multidisciplinary study of an agricultural area: the central ried of the Ill River. *Environmental Geology Water Science 20*: 125–137.

Böhlke, J.K. 2002: Groundwater recharge and agricultural contamination. *Hydrogeology Journal 10*: 153–179.

García-Linares, C., Martínez, M., Sánchez-Pérez, J.M. and Antigüedad, I. 2003: Biogeochemical functionality of a restored wetland (Basque Country, Spain). *Hydrology and Earth System Sciences 7(1)*, 109–121.

Gustafson, A. 1983: Leaching of nitrate from arable land into groundwater in Sweden. *Environmental Geology 5*: 65–71.

Ritter, W.F., Chirnside, A.E.M. and Scarborough, R.H. 1990: Soil nitrate profile under irrigation on coastal plain soils. *Journal of Irrigation and Drainage Engineering 116*: 738–750.

CHAPTER 3

Origin of nitrates in water inflows in Pb–Zn ore mines

J. Motyka and K. Różkowski
University of Science and Technology (AGH), Mickiewicza 30, 30-059 Cracow, Poland

ABSTRACT: In the Olkusz region Pb–Zn ore has been exploited for a few hundred years. Triassic carbonates hosting ore form a fissure-karstic aquifer. Under natural conditions the Triassic aquifer groundwater was good quality water. Deep mine drainage and the resulting changes in groundwater flow directions caused the activation of various pollution sources that have significantly influenced groundwater quality. This paper presents the spatial distribution of nitrates in Pb–Zn mine outflows in the Olkusz region. The most probable sources of nitrogen compounds determined in mine outflows are presented.

INTRODUCTION

The Olkusz region is a historical area of Pb–Zn ore exploitation. Ore occurs mainly within the dolomites of Middle and Lower Triassic age. During the last decades there were three active ore mines: 'Bolesław', 'Olkusz' and 'Pomorzany'. Since 1996 the 'Bolesław' mine has been closed; its drainage system however is still functioning.

In the Olkusz region, the Pb–Zn ore mines are among the most water-flooded mines in the world. According to the fissure-karstic character of the ore hosting rocks, single cavern outflow efficiency reaches dozens of $m^3 min^{-1}$, while total pumped water volume from all three mines, in relation to precipitation, can exceed $6\,m^3\,s^{-1}$ (Adamczyk and Motyka 2000). As a result of their fissure-karstic character, the dolomites and limestones are vulnerable to surface pollution infiltration. Pollutants can easily migrate for long distances in such types of aquifers.

Intensive drainage of the Triassic aquifer is connected with mining activity in the Olkusz region. Around the ore mines, wide spread regional depression cone occurs with depths ranging from 60 to 130 m, depending on the individual drainage centre. Natural groundwater flow directions are reversed on a regional scale and as a result different surface pollution sources appeared (Adamczyk, Motyka and Witkowski 2000). An indicator of anthropogenic groundwater contamination is the presence of nitrogen compounds, especially nitrates. Research into the mine water hydrochemical composition revealed increased concentrations of NO_3 ion in individual mine working outflows. The presented paper describes the origins of nitrates in Pb–Zn mine workings water inflows.

The work was conducted with support from the University of Science and Technology (AGH) as a part of the research programme No. 11.11.100.270.

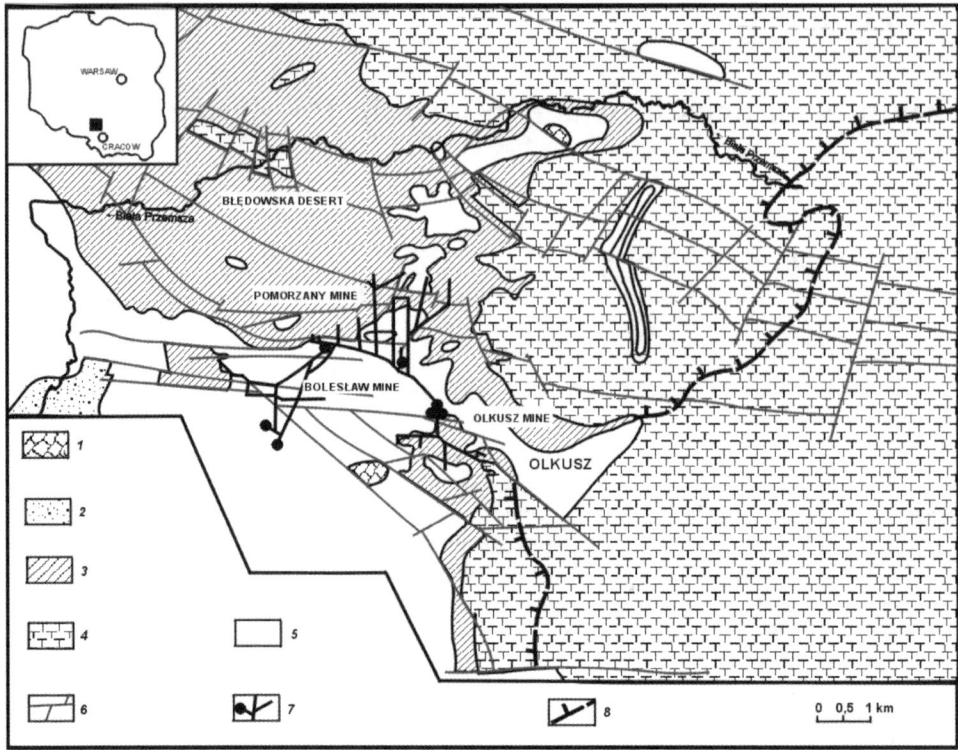

Figure 1. Geological map of the Olkusz ore district. 1. Devonian and Carboniferous, carbonates; 2. Permian, conglomerates; 3. Keuper, clays; 4. Upper Jurassic, limestones; 5. Lower and Middle Triassic, carbonates; 6. main faults; 7. main mine galleries; 8. range of the Upper Triassic under the Upper Jurassic.

HYDROGEOLOGICAL SETTING

The Olkusz Pb–Zn ore district belongs to the Cracow-Silessian Monocline built up of Triassic and Jurassic formations that are discordantly overlying folded and faulted Paleozoic basement (Figure 1). There are Quaternary, Jurassic, Triassic and Paleozoic aquifers in the hydrogeological profile of the Olkusz area (Figure 2). The Quaternary aquifer is of porous type, built up of fluvioglacial sands with gravel, debris or rarely dust, clay or boulder clay insertions. Jurassic, Triassic and Paleozoic (Carboniferous–Devonian) aquifers are of carbonate type, consisting of limestones and dolomites. The rocks are fissured, with karstic features such as caverns, karstic channels and breccia occurring. Jurassic and Paleozoic aquifers are of fissure-karstic type, while the Triassic aquifer, because of its high matrix porosity, is of fissure-karst-porous type.

The Quaternary aquifer is recharged mainly by precipitation infiltration, while under mine drainage conditions also by water infiltrating from the major Biała Przemsza River in the Olkusz region. The Jurassic aquifer, present in the eastern part of the Olkusz area (Figure 1), is recharged by precipitation infiltration and discharged through springs, natural watercourses, deep wells, as well as filtration to adjacent aquifers through zones of hydraulic connectivity.

BIAŁA PRZEMSZA

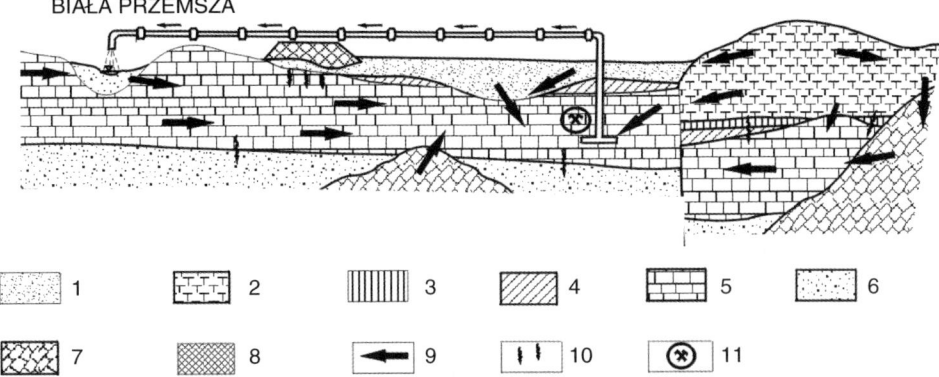

Figure 2. Scheme of groundwater circulation in the Olkusz area. 1. Quaternary, sands; 2. Upper Jurassic, limestones; 3. Middle Jurassic, marls; 4. Keuper, clays; 5. Lower and Middle Triassic, carbonates; 6. Permian, conglomerates; 7. Devonian and Carboniferous, carbonates; 8. tailings and dump sites; 9. flow direction; 10. percolation; 11. mine workings.

Triassic carbonates making up the Triassic aquifer are Pb–Zn ore bearing, which is exploited in the Olkusz mines. There are different pathways of recharge to the Triassic aquifer. At outcrops, precipitation infiltration predominates; a much smaller share is due to water discharging from watercourses, mainly from Biała Przemsza River. Water inflows occurring from the Quaternary aquifer through erosion window zones where direct hydraulic contact with Triassic aquifer occurs are of great significance. A very important source of recharge into the Triassic carbonates is due to water inflowing from the Upper Jurassic aquifer.

The following recharge pathways were confirmed (Figure 2):

● water flows through direct hydraulic contact zones between the Upper Jurassic limestones and the Triassic dolomites and limestones,
● water outflows from the Jurassic limestones into the Quaternary sands and also through erosion windows into the Triassic aquifer,
● water outflows from the Jurassic limestones into Carboniferous–Devonian carbonates and further on to the Triassic dolomites and limestones,
● water seepage from the Jurassic limestones into the Triassic carbonates through the Middle Jurassic marls.

The Carboniferous–Devonian (Paleozoic) aquifer built up of limestones and dolomites is situated deeper than all the aforementioned aquifers and, as a consequence, is the least studied aquifer. The aquifer is recharged by water infiltrating from Jurassic limestones at broad zones of hydraulic contacts in the eastern and southern parts of the Olkusz region.

GROUNDWATER QUALITY

The chemical composition of the Quaternary aquifer groundwater in the Olkusz Pb–Zn ore mine region is not well known. During the exploration stage for ore resources in the years 1957–1963, 67 groundwater samples were taken from wells and observation wells bored into the Quaternary sediments as well as from springs. Basic statistical parameters for the analysis are presented in Table 1.

Table 1. Basic statistical parameters of Quaternary aquifer water chemical composition (years 1957–1963).

Ion	Number of samples	Concentration (mg l^{-1})			Standard deviation (SD) (mg l^{-1})
		Min.	Max.	Med.	
Ca	55	7.9	277.2	86.7	56.5
Mg	65	0.9	129.0	15.3	23.2
Na	43	<0.5	118.0	14.3	21.5
K	42	<0.3	110.0	11.7	22.1
SO$_4$	62	8.0	508.9	69.8	82.5
Cl	62	0.19	284.0	23.1	42.6
NO$_3$	35	0.22	348.5	44.1	64.3
Fe	56	<0.01	2.0	0.15	0.31
Zn	46	<0.005	8.0	0.40	1.27

Table 2. Basic statistics of Upper Jurassic aquifer groundwater chemical composition (years 1996–1998).

Ion	Number of samples	Concentration (mg l^{-1})			Standard deviation (SD) (mg l^{-1})
		Min.	Max.	Med.	
Ca	33	69.9	226.3	116.3	36.78
Mg	33	1.26	20.1	5.66	4.97
Na	33	0.66	52.9	7.80	10.93
K	33	<0.5	34.1	3.73	6.48
SO$_4$	33	12.2	123.5	60.4	25.85
Cl	33	3.18	79.4	27.8	16.33
NO$_3$	33	<0.1	175.7	43.2	36.77
Fe	33	<0.01	0.093	0.012	0.021
Zn	33	<0.002	3.48	0.2	0.75

The data presented in Table 1 proves that already 30–40 years ago the Quaternary aquifer groundwater was already locally, strongly contaminated. This is revealed by the wide range of individual ion concentrations present. The contaminated waters are characterized by having increased nitrate, chloride and potassium concentrations, which indicates that already dozens of years ago groundwater pollution, mainly from domestic sewage and probably from municipal landfill site effluents, was occurring. The presence of anthropogenic pollution also influenced the diversity of water hydrochemical types. Natural uncontaminated waters were of Ca-HCO$_3$ or Ca-Mg-HCO$_3$ type, whilst contaminated waters were of multi-ionic types, that is, Ca-Mg-HCO$_3$-SO$_4$-Cl.

The groundwater chemical composition of the Upper Jurassic aquifer in the Olkusz region was recognized on the basis of sampled springs, dug wells, drilled wells and observation wells. Between the year 1996 and 1998, when water quality control studies were carried out, 53 water samples were taken. Basic statistics for the collected samples' chemical compositions are presented in Table 2.

Under natural conditions the Jurassic aquifer groundwater was of good quality. Because of the host rocks' geochemical character the hydrochemical type was Ca-HCO$_3$

Table 3. Basic statistics for the groundwater chemical composition of the Triassic aquifer (years 1957–1963).

Ion	Number of samples	Concentration ($mg\,l^{-1}$)			Standard deviation (SD) ($mg\,l^{-1}$)
		Min.	Max.	Med.	
Ca	130	17.2	184.3	65.4	20.43
Mg	130	2.6	67.5	19.4	10.88
Na	66	0.4	242.3	11.4	30.06
K	65	0.2	15.2	2.71	2.54
SO_4	148	7.4	420.0	51.9	63.5
Cl	151	0.6	344.3	14.3	30.97
NO_3	58	0.02	132.8	9.29	22.17
Fe	148	0.005	13.0	0.49	1.43
Zn	105	0.002	22.0	0.3	2.14

with mineralization occurring in the range 400 to $600\,mg\,l^{-1}$ and having low alkalinity ranging from 7.0 to 8.6. The Upper Jurassic fissure-karstic type aquifer is extremely vulnerable to surface pollution infiltration. Many years of sewage disposal mismanagement at the urbanized Upper Jurassic outcrops is the reason for significant groundwater quality degradation. As a result of the infiltration of domestic sewage and liquid manure into the Upper Jurassic aquifer, an increase in mostly nitrate and chloride concentrations has occurred. As a result of the excessive nitrate concentrations found in groundwater, a few water supply wells exploiting the Upper Jurassic aquifer groundwater were decommissioned. Operating groundwater intakes occasionally determine nitrate concentrations exceeding $70\,mg\,l^{-1}$ in sampled groundwater. Polluted Jurassic aquifer groundwater modifies to the hydrochemical types: $Ca-HCO_3-SO_4$, $Ca-SO_4-HCO_3$ and $Ca-HCO_3-SO_4-Cl$.

Triassic carbonate aquifer groundwater is of very good quality under natural conditions. During exploration for Pb–Zn ore resources in the Olkusz region between the years 1957 and 1963, 150 Triassic aquifer groundwater samples were collected. Basic statistics for the collected samples chemical compositions are presented in Table 3.

Before intensive mining drainage commenced, Triassic carbonate aquifer groundwater was of type $Ca-Mg-HCO_3$. However groundwater analyses results from the years 1957–1963 indicate that the Triassic aquifer groundwater quality was locally influenced by varied pollution sources. A clear influence was due to historical dumping sites resulting from Pb–Zn ore exploitation and processing. Mining wastes were disposed under various morphological conditions. Weathering and precipitation moving through these sites were most probably the reasons for local significant increases in sulphate, magnesium and calcium ion concentrations. Contaminated waters are of $Ca-Mg-SO_4-HCO_3$ or $Mg-Ca-SO_4-HCO_3$ types with a simultaneous increase in heavy metal concentrations. Significant increases in nitrate concentrations were also determined at Triassic carbonate outcrops in the vicinity of urbanized areas.

Mining drainage transformed hydrogeological conditions on a regional scale. The natural groundwater flow directions changed and Pb–Zn mine workings became drainage centres. The natural Triassic aquifer water table level decreased from 80 to 150 m around mine workings. As a result of the transformation of natural hydrogeological conditions, varied pollution sources were activated, the most important being: dumps and tailing

Table 4. Basic statistics for the outflows chemical composition at 'Olkusz' and 'Pomorzany' mine workings (2001).

Ion	Number of samples	Concentration ($mg\,l^{-1}$)			Standard deviation (SD) ($mg\,l^{-1}$)
		Min.	Max.	Med.	
Ca	145	53.5	994	241.9	224.96
Mg	145	10.5	2,500	203.6	397.04
Na	145	1.27	77.5	10.17	10.17
K	145	0.24	24.2	2.79	2.61
SO_4	145	44.6	12,000	1,187.2	2,038.04
Cl	145	5.16	96.99	18.84	12.42
NO_3	145	0.09	32.0	3.27	5.41
Fe	145	<0.005	24.0	1.52	3.05
Zn	145	0.0024	30.3	3.00	4.88

ponds, municipal landfill sites, liquid wastes from a paper factory, domestic sewage and agricultural contamination. As a result of the weathering of sulphide minerals, Triassic aquifer groundwater enriches in sulphates, calcium, magnesium and heavy metals (Table 4).

Depending on the proportion of natural groundwater similar to hydrochemical background, and the proportion of water with dissolved sulphide weathering products, various Triassic aquifer groundwater hydrochemical types are formed. The following hydrochemical types occur: $Ca-HCO_3$ or $Ca-Mg-HCO_3$, mixed $Ca-Mg-HCO_3-SO_4$ and $Mg-Ca-SO_4-HCO_3$, $SO_4-Mg-Ca$, sporadically SO_4-Mg. Locally, where municipal waste disposal leachates penetrate into the groundwater, chloride, sodium and potassium ions represent a large group of constituents.

Spatial distribution of nitrates in mine working water inflows

Spatial distribution of nitrates in Triassic aquifer groundwater was researched on the basis of water samples taken mainly from Pb–Zn mine gallery water outflows. Additionally sampled water originated from observation wells installed in Triassic carbonates. Nitrate concentrations were determined with a capillary electrophoresis system. On the basis of nitrate concentration studies a spatial distribution map was created (Figure 3).

Nitrates are introduced into the Triassic aquifer in various ways (Figure 4). In the south-eastern part of the Olkusz region, groundwater contaminated with nitrates originating from the Olkusz agglomeration domestic sewage infiltrates Triassic bedrock built by dolomites and limestones and filtrates further towards the 'Olkusz' mine galleries. Mine outflows posses characteristic nitrate concentrations ranging from about $0.5\,mg\,l^{-1}$ up to almost $20\,mg\,l^{-1}$ (Figure 3). The highest determined nitrate concentrations occur in the eastern part of the 'Olkusz' mine, that is from the groundwater flow contaminated by domestic sewage from the Olkusz agglomeration. Domestic sewage originating from highly urbanized areas migrating through fissured Triassic carbonates increased nitrate concentrations in water outflows in the south-eastern part of 'Pomorzany' mine, as well as the southern part of the 'Bolesław' mine (Figure 3).

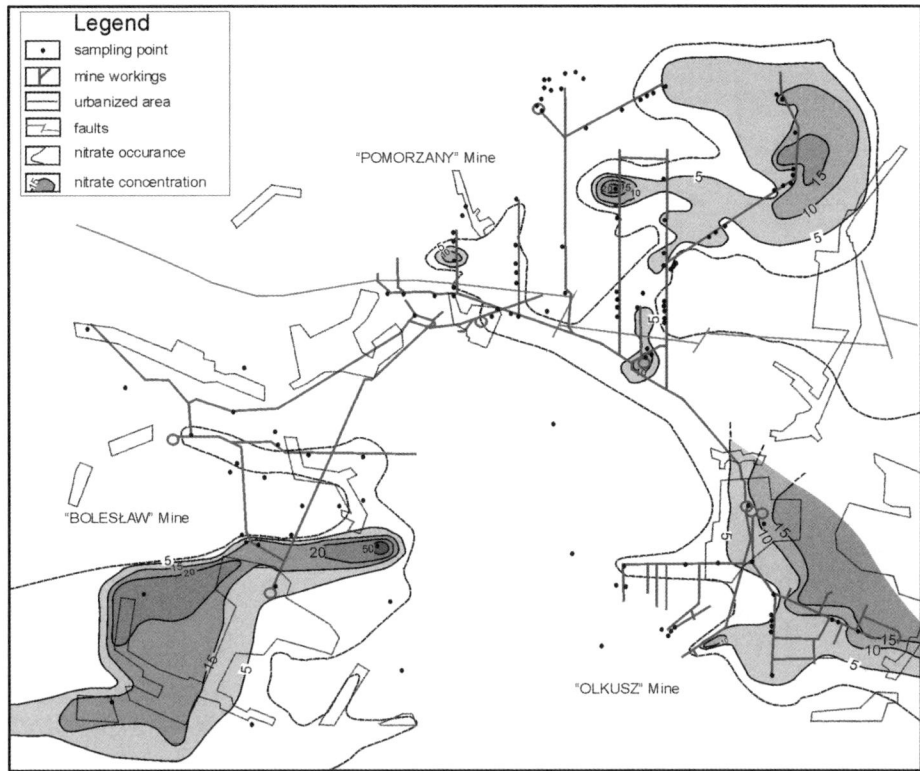

Figure 3. Contour map of nitrate concentrations [mg l^{-1}] within Triassic aquifer of the Olkusz.

In the eastern part of the Olkusz region the Triassic aquifer is hydraulically connected with Jurassic limestones. Various contacts between the aquifers can be differentiated into direct and indirect contacts (Wilk and Motyka 1977; Haładus, Motyka, Szczepański *et al.* 1978; Figure 4). Within hydraulic connectivity zones contaminated water from the Upper Jurassic aquifer filters into Triassic carbonates. Indirectly contaminated groundwater is drained by 'Pomorzany' mine workings. Individual water outflows in the eastern part of the mine are characterized by having nitrate concentrations exceeding 20 mg l^{-1} (Figure 3).

East of the 'Bolesław' mine, gossans were excavated in historical times. After the deposit was exploited, the southern sector of the excavation was filled with municipal wastes with no protection against effluent infiltration to Triassic carbonate bedrock being put in place. The main chemical components of the effluents are: chlorides, bicarbonates, sulphates, sodium, potassium, magnesium and ammonium. The ammonium ions exceed concentrations of 500 mg l^{-1}. The ammonium form of nitrogen is oxidized while infiltrating through an aeration zone, exchanging into stable nitrate (NO_3) under these conditions. In close proximity to the described municipal landfill site nitrate concentration exceeded 110 mg l^{-1}. Contaminated groundwater flow is heading towards the 'Bolesław' mine workings. Nitrate concentrations observed in the outflows located in the southern part of this mine exceed 15 mg l^{-1} (Figure 3).

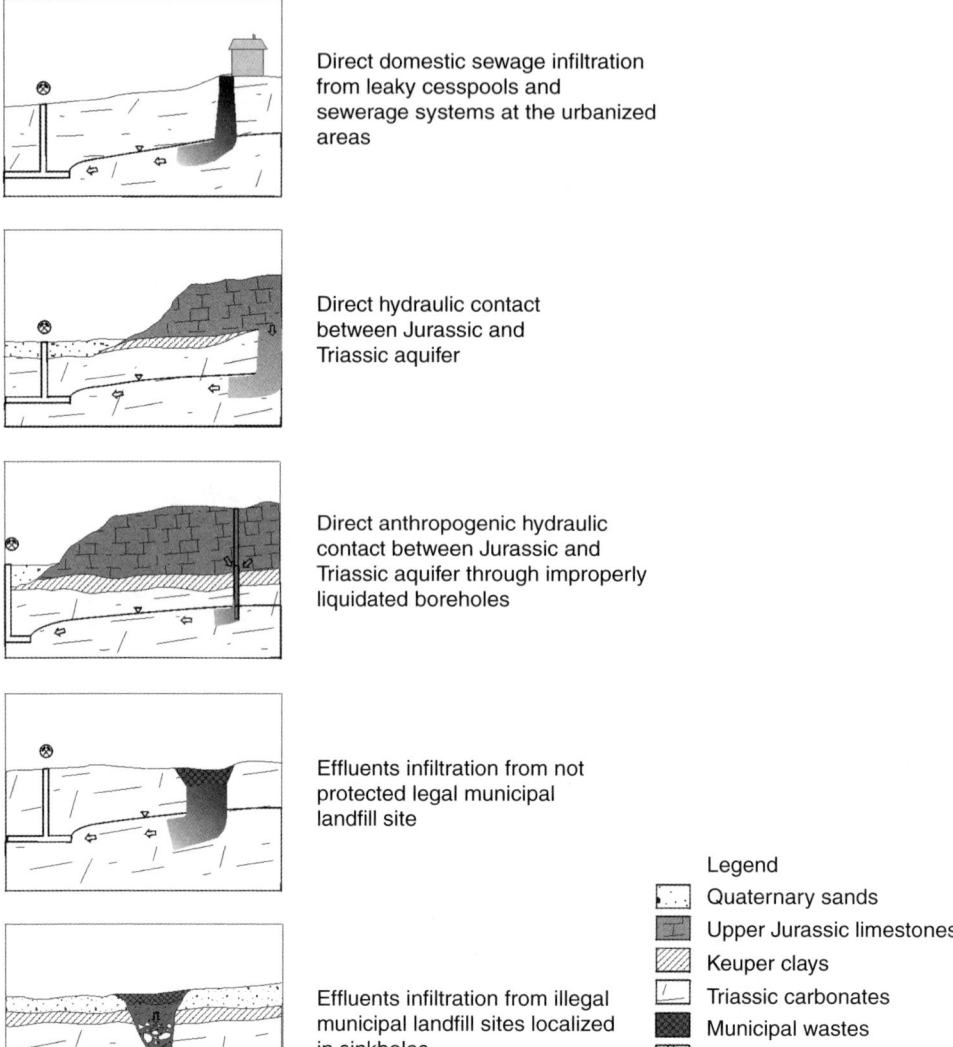

Direct domestic sewage infiltration from leaky cesspools and sewerage systems at the urbanized areas

Direct hydraulic contact between Jurassic and Triassic aquifer

Direct anthropogenic hydraulic contact between Jurassic and Triassic aquifer through improperly liquidated boreholes

Effluents infiltration from not protected legal municipal landfill site

Legend

	Quaternary sands
	Upper Jurassic limestones
	Keuper clays
	Triassic carbonates
	Municipal wastes
	Mine galleries

Effluents infiltration from illegal municipal landfill sites localized in sinkholes

Figure 4. Origin of increased nitrate concentrations in Triassic aquifer groundwater.

Locally, in individual outflows originating from the western part of 'Pomorzany' mine, nitrate concentrations are relatively high, that is, $10-20\,\mathrm{mg\,l^{-1}}$. The range of their distribution indicates a local pollution source of small size. It is very likely that nitrogen compounds penetrate Triassic carbonates as effluents from a variety of wastes, including municipal wastes illegally deposited in sink holes which formed above 'Pomorzany' mine workings more than 20 years ago. Despite levelling with filling material and recultivation, continued subsidence creates secondary morphological depressions, which are used for various wastes disposal, including municipal wastes.

CONCLUSIONS

At the initial phase of Pb–Zn deposits exploitation by 'Olkusz' and 'Pomorzany' mines, abstracted groundwater contained small quantities of nitrates, in the order of a few mg l^{-1}. Mine drainage, dewatering and the resulting drawdown cone altered natural groundwater flow directions. The main drainage centres have become mine workings. Various pollution sources contaminated drained and abstracted water with, among other things, nitrates. The present determined nitrate concentrations in the Olkusz ore district mines individually exceed 20–30 mg l^{-1}.

Groundwater polluted with nitrates affects mine water in various ways. Outflows from the eastern part of the 'Olkusz' mine, as well as from the western part of 'Bolesław' mine are contaminated with domestic sewage from urbanized areas located on Triassic carbonate outcrops. Increased nitrate concentrations in the eastern part of the 'Pomorzany' mine locally exceed 20 mg l^{-1}, is a result of recharge by Jurassic aquifer water with locally determined nitrate concentrations between 70–100 mg l^{-1}. Jurassic aquifer groundwater migrates to Triassic carbonates through the zones of direct and indirect hydraulic connections.

Nitrogen compounds present in the Olkusz ore district mine water originate from legal and illegal municipal landfill site effluents. Effluents from dumpsites which are legal but are constructed with no linings increase nitrate concentrations in outflows located in the south-eastern part of the 'Bolesław' mine. Increased nitrate concentrations locally exceeding 30 mg l^{-1}, observed in the western part of 'Pomorzany' mine, result from infiltrating municipal waste effluents. Wastes are illegally disposed of by the local community in sinkholes and depressions located over mine workings.

REFERENCES

Adamczyk, Z. and Motyka, J. 2000: Water inflow to lead and zinc mines in the Olkusz area (SW Poland). *Przegląd Geologiczny 48* (2): 171–175 (in Polish).

Adamczyk, Z., Motyka, J. and Witkowski, A. 2000: *Impact of Zn–Pb ore mining on groundwater quality in the Olkusz region*. Proceedings of the 7th IMWA Congress, Katowice-Ustroń, Poland, pp. 27–37.

Haładus, A., Motyka, J., Szczepański, A. and Wilk, Z. 1978: Forecasting of groundwater inflow into mines in fissured-karst horizons using the electric analogue simulation method. *Rocznik Polskiego Towarzystwa Geologicznego 48* (3/4): 559–587 (in Polish).

Wilk, Z. and Motyka, J. 1977: Contacts between water-bearing horizons (Olkusz ore mining region, near Cracow). *Rocznik Polskiego Towarzystwa Geologicznego 47* (1): 115–143 (in Polish).

CHAPTER 4

Nitrate in groundwater as an indicator of farmstead impacts on the environment

B. Sapek and A. Sapek
Institute for Land Reclamation and Grassland Farming at Falenty, 05-090 Raszyn, Poland

ABSTRACT: Nitrate concentrations in groundwater as an indicator of the environmental impact of all farm activities, particularly those in the farmstead and its vicinity, were studied using the results of several years' groundwater monitoring in the demonstration farms situated in different sites in Poland. All farmstead operations, mainly those connected with animal production, were shown to be an important source of nitrate and ammonium ions in ground and neighbouring surface waters. For that reason the nitrate, and even more, the dynamic system of ions — ammonium and nitrate in groundwater — is proposed as an indicator of the impact of the farmstead on the environment.

INTRODUCTION

The farmstead with its adjacent area is the integral part of each farm. There is a constant flux of matter between the farmstead and the rest of a farm (Sapek, B. 2002). It is known that agricultural activity under specific conditions could have a negative influence on the environment (Dewes 1995; Sapek, A. 1997; Schultz 1991; Schultz and Hradetzky 1999). Therefore, it is necessary to recognize the effect of the farmstead and its vicinity on the environment. Groundwater in agricultural areas is often polluted with nutrients, especially with nitrogen. Depending on redox conditions, the pollutant may be ammonium or nitrate nitrogen. Low nitrate concentration often indicates groundwater pollution with ammonium nitrogen.

The aim of this study was to estimate whether the nitrate concentration in groundwater could be an indicator of anthropomorphic processes associated with all farm activities, particularly of the impact of the farmstead and its vicinity on the environment.

MATERIALS AND METHODS

The effect of the farmstead and its vicinity on nitrate concentration in groundwater was estimated from the results of several years' groundwater monitoring in demonstration farms within the projects realized in cooperation with US Environmental Protection Agency ('Polish Agriculture and the Water Quality Protection' — PAWQP) in 1992–1996 and Swedish Institute for Environmental Protection ('Baltic Agriculture Runoff Program' — BAAP Project) in 1995–1997 and continued till now. Demonstration farms were situated in different sites in Poland — in Ostroleka region, and in Kujawsko-Pomorskie, Podlaskie and Mazowieckie provinces. The farms in Ostroleka region were situated in three small agricultural catchments and in the main represented local grassland farming and dairy production. Light sandy and shallow mineral–organic and organic soils prevailed there. In Kujawsko-Pomorskie province, all the farms were in one small catchment. Agriculture on rich soils, dominated by animal, mostly swine, production is the main activity of this region. Mixed crops and animal production (dairy and swine) characterized the demonstration farms in Mazowieckie province. In Podlaskie province, dairy production at demonstration farms on local mineral and organic soils predominated.

Water samples were taken from: i) control wells installed at selected monitoring sites, ii) operational farm wells and tap water, iii) non-operational farm wells, and iv) neighbouring ditches or streams. Samples were collected at monthly intervals. Concentration of $N-NO_3$ and $N-NH_4$ in water was determined calorimetrically using a TECHNICON auto analyser. Mineral forms of nitrogen in soil samples were determined in 1% K_2SO_4 extract using the same method.

RESULTS AND DISCUSSION

Variability of nitrate and ammonium concentrations in groundwater may indicate the type and intensity of inputs from different sources, named 'hot spots', of agricultural non-point pollutants (Sapek, B. 2002). Periodic quality control of water in demonstration farms during the PAWQP implementation was undertaken as 'educational monitoring' addressed to farmers (Sapek, B. 1997).

The results of ground and surface water quality control obtained during the PAWQP in 1993–1996 showed the highest water pollution with nitrate in places associated with animal production — in manure storage and in barns (Table 1). Mean values of nitrate

Table 1. Mean values of nitrate and ammonium nitrogen ($N-NO_3$, $N-NH_4$) concentration in groundwater from different places in demonstration farms of PAWQ Project in Ostroêka region in the 1993–1996.

Kind of solute	Values	Solute concentration ($mg\,l^{-1}$) in monitoring points									
		Meadows	Pastures	Clover reseeding	Grassland renovation	Arable land	By the manure storage	By the barn	Farm wells	Dithes	Outlet from water-shet
$N-NO_3$	Mean	2.9	5.0	3.8	5.0	4.5	25.1	18.4	10.6	3.9	5.2
	SD	5.0	7.6	10.1	7.2	4.3	37.7	21.0	15.8	7.2	11.5
	Max.	67.7	64.7	150.0	41.9	15.6	312	120.0	128.0	64.7	100.0
	n	541	462	316	61	37	342	212	282	272	151
$N-NH_4$	Mean	0.6	0.4	0.7	1.5	0.1	8.1	0.4	0.5	0.2	0.3
	SD	2.2	0.9	3.9	5.2	0.1	25.9	1.3	1.4	0.7	1.8
	Max.	33.3	9.8	65.9	39.9	0.81	250.0	14.8	12.6	9.8	21.0
	n	541	462	316	61	37	342	212	282	272	151

N – number of samples.
Standards for drinking water and for management using water: $50\,mg\,l^{-1}$ NO_3^- (11,3 $9\,mg\,l^{-1}$ $N-NO_3$); $1.5\,mg\,l^{-1}$ NH_4 ($1.16\,mg\,l^{-1}$ $N-NH_4$).
Polish standards of surface waters pollution for Ist classe of cleanlines: 5 and below $mg\,l^{-1}$ $N-NO_3$; 1 and below $mg\,l^{-1}$ $N-NH_4$.

concentration in water, expressed as a N-NO$_3$, were about twice as high as the standard (11.3 mg l^{-1} N-NO$_3$) (Rozporzadzenie 2000). It should be noted that maximum concentrations in these places were in the range of 100–300 mg l^{-1} N-NO$_3$. In spite of lower nitrate concentrations in other monitored sites, especially in clover reseeding, meadows and pastures, the maximum values exceeded the standard. Mean nitrate concentration in water from used farm wells was close to the standard, but maximum concentrations reached 120 mg l^{-1} N-NO$_3$ and this should be considered dangerous. According to Polish standards of surface water quality (Rozporzadzenie 1991), the mean nitrate values in ditches and in the outlets from catchments were within the limits for the first class water quality (Table 1).

Ammonium content in water is a potential source of nitrate. Under favourable redox conditions NH$_4^+$ (III) oxidizes to nitrate – NO$_3^-$ (V). For that reason, ground and surface water pollution with nitrate depends not only directly on its concentration, but also much more on the dynamic system of both ions. Measured ammonium concentrations in water (1.16 mg l^{-1} N-NH$_4$, on average) exceeded the standard value especially in such places as the manure storage or animal housing. Maximum concentrations of N-NH$_4$, which were found in groundwater under meadows, pastures, farm wells etc. (12.6–65.9 mg l^{-1}), confirm the possible risk of pollution (Table 1).

The results of ground and surface water quality monitoring performed in the BAAP project in 1999–2001 confirmed previous observations (Tables 1–3). Some differences between the results from similar 'hot spots' in the farmstead of demonstration farms in four Polish provinces reflected the effect of specific character and management levels of agricultural production on groundwater pollution with nitrate (Sapek, B. 2002).

Among the demonstration farms from three provinces, the highest and similar mean values of nitrate concentration in groundwater originating from manure storage were found in farmsteads from Mazowieckie and Kujawsko-Pomorskie provinces (14.3 and 14.1 mg l^{-1} N-NO$_3$), where mixed crops and animal (mainly swine) production dominated (Tables 2 and 3). It is noteworthy that the high ammonium concentrations in groundwater were recorded in monitoring sites similar to those already analysed in demonstration farms of Podlaskie province, where dairy production on local mineral and organic soils predominated. Observed groundwater pollution from places of animal waste storage is confirmed by the results of Dewes (1995) who studied nitrate and ammonium concentrations in manure heap leachates. The highest leachate concentration that he found was 1,139 mg l^{-1} N with 66% as N-NH$_4$ and only 4% as N-NO$_3$. In similar investigations, Schultz (1991) found the N-NH$_4$ concentration in groundwater between 0.023 and 300 mg l^{-1} but only 0.003 mg l^{-1} of N-NO$_3$. The maximum concentration of ions in groundwater under the field where the manure stored was 51.5 in the case of N-NH$_4$ and 221 mg l^{-1} in the case of N-NO$_3$ (Schultz and Hradetzky 1999).

Other monitored sites, described in Tables 1 and 2 as 'other places in the farmstead', were situated on down gradient from the farmstead, in the vicinity of a stream or a ditch or at the animal housing etc. They represented mostly groundwater polluted with ammonium (mean values between 4.7 mg l^{-1} N-NH$_4$ in Mazowieckie and 8.9 mg l^{-1} N-NH$_4$ in Podlaskie province) (Tables 2 and 3). As underlined earlier, this high ammonium concentration is a potential source of groundwater pollution with nitrate. Groundwater under the meadow at the demonstration farm in Podlaskie province was extremely polluted with nitrate (mean value 22.4 mg l^{-1} N-NO$_3$). It can be assumed that these high nitrate concentrations were a consequence of mineralization processes that were going on in the

Table 2. Mean values of nitrate and ammonium nitrogen (N-NO$_3$, N-NH$_4$) concentration in groundwater from different places in demonstration farms of BAAP Project in Mazowieckie and Podlaskie province in the 1999–2001.

Province	Kind of solute	Values	Solute concentration (mg l^{-1}) in monitoring points								
			Tap water	Farm wells (not used)	By manure storage	Others places on farmstead	Meadows	Pastures	Dithes before farm	Dithes after farm	Outlet from watershet
Mazowieckie	N-NO$_3$	Mean	0.62	30.1	14.3	6.8	9.0	3.5	3.9	3.4	–
		SD	1.41	27.7	15.4	14.9	20.5	2.7	7.6	2.6	–
		Max.	9.2	88.8	46.7	67.7	67.0	8.7	47.5	9.9	–
		n	40	39	27	35	10	10	38	36	–
	N-NH$_4$	Mean	0.14	0.48	7.8	4.7	0.5	0.21	2.8	0.9	–
		SD	0.13	0.71	16.5	8.3	1.3	0.18	8.3	1.9	–
		Max.	0.5	3.4	83.7	41.0	4.1	0.6	50.0	8.1	–
		n	40	39	27	35	10	10	38	36	–
Podlaskie	N-NO$_3$	Mean	1.3	10.3	1.2	1.7	22.4	1.1	4.2	3.3	4.2
		SD	3.8	6.4	2.4	2.6	68.6	1.7	4.9	3.4	5.1
		Max.	15.4	27.0	12.7	14.0	240.0	4.5	24.4	13.5	18.0
		n	44	44	30	47	12	6	29	22	14
	N-NH$_4$	Mean	0.64	0.35	12.3	8.9	0.6	1.2	1.9	0.69	0.41
		SD	0.33	0.57	19.5	16.9	1.6	1.1	2.3	0.89	0.50
		Max.	1.3	3.3	84.0	77.0	5.7	2.4	7.8	3.6	1.5
		n	44	44	30	47	12	6	29	22	14

Description as in Table 1.

Table 3. Mean values of nitrate and ammonium nitrogen (N-NO$_3$, N-NH$_4$) concentration in ground and surface water from different places in demonstration farms of BAAP Project in Kujawsko-Pomorskie province in 1997–2001.

Kind of solute	Values	Solute concentration (mg l^{-1}) in monitoring points				
		Tap water	Farm wells (not used)	By the manure storage	Others places at farmstead	Ditches
N-NO$_3$	Mean	0.55	25.4	14.1	21.2	5.7
	SD	0.85	15.4	22.1	31.8	6.9
	Max.	6.8	66.4	93.0	109.0	25.6
	n	96	97	59	38	109
N-NH$_4$	Mean	0.14	0.23	1.0	5.5	0.30
	SD	0.30	0.74	1.7	8.9	0.43
	Max.	1.9	7.3	8.2	42.8	2.7
	n	96	97	59	38	109

Description as in Table 1.

organic soils that dominate there. A similar phenomenon was considered by Nadany and Chrzanowski (this volume) on the Kuwasy object, representing dewatered peatlands that are managed mostly as permanent meadows and pastures.

The demonstration farms of three provinces under study now have a piped water supply and farm wells are no longer used for drinking purposes. The mean nitrate concentrations in water from these wells exceeded the allowable standard value (Tables 2 and 3). Particularly polluted with nitrate were the farm wells from demonstration farms in Mazowieckie and Kujawsko-Pomorskie provinces. In this region the animal production, especially swine, is more intensive than in Podlaskie province. The negative influence of factors such as the type and intensity of agricultural production and the manner of animal waste storage on the water quality from farm wells was discussed by Ostrowska, Plodzik, Sapek *et al.* (1997). Nitrate and ammonium concentration in piped water from demonstration farms under study were consistent with standard values.

Generally, surface waters in ditches and streams flowing in the vicinity of farmsteads represented a first class water quality with reference to nitrate and ammonium concentration. However, having in mind the dynamics of the $NH_4^+ \Longleftrightarrow NO_3^-$ system, the maximum concentrations of analysed ions may still pose the risk of water pollution with nitrate (Tables 2 and 3). This phenomenon could be explained in the light of monitoring groundwater from the control wells (GT5) installed by the ditch, on the course of water flow from a farmstead (Figure 1). Nitrate concentration in 1997 was in the range of 20–50 mg l^{-1} N-NO$_3$, but ammonium concentration was very high – about 20 mg l^{-1} N-NH$_4$. Annual precipitation in this year was 661.8 mm. In 1999 precipitation was only 579.6 mm and nitrate concentration in groundwater from this control well increased to about 100 mg l^{-1} N-NO$_3$ but ammonium concentration under these conditions was very low.

The sources of nitrate leaching to groundwater and later to surface water are the soils which are excessively enriched in mineral nitrogen (N-min = N-NH$_4$ + N-NO$_3$) in places named 'hot spots' here. In the same places as the control well GT5, the N-min content, expressed in kg ha^{-1}, in the soil profile to the depth of 200 cm was about 580 kg ha^{-1} in

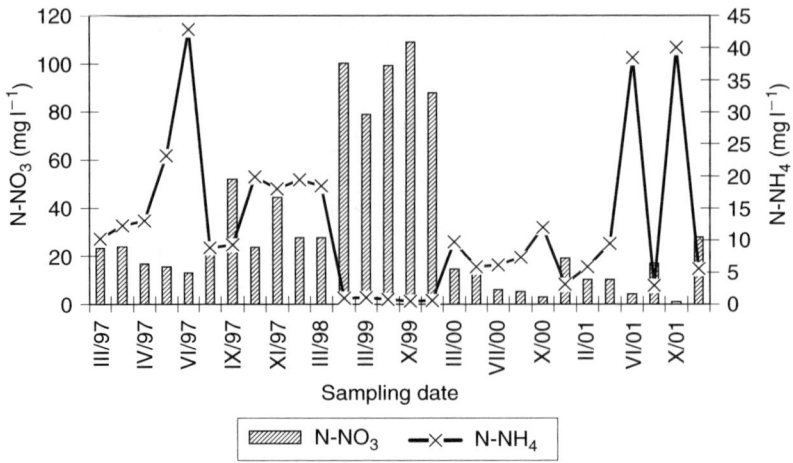

Figure 1. Nitrate (N-NO$_3$) and ammonium (N-NH$_4$) concentrations in groundwater in the place by the ditch in the vicinity of farmstead (GT5).

1997 (Figure 2). This year nitrates were the main N form to about 120 cm depth and dominated below the ammonium form. This would indicate anoxic conditions in deeper parts of the profile. In 1999, when precipitation was remarkably lower in comparison to 1997, the N-min content increased to about 880 kg ha^{-1}, and nitrates dominated in an upper 80 cm surface layer, acting as a potential source of pollution for the ditch. The study by Dewes and Schmit (1994) showed that N-NH$_4$ accumulation in the soil below a long-term manure store differed from the adjacent field only in 0–15 cm soil layer. Higher N-NO$_3$ content appeared in the deeper soil layer.

Results presented in Tables 1–3 on nitrate and ammonium concentration in water from different potential 'hot spots' at the farmstead demonstrate that places associated with animal production, situated near the manure pad, slurry tank, barn etc. are potentially at risk of soil and groundwater pollution. Proper management of animal waste, particularly the construction of roomy and compact manure pads, slurry and urine tanks as well as all farmstead operation that could limit nitrogen losses, would effectively decrease this risk (Sapek 2002). Changes of nitrate concentration in groundwater from the control well (GT4) after constructing a new manure pad in one of the demonstration farms could be an example of such environmental friendly activities (Figure 3). This concentration, ranging in 1997 between 60 and 90 mg l^{-1} N-NO$_3$, decreased to below 10 mg l^{-1} N-NO$_3$ in 2001. High N-NH$_4$ concentrations that appeared between April 2000 and October 2001 deserve special consideration. This phenomenon could result from the 'memory' of soil saturated with mineral nitrogen in this place.

Some additional information on nitrate and ammonium behaviour in ground and certain surface waters shows significant correlation between concentrations of soluble compounds (Table 4). One of more interesting is the correlation between nitrate and phosphorus in groundwater in the vicinity of the manure store, which suggests an 'animal' source of pollution. The lack of significant correlations for 'others places at farmstead' probably resulted from a high differentiation of pollution sources. Noteworthy are the correlations between nitrate, ammonium and other solutes in the groundwater obtained for meadows and pastures. Significant positive correlation coefficient between N-NO$_3$ and

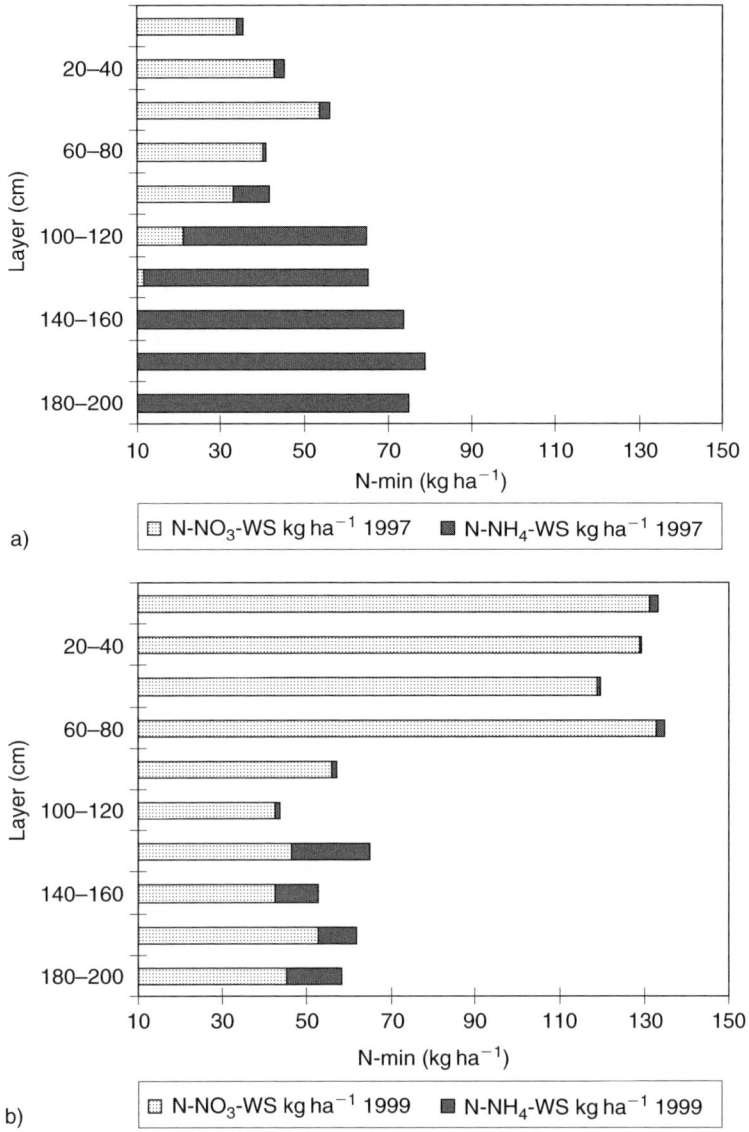

Figure 2. Mineral nitrogen content (N-min = N-NH$_4$ + N-NO$_3$) in the soil profile situated by the ditch in the vicinity of farmstead: a) in 1997, b) in 1999 (GT5).

N-NH$_4$ concentration in water under meadows reflects the nitrogen fertilization effect on the one hand, and the influence of mineralization processes on groundwater quality on the other. Correlation between N-NH$_4$ and K, Na, Cl and Mg concentrations in water confirms the effect of fertilization on water quality. On pastures, correlation between N-NH$_4$ and K concentration is evident that urine, especially on pastures with high animal density, is the source of groundwater pollution. As was shown by Dewes (1997), the potassium leaching from animal waste stores is an important threat to groundwater quality.

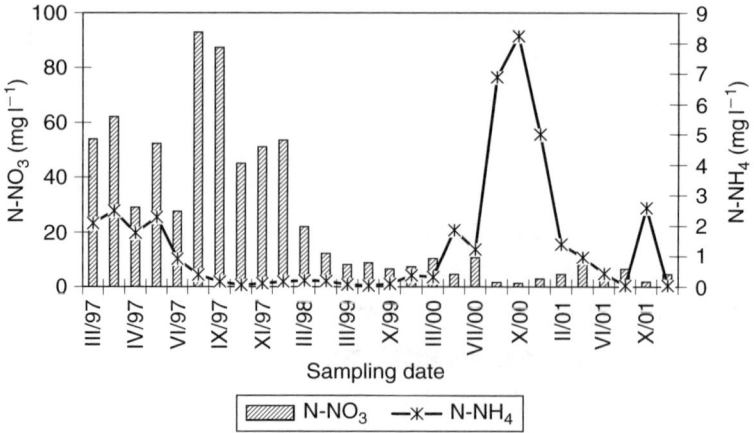

Figure 3. Nitrate (N-NO$_3$) and ammonium (N-NH$_4$) concentration in groundwater after construction of new manure pad in 1997 (GT4).

Table 4. Significant correlation coefficients between the solute concentration in water from the farmstead and its vicinity.

No.	Monitoring point	n	Elements						
			N-NH$_4$	P	Cl	Na	K	Mg	Ca
n.s.	Tap water	209	N-NO$_3$ n.s.	n.s.	n.s.	n.s.	n.s.	n.s.	n.s.
			N-NH$_4$ –	n.s.	n.s.	0.46**	n.s.	n.s.	n.s.
2.	Farm wells not used	209	N-NO$_3$		0.64**	0.62**		0.27*	0.49**
			N-NH$_4$ –	n.s.	n.s.	n.s.	n.s.	n.s.	n.s.
3.	By the manure storage	135	N-NO$_3$ n.s.	0.42**	–	0.30**	n.s.	n.s.	−0.31**
			N-NH$_4$ –	n.s.	n.s.	n.s.	0.28**	0.28**	n.s.
4.	Other places at farmstead	149	N-NO$_3$ n.s.	n.s.	n.s.	n.s.	n.s.	0.22*	n.s.
			N-NH$_4$ –	n.s.	n.s.	n.s.	n.s.	n.s.	n.s.
5.	Silage silos	31	N-NO$_3$ n.s.	n.s.	n.s.	0.38*	n.s.	n.s.	0.36*
			N-NH$_4$ –		n.s.	n.s.	n.s.	n.s.	n.s.
6.	Meadows	28	N-NO$_3$ 0.76**	n.s.	n.s.	0.91**	n.s.	n.s.	n.s.
			N-NH$_4$	n.s.	0.80**	0.93**	0.61**	0.35*	
7.	Pastures	19	N-NO$_3$ n.s.	n.s.	0.67**	0.62**	0.60**	n.s.	n.s.
			N-NH$_4$ –	n.s.	n.s.	n.s.	0.51*	n.s.	n.s.
8.	Ditch (before farm)	80	N-NO$_3$ n.s.	n.s.	n.s.	n.s.	n.s.	n.s.	0.34*
			N-NH$_4$ –	n.s.	0.23*	0.38**	0.25*	n.s.	n.s.
9.	Ditch (behind farm)	67	N-NO$_3$ −0.25*	n.s.	n.s.	n.s.	n.s.	n.s.	n.s.
			N-NH$_4$ –	n.s.	0.68**	0.65**	0.73**	0.64**	n.s.
10.	Ditch[1]	109	N-NO$_3$ −0.24*	n.s.	0.23*	−0.32**	n.s.	n.s.	0.38**
			N-NH$_4$ –	n.s.	0.22*	n.s.	0.22*	n.s.	n.s.
9.	Catchment	14	N-NO$_3$ n.s.	n.s.	n.s.	n.s.	n.s.	n.s.	n.s.
			N-NH$_4$ –	n.s.	n.s.	n.s.	n.s.	n.s.	n.s.

* Significance at $\alpha = 0.05$, ** significance at $\alpha = 0.01$; n.s. – correlation coefficient not significant.
[1] values concerning only Kujawsko-Pomorskie.

The relation of the concentration between the two ions nitrate and ammonium in water taken from the ditch depended on redox conditions at the particular time. The negative correlation between these ions confirmed this phenomenon (Table 4).

CONCLUSIONS

From the results and discussion presented in this paper, the effect of farmstead operations on the level of nitrate and ammonium concentrations in groundwater seems unquestionable. All types of activity associated with animal production such as animal waste storage, pasture and fodder production (green fodder and silage) pose a threat of water pollution with mineral nitrogen and in particular with nitrates. All activities in the farmstead and its vicinity are anthropomorphic processes that have, as we have shown, important impacts on the concentration of the two ions ammonium and nitrate in ground and surface waters. For that reason nitrate, and even more, the dynamic system of ammonium and nitrate ions in groundwater, is proposed as an indicator of farmstead impact on the environment.

REFERENCES

Dewes, T. 1995: Nitrogen losses from manure heaps. Nitrogen Leaching in Ecological Agriculture, 309–317.

Dewes, T. 1997: Zusammensetzung und Eigenschaften von Sickerwasser aus Stallmiststapeln. *Zeitschrift fuer Pflnzenernährung. Bodenk. 160*: 97–101.

Dewes, T. and Schmit, L. 1994: Deposition von Stickstoff und Kalium aus Stallmissststapeln in Bšden unter langjähring genutzen Mistalplätzen. *Agrobiol. Res. 47* (2): 115–123.

Ostrowska, E.B., Plodzik, M., Sapek, A., Wesolowski, P. and Smoron, S. 1997: Water quality in the farm wells (in Polish). Falenty, Instytut Melioracji i Użytków Zielonych. *Materiały Seminaryjne 39*: 161–168.

Rozporzadzenie Ministra Ochrony Środowiska, Zasobów Naturalnych i Leśnictwa z dnia 5.11.1991 r. Dz.U. nr 116 poz. 503.

Rozporzadzenie Ministra Zdrowia z dnia 4 września 2000 r. Dz U. nr 82 poz. 937.

Sapek, A. 1997: *The effects of agriculture on water quality: A Polish perspective.* Baltic Basin Agriculture and Environment Series. Ames: Centre for Agricultural and Rural Development. Report 97-BB, 6: 1–22.

Sapek, B. 1997: *Monitoring of mineral nitrogen in a farm.* In: *Sustainable Agriculture and Rural Area Development.* Activity of Working Group Reports and Conference Proceeding. Warsaw, 8–10 October 1996. IMUZ Falenty, pp. 122–129.

Sapek, B. 2002: *The impact of farmstead operation on groundwater quality.* In: J. Steenvorden, F. Claessen and J. Willems (eds): *Agricultural effect on ground and surface waters: Research at the edge of science and society.* Proceedings of a symposium held at Wageningen, October 2000. IAHS Publ. 273: pp. 125–130.

Schultz, L. 1991: Einfluss der Lagerung von Wirtschaftsdünger auf Gewässer. *Der Förderungsdienst/Beratungsservice 39* (3): 16–20.

Schultz, L. and Hradetzky, R. 1999: *Grundwasserschutz bei der Lagerung von Stallmist am Feld. Wasserwirtschatskataster*, Wien, Feb. 1999. 1:1-A-74.

Section 2

Experimental investigations of nitrates in groundwater

CHAPTER 5

Spatial distribution of nitrate in Cenozoic sedimentary aquifers controlled by a variable reactivity system

R. Eppinger
Ministry of Flanders, Environment, Nature, Land and Water Administration — Water Division, Brussels, Belgium

K. Walraevens
Laboratory for Applied Geology and Hydrogeology, Ghent University, Ghent, Belgium

ABSTRACT: Once nitrate-contaminated water has entered the aquifer below the root zone, microbiologically catalysed nitrate reduction in groundwater is mainly controlled by the reduction capacity of the sediment. Three main sedimentary energy sources (electron donors) to nitrate-reducing micro-organisms are known: organic matter, sulphides (pyrite) and Fe^{2+}-bearing minerals. Reactivity tests of Flemish sandy sediments of Cenozoic age obtained from different survey sites show that the presence of organic matter and (or) pyrite determines whether nitrate reduction occurs in groundwater. The total reduction capacity of the sediments contributes to determining the 'depth sensitivity' of the aquifers towards nitrate contamination, and allows a forecasting of potential pollution problems. Next to biological and chemical boundary conditions that must be present for running denitrification processes in nitrate-contaminated aquifers, it should be stressed that also the physical accessibility of nitrate molecules and nutrients to micro-organisms controls the occurrence of nitrate reduction. Not all reactive matter is available to denitrifiers. It has been observed that fine-grained layers with potentially high reduction capacity do not have a barrier function to nitrate contamination, but represent long-term pollution sources of low effective reactivity.

INTRODUCTION

Nitrate pollution of shallow aquifers is still one of the main problems that endanger drinking water supply in rural areas with a high impact of agricultural fertilization. Five test sites, spread over the whole of Flanders, each with 8 multi-level wells, have been installed in the scope of a project of the Flemish government (AMINAL) for detailed research about mechanisms of nitrate pollution and nitrate removal in Flemish sandy aquifers. The sites are located in different hydrogeologically homogeneous zones. These zones, mapped on a scale of 1:50,000, indicate areas of Flanders with comparable hydrodynamical and hydrogeochemical boundary conditions of the phreatic aquifers. Five of the most important zones have been chosen for the installation of the test sites.

   Next to water sampling of the filters, a cored drilling has been executed for each test site to obtain *in-situ* sediments for a detailed research of the sediment parameters, such as total reduction capacity (TRC) as a common parameter for all reduced substances in the aquifer, total organic carbon contents (TOC), mineralogy and grain-size distribution. Also, pore water samples have been taken to compare ion concentrations with sediment reactivity. The main objective is to understand the principles of nitrate reduction processes in Flemish aquifers for developing a nitrate specific model used for groundwater protection.

METHODS

Core sediments of each survey site have been stored in nitrogen gas-filled bags. Complete ionic balances have been made for all obtained pore water samples by executing chemical analyses according to Standard Methods (APHA 1998).

Determination of the total reduction capacity

The basic reactivity of the sediment can be determined by analysing the total reduction capacity (TRC), a term introduced by Pedersen, Bjerg and Christensen (1991). The TRC-value quantifies the total amount of reduced substances in the aquifer sediment, generally composed of organic carbon, reduced sulphur compounds (mainly sulphides) and Fe^{2+}-bearing minerals. The TRC-method is very sensitive and efficient, especially if there are only small amounts of reduced substances present in the sediments. Small quantities of organic matter for example are difficult to detect by conventional methods. In such a case the significant reactivity can be expressed by the TRC. However, the term 'total reduction capacity' should not lead to a wrong impression. Some of the organic compounds are low reactive and difficult to oxidize by chemical methods. The consequence is that not 100% of the organic matter has been detected by the TRC-method. The oxidation rate measured varies between 85% and 100%, depending on the composition of the organic substances. One should conclude that the low reactive organic nutrients that form the rest reactivity are not available to nitrate-reducing micro-organisms.

The total reduction capacity (TRC) of the sediment has been measured with a potassium-dichromate method modified after Mebius (1960) and Pedersen, Bjerg and Christensen (1991). Wet sediment samples are oxidized by boiling them in a reagent mixture of potassium dichromate and sulphuric acid–silversulphate in an open-reflux system. The oxygen consumption is proportional to the loss of potassium dichromate by reduction, measured by titration with ammonium-iron(II)-sulphate (APHA 1998). The oxygen consumption can be easily transferred in electron equivalent concentrations (meq) to be comparable with other measurements.

Determination of total organic carbon (TOC)

A TOC-analyser Solid-Sample-Module, SSM-5000A (Shimadzu) has been used for the determination of the concentrations of total organic carbon present in the analysed sediments. With this method it is not possible to distinguish between different organic compounds and their different reactivity, for example between lower reactive lignin or tannin and freshly deposited humic and fulvic acids. The measured value simply shows the common amount of organic carbon in the sediment that has been combusted. Van der Grift, Griffioen and Van Buizen (1999) concluded that not only the composition of the organic substance influences the effective reactivity but also the particle size of it. With increasing particle size, the reactivity of the sediment increases.

On the one hand, the inaccuracy of the TOC-method for values below $100\,meq\,kg^{-1}$ limits a precise measurement in low reactive sediments, while on the other hand the method allows an overview to be obtained about the relation between inorganic and organic substances in the sediment that could participate in nitrate reduction processes.

Determination of pyrite

Next to organic carbon, sulphides are the most important energy source to nitrate-reducing micro-organisms such as *Thiobacillus denitrificans*. Pyrite, in particular, is the most common sulphide mineral to be found in shallow marine sedimentary deposits and as post-sedimentary formed secondary mineral in organic rich layers. The concentrations of other sulphide minerals in shallow sedimentary deposits in Flanders are negligible in comparison to pyrite.

A chromium(II)-chloride method of Canfield, Raiswell, Westrich *et al.* (1986) has been modified for the analysis of the pyrite contents in the sediment and combined with a method of Landers, David and Mitchell (1983) to determine other sulphur compounds. According to Heron, Crouzet, Bourg *et al.* (1994) the combination of these two methods leads to most precise results for the determination of iron sulphides in sediments.

After drying the sediments under oxygen-free conditions, a homogenized sample is filled in a digestion–distillation apparatus to be boiled in a mixture of hypo phosphoric acid, hydroiodic acid and formic acid. During boiling, the apparatus is purged with nitrogen gas. All monosulfides, acid volatile sulphides (AVS), organic sulphur compounds and sulphates are reduced to H_2S, transported by the gas stream and trapped in a zinc acetate solution. Here zinc sulphide is formed that can be determined spectrophoto-metrically (670 nm) after adding ammonium iron(II) sulphate and p-aminodimethyl-anilinesulfate to the trapping solution (Landers, David and Mitchell 1983).

All sulphur compounds except pyrite have been removed from the sediment by the method of Landers, David and Mitchell (1983). After addition of ethanol, concentrated HCl and chromium(II) chloride the sediment residue in the digestion flask is boiled once again under nitrogen purging. This time H_2S is formed due to the dissolution of pyrite. Also zinc sulphides are formed here in a second flask with zinc acetate trapping solution and can be determined spectrophotometrically. In general the method of Canfield is very selective and dilutes only monosulphides and disulphides.

Analyses with the combined methods, that allow the separate determination of disul-phides, have shown that the amount of other sulphur compounds in the analysed sediments were negligible, so that for the total reactivity only the pyrite concentrations were relevant.

In Table 1 the variation of the charge equivalent concentrations analysed by the methods described here are shown for the oxidized and reduced layers of the aquifers of the five test sites.

Table 1. Variation of the reduction capacity in oxidation and reduction zones of the different survey sites.

Survey site	Redox conditions	TRC (meq kg^{-1})		TOC (meq kg^{-1})		Pyrite (meq kg^{-1})	
		Range	Average	Range	Average	Range	Average
Adegem	Oxidation zone	83–180	120	100–170	120	2–8	5
	Reduction zone	210–81.900	10.200	150–78.000	17.000	11–510	150
Kampenhout	Oxidation zone	91–690	330	63–530	250	0–110	37
	Reduction zone	1.040–1.110	1.070	400–440	420	770–820	800
Mol	Oxidation zone	49–86	60	91 120	110	0–2	0
	Reduction zone	640–1.580	1.020	500–2.370	1.350	5–100	33
Torhout	Oxidation zone	180–590	370	270–330	300	1–350	100
	Reduction zone	220–1.140	610	97–360	230	47–1.090	500
Peer	Oxidation zone	27–110	56	72–170	110	0–1	0
	Reduction zone	390–1.780	800	450–2.070	930	5–220	80

RESULTS

Pore water analyses show unacceptably high nitrate contamination of the shallow parts of the aquifers for four of the five chosen test sites (Figures 1–3). Low concentrations were only found at the test site Mol due to very shallow reactive organic deposits in the aquifer and a limited application of fertilizers on the neighbouring fields.

In correlation to a strong increase of the reduction capacity of the sediment nitrate is nearly completely removed from deeper parts of the aquifers. The nitrate reduction zones occur in different depths for different aquifers depending on the type and oxidation status of the sediment as well as on the nitrate input (see Figures 1–3).

Pyrite forms the main electron donor source to nitrate-reducing micro-organisms for the test sites of Torhout and Kampenhout (Figure 2). Nitrate is reduced by pyrite oxidation indicated also by a significant increase of the sulphate load.

$$2FeS_2 + 6NO_3^- + 2HCO_3^- \rightarrow 3N_2 + 2FeOOH + 4SO_4^{2-} + 2CO_2 \qquad (1)$$

The shallow aquifer systems of Torhout and Kampenhout are built by marine deposits of the Lower Eocene age. The genesis explains the presence of higher pyrite concentrations here. The existence of pyrite at the Torhout site is probably based on a secondary process, indicated by the peak concentrations of pyrite in the upper reduction zone (Figure 2a). Iron sources for the post-sedimentary forming of pyrite are the high glauconite contents in the Sands of Egem. Sulphides are delivered by the degradation of organic matter and the reduction of sulphates.

Organic matter is the main electron donor source to the micro-organisms for the site of Adegem. The high reduction capacity is completely controlled by the high organic contents of quaternary peat layers between 7 and 8 m (see semi-logarithmic scale, Figure 1a). Denitrification is also indicated by an increase of the bicarbonate contents (pH-dependency, see reaction 2). A paleo soil at the top of the Miocene Diestian sands of the site of Peer forms the basic nutrient to nitrate removal. Also an increase of the bicarbonate concentrations here has been observed (see Figure 3).

$$5C_{org} + 4NO_3^- + 2H_2O \rightarrow 2N_2 + 4HCO_3^- + CO_2 \qquad (2)$$

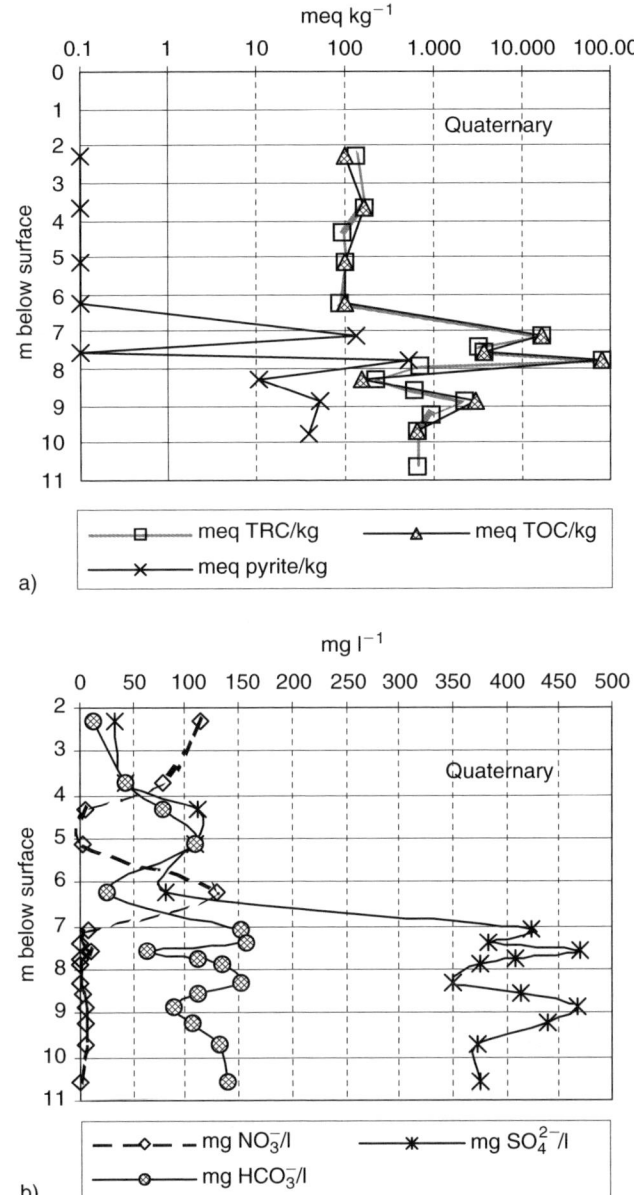

Figure 1. Test site Adegem: a) reduction capacity dominated by TOC and b) change of main anion concentrations in pore waters with increasing depth.

However, a clear change in sediment reactivity at the border between oxidation and reduction zone is observed for the sites of Adegem (Figure 1), Peer (Figure 3) and also Mol. Therefore still higher concentrations of organic matter and pyrite in the shallower part of the aquifers of Kampenhout and Torhout do not result in a lowering of the nitrate content (Figure 2). Here a kind of transition zone exists where only the part of the

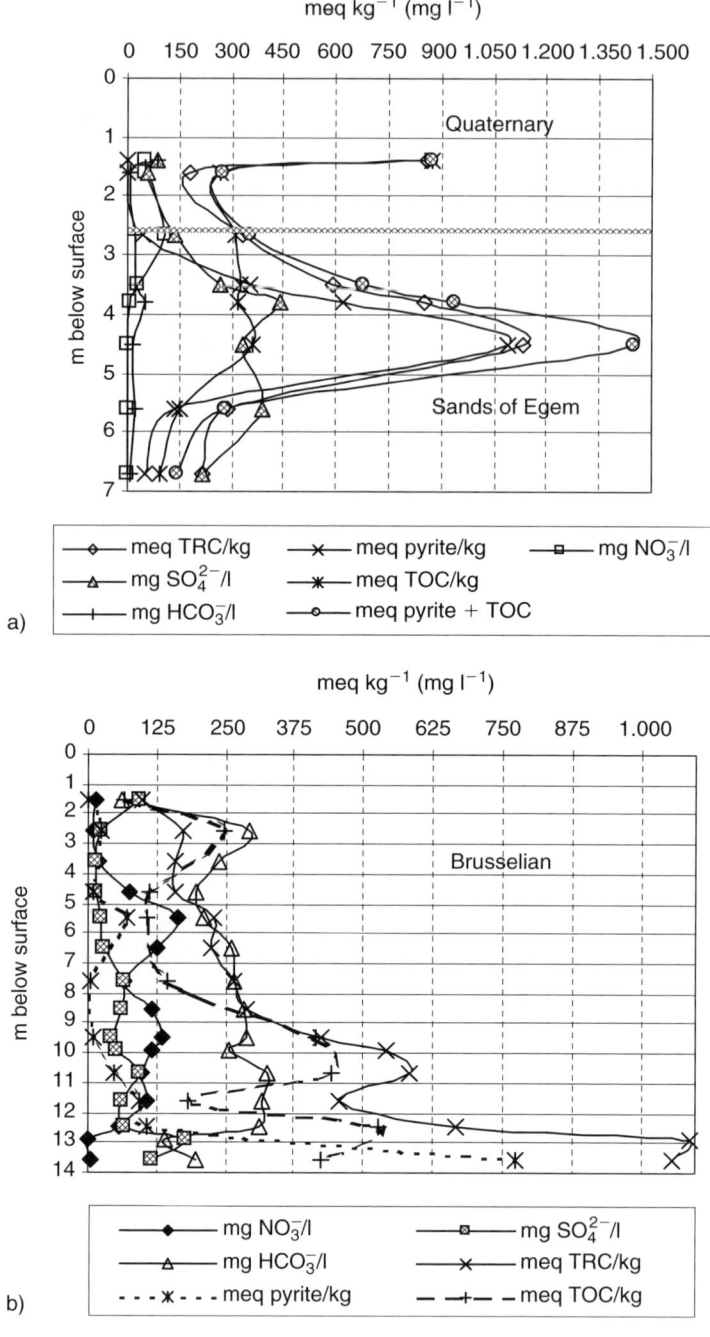

Figure 2. Comparison of reduction capacity and main anions in pore waters for the test sites a) Torhout and b) Kampenhout.

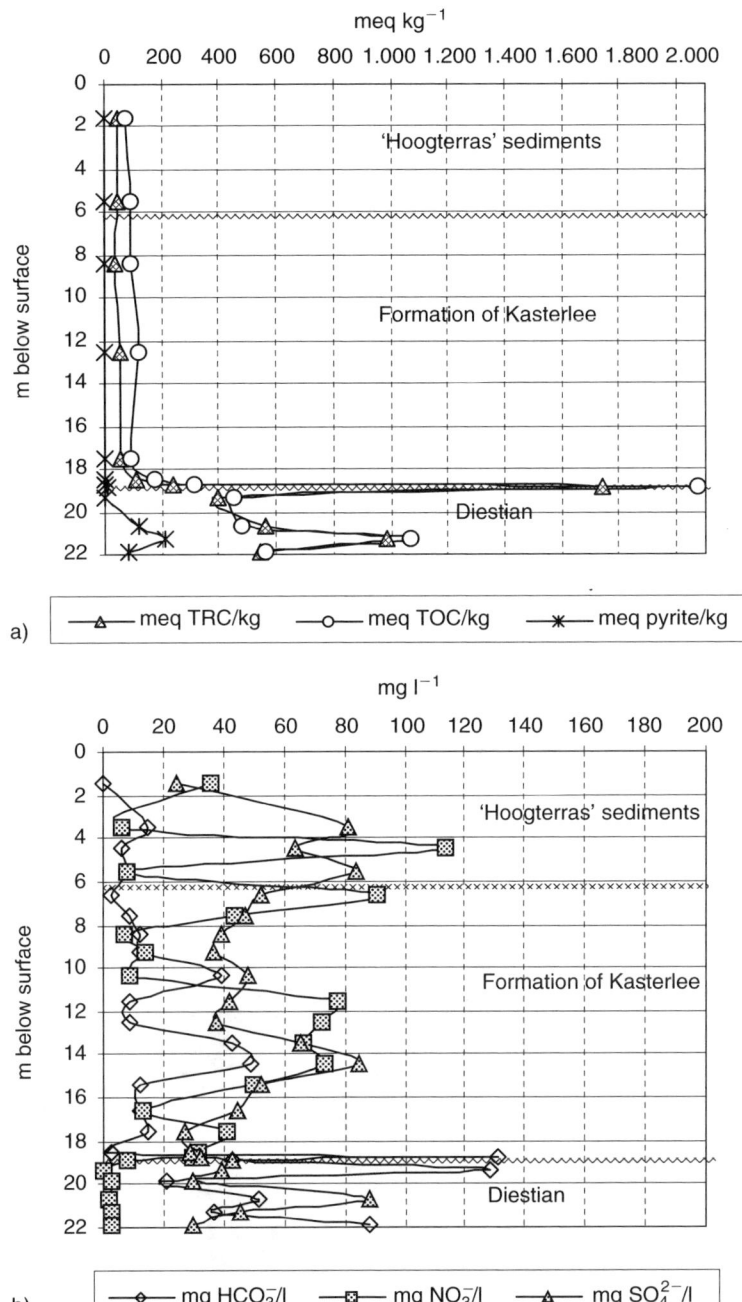

Figure 3. Test site Peer: a) comparison of TRC, TOC and pyrite charge equivalents and b) anion concentrations in pore waters.

sediment in contact with preferential flow paths is involved in reduction processes and gets oxidized. An indication for the presence of the transition zone for these two test sites is given by the range of reactivity measurements in the oxidation zone as shown in Table 1. A part of the reactive matter must be physically inaccessible to the micro-organisms due to fine-grained calcareous material in the Brusselian sediments of Kampenhout and clayey material in the Sands of Egem of Torhout. The micro-organisms are not able to reach all reduced compounds because of pore spaces smaller than their body size. Furthermore the negative particle load of, for example, clay minerals will lead to repulsive forces. Figure 4 gives an explanation of how the inaccessibility due to the particle size has to be understood.

Next to the physical accessibility, nitrate reduction is limited by the effective reactivity of the reducible substances. Low reactive matter forms a kinetically insufficient energy source to nitrate reducing micro-organisms. In good flushed sands without a fine-grained matrix practically no pyrite is found in the oxidation zone of the phreatic aquifers (see Adegem, Peer and Mol in Table 1). This electron donor is easily oxidized in conditions of good accessibility. Therefore a rest reactivity has been measured in the oxidation zones of all test sites due to the presence of low reactive organic matter (see Table 1 and Figures 1–3).

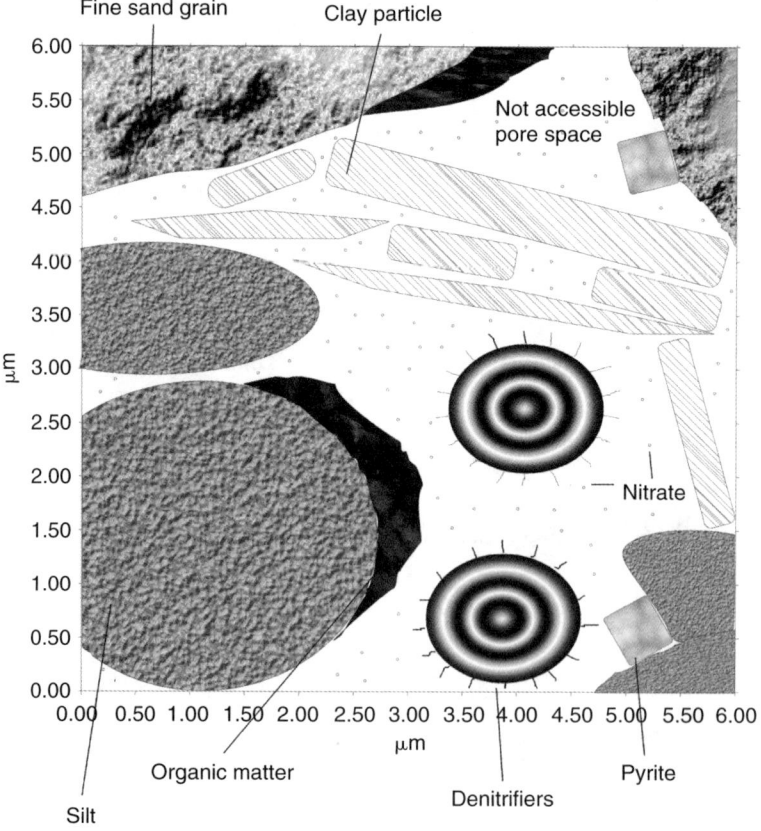

Figure 4. Accessibility of nitrate and nutrients to micro-organisms.

CONCLUSIONS

1) Nitrate reduction in groundwater is almost completely controlled by the presence of organic matter and (or) pyrite for the observed survey sites. They occur everywhere as a major part of the total reduction capacity.

   a) Organic matter determines the denitrification in younger deposits of Quaternary age or continentally influenced Tertiary layers as shown for the Flemish valley site of Adegem (main reduction level 7.10 m) and the Plio-Miocene deposits of Peer (main reduction level at 18.9 m).

   b) Pyrite is found to be the most important electron donor for nitrate reduction processes for older sedimentary deposits of Tertiary age of marine origin, such as for the sites of Kampenhout and Torhout (main reduction levels at 12.8 m and 3.8 m).

2) Other Fe^{2+}-bearing minerals that are potential electron donors, like glauconite and siderite, have only an impact as iron source for a post sedimentary forming of pyrite and other iron sulphides, but have no importance to nitrate reduction itself.

3) Next to the quantity of nitrate input, nitrate reduction in groundwater is determined by the sediment type, grain-size distribution (physical accessibility), effective reactivity and the advection of nitrate.

4) The 'depth sensitivity' of aquifers against nitrate contamination is clearly related to the reactivity and the oxidation state of the sediment. Strongly oxidized sediments such as the Formation of Kasterlee (Peer) can result in nitrate contamination to large depths in the aquifer. Protection zones have to be created.

5) The inaccessibility of nitrate to micro-organisms in fine-grained sediments could lead to long-term pollution sources and limits the barrier function of such kind of sediments.

6) The total amount of reducible substances in sedimentary deposits as nutrient source to nitrate reducing micro-organisms consists of three different parts:

   a) easily accessible reactive matter,

   b) physically not accessible reactive matter,

   c) low reactive organic matter as insufficient energy source.

REFERENCES

APHA, AWWA, WPCF 1998: *Standard methods for the examination of waste and wastewater.* 20th edition. APHA, Washington.

Canfield, D., Raiswell, R., Westrich, J.T., Reaves, C.M. and Berner, R.A. 1986: The use of chromium reduction in the analysis of reduced inorganic sulfur in sediments and shales. *Chemical Geology* 54: 149–155.

Eppinger, R. 1994: Hydrogeologische und hydrogeochemische Untersuchungen im Bereich von Volkegem (Hydrogeologic and Hydrogeochemistric research in the Volkegem region, Belgium) (in German with English and Dutch summary). Thesis. Kiel: Geological institute, University of Kiel, Germany.

Eppinger, R. and Walraevens, K. 1999: Mobility and removal of nitrate and their relationship to groundwater protection zones. Final report. EU-project: N°ENV4-CT96-5031 (DG12-ASAL). Ghent University, Ghent.

Heron, G., Crouzet, C., Bourg, A.C.M. and Christensen, T.H. 1994: Speciation of Fe(II) and Fe(III) in contaminated aquifer sediments using chemical extraction techniques. *Environ. Sci. Technol.* 28: 1698–1705.

Landers, D.H., David, M.B. and Mitchell, M.J. 1983: Analysis of organic and inorganic sulfur constituents in sediment, soils and water. *Intern. J. Environ. Anal. Chem.* 14: 245–256.

Mebius, L.J. 1960: A rapid method for the determination of the organic carbon in soil. *Anal. Chim. Acta* 22: 120–124.

Pedersen, J.K., Bjerg, P.L. and Christensen, T.H. 1991: Correlation of nitrate profiles with groundwater and sediment characteristics in a shallow sandy aquifer. *J. Hydrol.* 124: 263–277.

Van der Grift, B., Griffioen, J. and Van Buizen, H. 1999: Een geïntegreerd transportmodel voor grondwaterkwaliteit, Deelrapport C: Bepaling van het reactief vermogen van bulk organisch materiaal in zandige sedimenten met behulp van de micro-oxymax (An integrated transport model for groundwater quality, report part C: Determination of the reactivity of bulk organic matter in sandy sediments by the micro-oxymax) (in Dutch). NITG 99-71-B. Nederlands Instituut voor Toegepaste Geowetenschappen TNO. Delft and KIWA: Nieuwegein Netherlands: p. 42.

CHAPTER 6

Field investigation of nitrogen in unsaturated and saturated zones

W. Mioduszewski[1], M. Fic[2] and A. Zdanowicz[1]

[1]*Department of Water Resources IMUZ Falenty, 05-090 Raszyn, Poland, fax: +48 22 6283763*
[2]*Department of Sanitation IMUZ, 05-090 Raszyn, Poland, fax: +48 22 6283763*

ABSTRACT: The study of nitrogen transport in the soil water (unsaturated zone) and groundwater (saturated zone) has been conducted for two years on the experimental plot in Falenty. From the unsaturated zone, the soil water was extracted from ceramic cups installed at depths of 0.3, 0.6, 0.9 and 1.2 m below soil surface. Multilevel piezometers were installed in two places at depths of 3.0, 6.0 and 9.0 m below soil surface. Additionally 12 micropiezometers (own construction), with very small filters for water sampling from saturated zone were installed at depths of 3.0 and 5.0 m. Water was sampled every two 2–3 weeks and chemical analyses undertaken.

The age of water extracted from piezometers was also measured using the tritium method.

Nitrate content in the soil waters (unsaturated zone) varied substantially, frequently being higher than in the saturated zone, suggesting that pollution of groundwater in this area originates from water feeding from adjacent areas (housing estate and farmland). Despite these substantial differences, it was possible to observe seasonal changes in soil waters. The lowest nitrate content is registered usually during the summer.

INTRODUCTION

Human economic activity and agricultural use of nutrient compounds from fertilization, is posing a serious threat of pollution to groundwater. Agricultural use of fertilisers in infiltration areas is particularly detrimental, due to its impact on the water that percolates directly down to the aquifer. Numerous studies and measurement data show that water in shallow farm wells drawn from the unconfined aquifer contains excessive nitrate amounts that make it an unfit source of potable water. Much better quality characterises water found in confined layers insulated from the ground surface, lying at a greater depth and retained over a longer period (Bowen 1986; Fetter 1999).

A range of phenomena and processes associated with nitrogen change in waters percolating down the unsaturated and saturated zones still require elucidation by research. For this purpose, a detailed study of nutrient content in soil and groundwater was carried out at an experimental plot set up in the catchment of a small stream 'Stawy Raszyńskie' at Falenty. Complementary data were derived from a detailed study of the geology and land use affected in the catchment as well as by measuring parameters possibly influencing nutrient content. Fieldwork done at the experimental plot at Falenty forms part of an international research project (INCO-COPERNICUS no

IC15-CT98-0131) entitled 'Elaborating the methods for analysing the impact of changes in land use on groundwater quality'. The present paper reports on the section of the project related to field work.

DESCRIPTION OF THE CATCHMENT AND THE EXPERIMENTAL PLOT

The experimental field plot lies in the catchment of a small stream at Falenty near the fish ponds (the Raszyn Ponds Catchment) (Figure 1). Detailed analyses using historical records and results of previous research helped to develop an understanding of the geological structure of the study area and to identify its hydrogeology. In the immediate vicinity of the stream (site of the experimental plot), is an area of permeable soil (coarse and medium-grained sand) characterised by a high hydraulic conductivity, in the order of 20–30 m/day. In the direction of the watershed zone, there are poorly permeable soils (Figure 2), with a hydraulic conductivity of less than 5 m/day (Mioduszewski 2001).

A detailed study and analysis was made of the impact of land use in the catchment and on the direction of the groundwater flow to the experimental plot over the recent

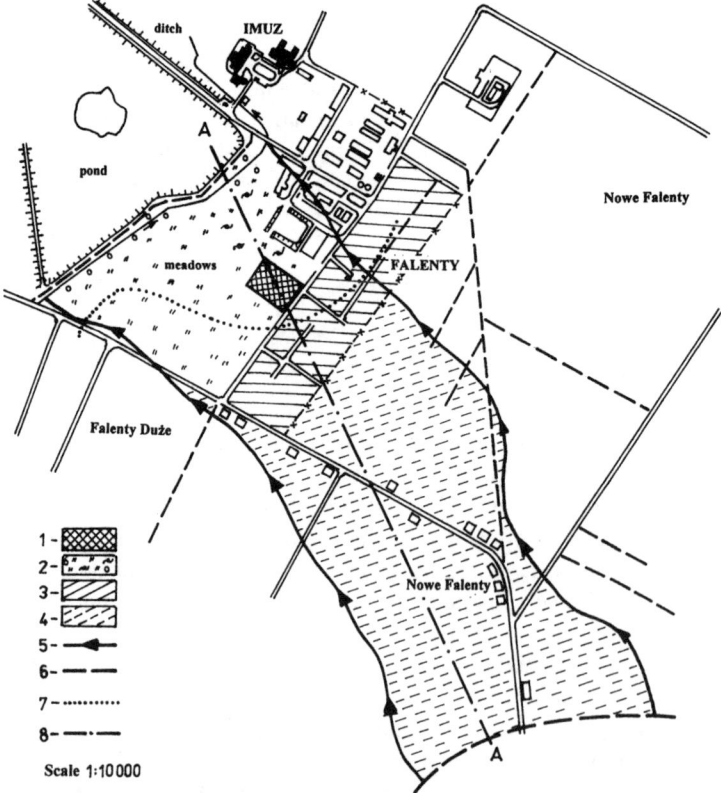

Figure 1. The water catchment with the field plot. 1 – experimental plot, 2 – meadows, 3 – settlements (houses), 4 – arable land and orchades, 5 – flowlines, 6 – groundwater divide, 7–border of old valley, 8 – cross-section.

10-year period (1991–2000). This helped to describe the entire recharge area possibly supplying the aquifer, with its impact on the quality of groundwater feeding the experimental plot.

From the perspective of land use, the study catchment may be distinguished into the following three areas:

- Area I – downstream part of the catchment, containing the experimental plot, under grassland, it yields – depending on the year and management – ranging between 700 and 1200 kg/ha; with fertiliser application, registering 50 kg of nitrogen per hectare.
- Area II – a developed stretch of the catchment upstream of the grassland. An area of a recent (1990–2000) housing development (cottages), its effluent draining to central sewage system, while rain water from the paved area drains to a road ditch (Figure 1). Although the development is equipped with a central water supply system, some of the cottages also draw their water individually from the depth of ca. 15–20 m.
- Area III – upstream of the residential development, under crop cultivation of varying intensity. Field plots range between 0.5 and 3.0 ha forming a mosaic of crop fields (maize, wheat, potatoes) and vegetable gardens (cabbage, onion, carrots), there are also small tracts that periodically are not under any type of cultivation. The level of fertiliser application varies considerably ranging from several kilograms to over 100 kg of nitrogen per hectare, depending on the field and the year. The average nitrogen load introduced with organic and mineral fertiliser in Area III ranged from 34 kg N/ha to 79.3 kg N/ha in 1995 and 1998 respectively (averaging 50 kg N/ha over the 10-year period).

The analysis of geology and land use in the study catchment helped to determine possible sources of pollution of groundwater within the boundaries of the experimental plot.

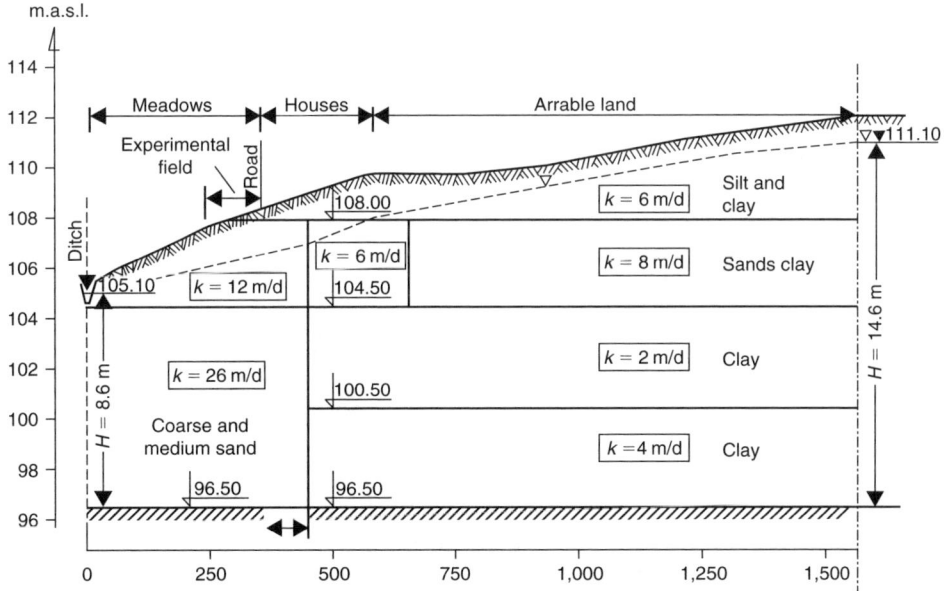

Figure 2. Schematization of the geological structure in cross-section; k – hydraulic conductivity, x – hydrogeological window.

These may include fertiliser application in the grassland surrounding the experimental plot (Area I) and inflow from areas II and III of the catchment, i.e. the residential and the crop and vegetable cultivation areas, respectively.

The results of the geological study indicated possible additional recharge of the upper aquifer from so-called distant water circulation. Apparently the impermeable strata underlying the sandy formations contain a hydrogeological window (Lieber and Mioduszewski 2002), which additionally feeds the upper aquifer.

MEASURING DEVICES AND STUDY METHODS

The plan of the experimental plot and the siting of the measuring devices are shown in Figure 3. Figure 4 shows the spacing of piezometers and micropiezometers in the cross-section. The installed monitoring system comprises the following elements:
- Measurements of solar radiation and wind velocity are provided by the meteorological station of the Institute of Meteorology and Water Management which is situated some 6 km from the experimental plot.

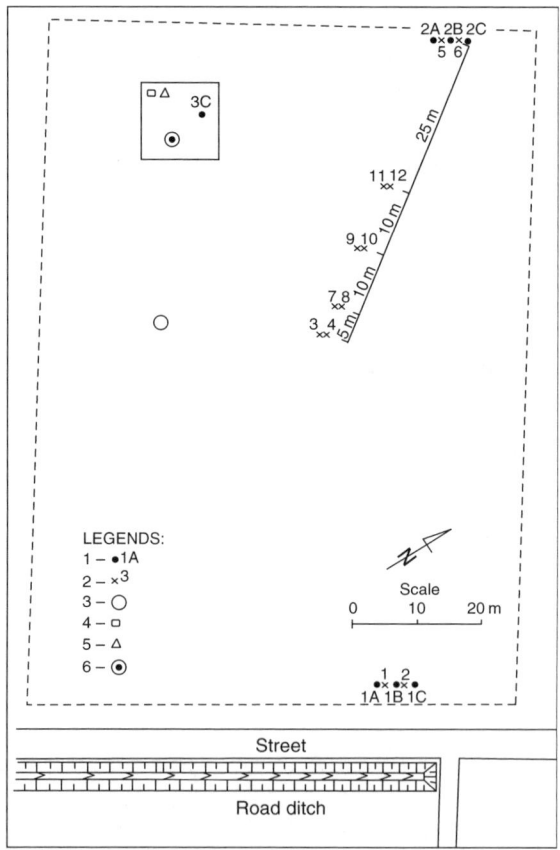

Figure 3. Measurement equipment on the field plot. 1 − piezometers, 2 − micropiezometers (M), 3 − drainage well, 4 − soil moisture measurement, 5 − ceramic cups, 6 − meteorological station.

- Piezometers, to measure the depth of the groundwater table and take water samples from the saturated zone. The piezometers in two points are fitted with filters installed at the depth of 3.0, 6.0 and 9.0 m below the ground surface (Figure 4). The piezometers are made of plastic pipe with a nylon filter, 1 m in length. An additional piezometer was installed at the meteorological station.
- Micropiezometers, to take water samples from the saturated zone. Small filters were designed and installed, in the form of a perforated tube 1 cm in diameter, 10 cm long, wrapped in nylon mesh. Water samples are taken using a vacuum pump attached to the micropiezometer by means of a plastic pipe, 3 mm in diameter. The micropiezometers are set up at the depth of 3 and 5 m (Figure 4). They make possible 'point water' sample-taking for chemical analysis without interfering with the movement of the groundwater.
- A meteorological station, to measure the parameters: atmospheric precipitation, air and soil temperature at the depth of 15 and 30 cm, and soil moisture. All measurements are made daily, according to standards followed in Poland.
- Electrodes, to measure soil moisture, installed at 0.1, 0.3, 0.6 and 1.2 m below the ground surface. Soil moisture measurements are made 2–3 times a week using the FOM field meter produced by the Institute of Agrophysics of the Polish Academy of Science (Malicki 1999).
- 'Ceramic cups', to take samples of soil-water from the unsaturated zone, installed at the depth of 0.3, 0.6, 0.9 and 1.2 m below the ground surface. Water samples are taken using a special suction pump.

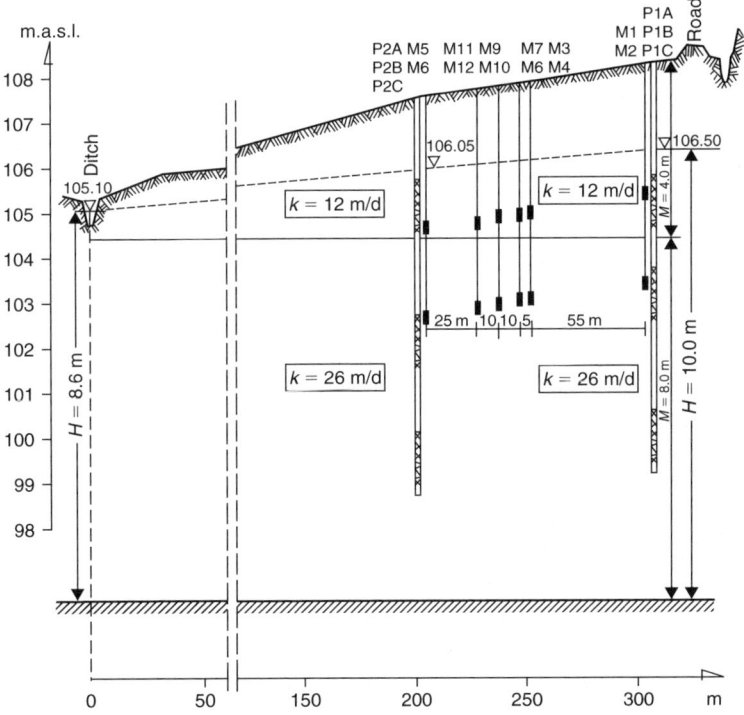

Figure 4. Piezometers and micropiezometers in cross-section. k – hydraulic conductivity, P – piezometers, M – micropiezometers.

Water samples for chemical analysis are taken from all piezometers and micropiezometers, and additionally, for comparison, from a nearby outlet of a tile drainage system (Figure 3) and a farm well which draws its water from the deeper level of the aquifer. Samples are taken in general once a month, but from micropiezometers, more frequently. Systematic study of water quality covers measurements of the level of pH, electrical conductivity, the content of oxygen, sulphates, nitrates and nitrites (Fetter 1999). In the next section of this paper, only nitrate measurements are discussed because this is the chemical compound that, in the study area, has the greatest bearing on water quality.

ANALYSIS OF MEASUREMENT RESULTS

Measurements of groundwater levels and soil moisture indicate evident seasonal fluctuation. Highest water table levels and soil moisture were registered in February, while the lowest values of these parameters were noted during the summer. This is a pattern typical for Polish conditions, resulting from the relationship between precipitation and evapotranspiration. Soil moisture in the upper layers (0.3 and 0.6 m below the groundsurface) depends largely on precipitation, while at 0.9 and 1.2 m below the groundsurface this influence apparently wanes. This shows that groundwaters (saturated zone) are fed primarily during non-growing season. The same period presumably also registers the inflow of the chemical compounds to the aquifer. Highest nitrate content in the soil-water (unsaturated zone) was registered outside the growing season, reaching up to $30 \, \text{mg} \, \text{l}^{-1}$ at the depth of 0.6 and 0.9 m, while at 0.3 and 1.2 m below the surface it was slightly lower (Figure 5A). Low nitrate content at the depth of 0.3 m probably results from the washing of the soil profile and nitrogen uptake by the plants. Less obvious is the reason for the lower nitrate content (below $10 \, \text{mg} \, \text{l}^{-1}$) registered at the depth of 1.2 m throughout the year, suggesting low nitrate loads are introduced into the phreatic zone in the study area. Nitrate content in the waters of the saturated zone varied substantially, frequently being higher than in the unsaturated zone, suggesting that pollution of groundwater in this area originates from water feeding from adjacent areas (housing estate and farmland).

It was instructive to compare data from piezometer and micropiezometer samples. Differences in nitrate concentration result from the fact that in the former, the water sample (after first pumping off at least three times the volume of the piezometer pipe), represents a certain, not easily assessed mean value from the area surrounding the piezometer filter with a length of 1 m, while water taken from the micropiezometer represents a point sample which describes the quality of the groundwater found in the immediate vicinity of a small filter (a small volume of water is pumped off in order to take a sample). Therefore subsequent analysis takes into account only the study of water samples taken from micropiezometers.

Figures 5B and 5C show the results of measurements of nitrate concentration in water samples taken from micropiezometers situated in the central section of the experimental plot. These micropiezometers are spaced at a slight distance from each other, at two different depths, i.e. 3 m and 5 m (Figures 5B and 5C respectively). A very high variability of nitrate concentration was observed. It is hard to explain such a substantial and rapid fluctuation of nitrate content in groundwater. In some cases, over a month, nitrate content changed by $20-30 \, \text{mg} \, \text{l}^{-1}$. Despite these substantial differences, it was possible to

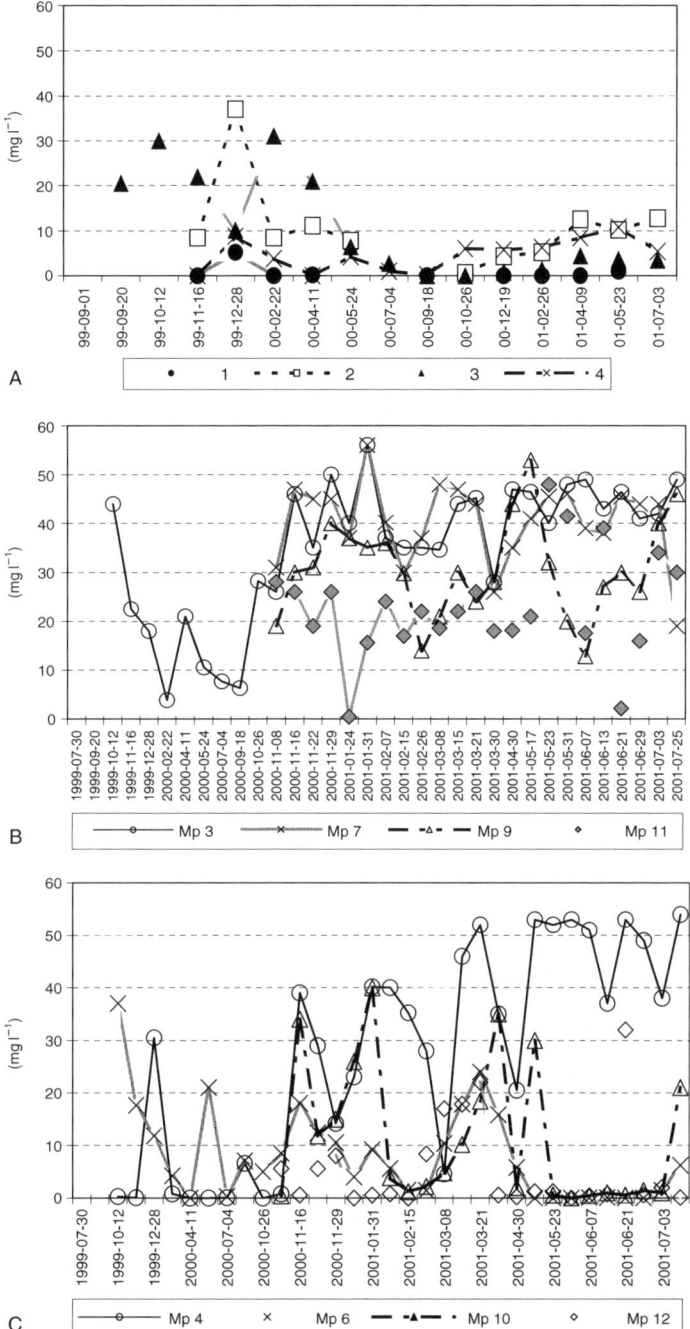

Figure 5. Measurement of nitrate concentration A – in soil-water; 1 – 0.3 m, 2 – 0.6 m, 3 – 0.9 m, 4 – 1.2 m below the ground surface; B – at 3 m below the ground surface (Mp–micropiezometer number); C – at the depth of about 5 m below the ground surface (Mp – micropiezometer number).

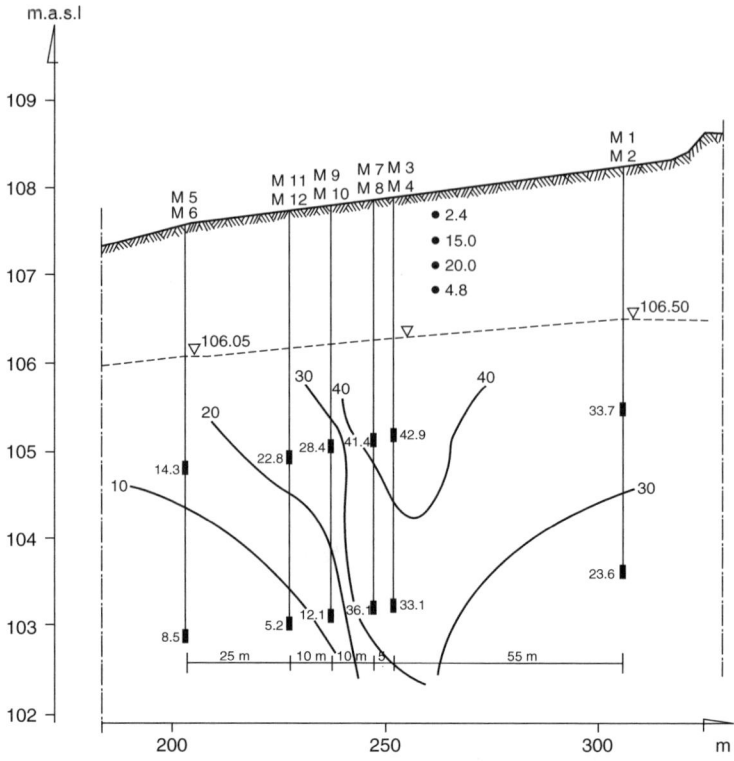

Figure 6. The average concentration of nitrate (mg l^{-1}) (lines of equal values) – in cross-section A–A. Measurements for samples taken from micropiezometers and ceramic cups.

observe, in waters of the saturated zone, a seasonal trend. The lowest nitrate content is registered usually during the summer.

Water quality at the outlet from the tile drainage system was relatively stable. Nitrate content ranged between 6–12 mg l^{-1}, with a mean value of ca. 10 mg N/l and was lower than the concentration in groundwater from the uppermost level of the aquifer.

Substantially lower nitrate content was observed at a greater depth. Water drawn from the deep farm well and piezometers at the depth of 9 m is practically nitrate free (concentrations close to zero). Figure 6 gives the mean arithmetical nitrate content calculated on the basis of the entire observation period (data from micropiezometers only). Isolines indicating similar concentrations are also shown. Substantial spatial variation is evident. It may be judged from the shape of the isolines that there is an influx of polluted waters from the direction of the housing estate (Area II) and areas under intensive agriculture (Area III). There is little impact on groundwater quality from waters percolating through from the unsaturated zone.

Low nitrate content in deeper lying water in the unconfined aquifer suggests that this layer is fed by 'pure' water inflowing from the confined aquifer through the presumed hydrogeological window. The presence of this additional recharge is evidenced by the study of the age of the water (measurements of tritium content made at the Institute of Nuclear Physics at the Mining and Metallurgy University). This suggests that water found at the depth of 9 m is older than water sampled at 3 m below the surface.

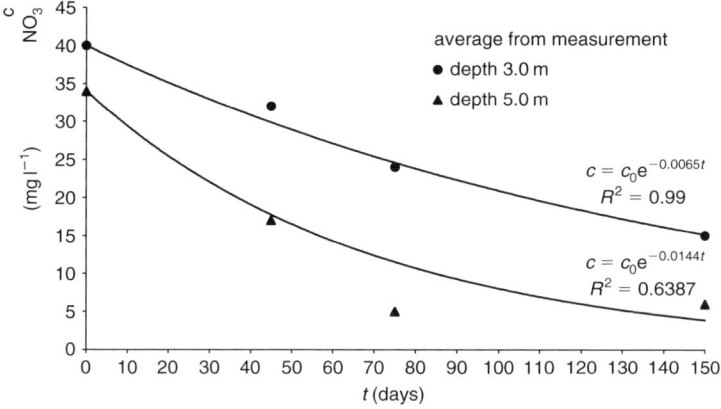

Figure 7. Decomposition NO_3, measurement from micropiezometers.

Changes in concentrations in the upper level of the aquifer (micropiezometers 3, 7, 9, 11, 5) may result also from denitrification processes ongoing in the saturated zone. This is because the soil contains small amounts of organic material (0.8–1.2%), and the time taken by water to flow from micropiezometer 3 to 5 is several months.

Figure 7 represents a diagram of nitrate concentration change over time. Numerical models MODFLOW and FLOTRANS were used to investigate the rate of nitrate migration between individual micropiezometers. It was assumed that change in nitrate concentration between individual micropiezometers results only from the rate of migration of this chemical compound. This assumption is supported by results from modelling (Lieber and Mioduszewski 2002). Figure 7 shows mean nitrate values registered over the observation period (a year). Taking the measurement from micro-piezometer no. 3 as initial, $t = 0$, successive readouts were made after a suitable lapse of time suggested by the rate of flow from the starting point.

While the described study should be viewed as preliminary, it has helped to define the processes responsible for the lowering of the nitrate content in the phreatic zone. Over the filtration distance of ca. 60 m (filtration time – 150 days), the concentration of NO_3 decreased by 50%. The research should be repeated in different hydrological and soil conditions, as well as over a longer observation period.

CONCLUSIONS

The field study was intended to determine nitrate content in soil and groundwater. It helped to determine the marked complexity of processes at work during water movement in a porous media. Even in confined layers, the condition and pollution of groundwater depends not only on the manner and intensity of land use directly over the aquifer, but additionally, and often principally, on the quality of groundwater flowing from adjacent areas. In the studied case, substantial impact was exerted by water flowing from the deeper levels of the aquifer. Moreover, study results indicated the decrease in nitrate content caused by processes taking place in the aquifer.

REFERENCES

Bowen, R. 1986: *Groundwater*. Elsevier, London, p. 427.
Lieber, A. and Mioduszewski, W. 2002: Numerical modelling of groundwater flow for an aquifer in a small river catchment. *Journal of Water and Land Development 6*: 19–27.
Mioduszewski, W. 2001: *The study of the nitrogen in saturated and unsaturated zones (in Polish). Wiadomości Melioracyjne i Łąkarskie*, str, pp. 192–195.
Malicki, M.A. 1999. *Methodical questions of monitoring of water status in selected biological materials (in Polish). Acta Agrophysica 19*: p. 108.
Fetter, C.W. 1999. *Contaminant Hydrogeology*. Prentice Hall, New York, pp. 689.

CHAPTER 7

Nitrate in groundwater from renaturalized and managed peatlands

P. Nadany
Institute for Land Reclamation and Grassland Farming at Falenty, PL-05-090 Raszyn, Poland

S. Chrzanowski
Biebrza Experimental Station, PL-19-200 Grajewo, Poland

ABSTRACT: Results of a study concerning the variability of nitrate and ammonia nitrogen concentrations in groundwater of peatlands as the effect of different dewatering degree and management systems are presented. The investigation was conducted in three different sites on the area of peatlands of the Biebrza River Valley in south-eastern Poland. All the sites were divided into zones of advancing bush and tree succession or ground cover. During field exploration, the quality of ground, surface and drinking water were monitored and changes of groundwater levels were recorded. The results of the investigation indicate a variation of nitrate nitrogen (N-NO$_3$) concentration in groundwater between the sites, a relationship between ground cover and N-NO$_3$ concentration and excessive concentrations of ammonium nitrogen in ground and surface water.

INTRODUCTION

Eutrophication and related problems caused by excessive concentrations of nitrogen and other nutrients is a serious environmental problem throughout the world. Significant loads of nutrients may be stored in peat soils of peatlands. Peatlands are the transitional lands between terrestrial and aquatic systems and improper water and land management can easily disrupt their function as natural biological filters of water. Peat soils contain up to 5% (of dry matter) of nitrogen and are recognized as nitrate-rich soils. The dewatering of theses soils some changes in the natural equilibrium in the natural habitants can occur resulting and the peatland may become a potential source of pollutants. Decrease of moisture in peat accelerates oxidation of organic matter and peat soil degradation. Recent literature and current studies strongly indicate that large amounts of nitrogen and other nutrients can be released from drained and degraded peat soils. The investigations presented below show the influence of human activity on increasing mineral nitrogen concentrations in groundwater.

The aim of this study is to determine the variability of nitrate and ammonium nitrogen concentrations in groundwaters of peatlands characterized by different management, dewatering scale and the advancement of moorsh-forming processes — the result of peat soil degradation.

STUDY AREA

The study area is situated in the north-eastern part of Poland in the Biebrza River Valley. The Biebrza River Valley is the greatest complex of peatlands in western and central Europe. To protect this precious ecosystem, Biebrza National Park (BNP) was created in 1993 with an area of nearly 60,000 ha. Although the valley is listed among the most clean and unspoilt sites in Poland, it is not free from problems. Human activity on this territory has resulted in the lowering of the groundwater level during the last century. The lowering of groundwater levels and the cessation of extensive agriculture on this territory has increased the degradation of peat soils, and allowed the succession of bushes and trees. Growth of bushes and trees is an additional factor that accelerates peat soil degradation and the release of mineral nitrogen to groundwaters. These effects justify increased concern for saving the natural landscape and quality of water resources there.

The investigations were carried out on three different sites chosen for their advancing anthropogenic impact. These sites were called Grobla Honczarowska, Gugny and Kuwasy.

The Grobla Honczarowska is located on peatlands known as Bagno Ławki. This is one of the most natural sites in the southern part of the BNP. Due to its location in the lowest basin of the Biebrza River Valley, declining groundwater levels did not exert such pronounced effects there as on others studied sites. The site is periodically flooded by water from the Biebrza River and the advancement of peat degradation is insignificant. The largest part of this site is open country covered with wetland vegetation surrounded by trees and bush zone, formed as a result of ecological succession.

The Gugny site is situated in low basin of the Biebrza valley in the vicinity of Gugny village. Before the establishment of the BNP, this area was used as an extensive permanent grassland (meadows and pastures). During later years, agricultural activity was stopped and the site is now exposed to the renaturalization process. It is situated in the flooding zone of the Biebrza River but floods are not as intensive and long as on Grobla Honczarowska. Therefore, peat soils are more dewatered and the advanced mineralization process is observed. Peat profiles contain a level of moorsh to the depth of 0−20/30 cm. Just as with the above this is an open ground area and succession of bushes and trees is observed there.

The Kuwasy site is situated in the buffer zone of the BNP in the catchment of Kuwasy Canal. Due to natural conditions and management practices, this area is the most degraded in comparison with other investigated sites. During the last century, the site was subject to changes made by intensive human activity. The area is under intensive management as permanent grasslands (meadows and pastures) used by local farmers and the Biebrza Experimental Station. Non-grassland areas are usually covered with trees. For dewatering and irrigation purposes, the entire area has a drainage system and canal networks equipped with weirs; and therefore groundwater level fluctuation is artificial. Agricultural activity, improper land management and excessive dewatering of peat soils are the reasons why peat degradation is occurring and peat soils are most degraded and transformed at this site in comparison with other sites in the BNP. The thickness of moorsh layer which reaches 5−40 cm is greater than in the Gugny site.

RESEARCH DESCRIPTION AND METHODS

The study started in May 2000 and continued until May 2003. During field exploration, the quality of ground, surface and drinking water were monitored and changes of groundwater levels were recorded. Groundwater samples were taken from installed plastic wells, surface water from ditches and canals on the monitored site, and drinking water from dug and deep farm wells situated in villages near the investigated areas. To analyse the relationship between nitrogen concentration and succession of bushes and trees, all sites were divided into zones of advancing succession. Three zones were distinguished on Grobla Honczarowska: open ground, sites with active succession and sites overgrown by trees. At Gugny, two zones were distinguished: open ground and sites with active succession. In the catchment of Kuwasy Canal, open ground and areas covered by trees were distinguished.

Eleven plastic wells were installed on Grobla Honczarowska, eight on Gugny and eleven on Kuwasy. Groundwater levels were measured periodically in the wells. In intensively managed agricultural areas, where peat soils are most dewatered and transformed, sampling was undertaken once a month, and on the areas of the BNP of nearly natural character, once every 6 weeks. $N-NO_3$ and $N-NH_4$ were determined using Automatic Flow Analysis. Determination of $N-NO_3$ was based on the method of cadmium reduction (Searle 1984) and determination of $N-NH_4$ was based on modified reaction of Berthelot (Krom 1980, Searle 1984). In total 1,056 samples were processed between May 2000 and December 2001. Statistical analysis was made with the Statistica software. All statistical tests were run for significance level – $\alpha = 0.05$ and $\alpha = 0.01$.

RESULTS

Variations of groundwater levels strongly indicate that in more frequently drained areas, the degradation of peat soils is more advanced than on areas of more natural conditions. Furthermore, in the intensively managed catchment of Kuwasy Canal, the groundwater level was most variable in comparison with other areas in the BNP (Table 1).

Results showed high variability of nitrate ($N-NO_3$) and ammonium nitrogen ($N-NH_4$) concentrations in ground and surface water. The differences were observed both between and within the studied sites (Table 2). High $N-NO_3$ and $N-NH_4$ variability, seen in high standard deviations (Table 2), suggests that mechanisms of $N-NO_3$ and $N-NH_4$ cycling were also diversified. As a consequence of different site conditions, sites were divided into zones of different peatland cover and type of management (Table 3).

The results obtained provided some information on nitrogen behaviour in soil in relation to site conditions but could not be processed by statistical data analysis. High variability of $N-NO_3$ and $N-NH_4$ together with the impossibility of performing enough field measurements meant that the distribution of analysed compounds in collected

Table 1. Groundwater levels on the study area.

Sites	Groundwater below ground surface			
	Mean	SD	Min.	Max.
Grobla Honczarowska	24	23	0	94
Gugny	39	27	0	96
Kuwasy	54	21	0	113

SD – standard deviation.

Table 2. Nitrate nitrogen (N-NO$_3$) and ammonium nitrogen (N-NH$_4$) concentrations in groundwater and surface waters.

Sites	n	N-NO$_3$ (mg l^{-1})		N-NH$_4$ (mg l^{-1})	
		Mean	SD	Mean	SD
Groundwater					
Grobla Honczarowska	122	1.01	2.17	0.80	1.78
Gugny	83	1.28	2.38	1.6	2.76
Kuwasy	181	9.83	21.38	2.29	4.95
Surface water					
Grobla Honczarowska	100	0.87	2.23	0.69	1.41
Kuwasy	139	2.45	22.01	0.37	0.52

SD − standard deviation; *n* − number of samples.

Table 3. Nitrate nitrogen (N-NO$_3$) and ammonium nitrogen (N-NH$_4$) concentrations in groundwater in relation to ground cover and land management.

Ground cover and land management	n	N-NO$_3$ (mg l^{-1})		N-NH$_4$ (mg l^{-1})	
		Mean	SD	Mean	SD
Grobla Honczarowska					
Open ground	82	0.45	0.99	0.76	1.96
Sites with proceeding succession	21	1.69	2.61	0.59	1.20
Area covered by trees	11	2.8	4.22	1.44	1.7
Gugny					
Open ground	70	1.14	1.85	1.66	2.9
Sites with proceeding succession	15	1.80	4.0	1.17	1.71
Kuwasy					
Open ground	170	5.03	12.05	2.28	5.02
Birch − alder forest	17	54.76	35.36	1.59	3.37

SD − standard deviation; *n* − number of samples.

Table 4. Site effect of nitrate nitrogen (N-NO$_3$) and ammonium nitrogen (N-NH$_4$) variability.

Sites	N-NO$_3$		N-NH$_4$	
	Gugny	Kuwasy	Gugny	Kuwasy
Grobla Honczarowska	is.	*	is.	is.
Gugny	−	is.	−	is.

* Significant differences; is. − insignificant differences.

material was different from their distribution in the environment. Because of this inconsistency, nonparametric statistical data analysis was used. N-NO$_3$ and N-NH$_4$ variability in samples of water between and within the sites was characterized by the Wald-Wowitz test (Stanisz 1998) (Tables 4 and 5).

Table 5. Nitrate nitrogen (N-NO$_3$) and ammonium nitrogen (N-NH$_4$) variability in relation to ground cover and land management on Grobla Honczarowska.

Ground cover and land management	N-NO$_3$ – Ground cover and land management		N-NH$_4$	
	Sites with active succession	Area covered by trees	Sites with active succession	Area covered by trees
Open ground Gugny	is.	*	is.	is.
Open ground Kuwasy	is.	–	is.	is.
Open ground	is.	**	is.	is.

* Significant differences; ** highly significant differences; is. – insignificant differences.

Table 6. N-NO$_3$ and N-NH$_4$ concentration in samples of drinking water.

Well	n	N-NO$_3$		N-NH$_4$	
		Mean	% above standard*	Mean	% above standard*
Dug wells					
1	12	26.58	135	0.13	
2	11	14.42	28	0.45	
3	16	18.05	60	1.37	10.1
4	13	4.99		0.17	
5	12	19.87	76	1.61	29.9
6	11	26.01	130	1.91	54.1
7	11	6.65		0.67	
Deep wells					
1	12	0.43		0.45	
2	10	0.43		0.09	
3	16	2.38		0.76	
4	15	1.46		0.88	
5	13	1.40		0.20	
6	12	17.94	59	1.47	18.6
7	16	0.61		1.44	16.2

Polish standards for drinking water quality – 50 mg NO$_3^-$/l and 1.5 NH$_3$ mg/l (N-NO$_3$* 4.43 = NO$_3^-$; N-NH$_4$* 1.21 = NH$_3$); n – number of samples.

N-NO$_3$ and N-NH$_4$ concentration in samples of drinking water taken from dug and deep wells in nearby farms is presented in Table 6.

DISCUSSION

In many sites of the Biebrza River Valley, lowering of groundwater level is observed as a result of land drainage. The Biebrza River regulations and other factors had an adverse effect on soil conditions in the studied area. According to general analysis of the impact of drainage and long-term intensive utilization of grasslands on organic soils, the primary effect of land reclamation is the alteration of soil processes. Dewatering, excessive aeration and microbiological processes are the major reasons for decomposition and loss of the soil organic matter and nitrogen (Okruszko 1993). This study confirms the impact of dewatering on the advancement of peat soil degradation and moorsh forming processes. Current literature and studies carried out in different sites indicate the potential threat of releasing significant amounts of mineral nitrogen from dewatered and degraded peat soils. The study on nitrogen mineralization rate in peat soils of the depression sink in the Bełchatów strip mine showed that intensively dewatered peat soil could release up to 500 kg $N-NO_3$ per hectare (Sapek, A., Sapek B. and Gawlik 1990). High concentrations of $N-NO_3$ in the agriculturally managed site at Kuwasy should be expected based on the fact that peat soils are most transformed there. Indeed, $N-NO_3$ concentration in samples of groundwater from the Kuwasy site was significantly higher than in groundwater from the Grobla Honczarowska site where the mean concentration of this compound was very low. The situation is different in the case of $N-NH_4$. Increased concentration of $N-NH_4$ is able to accelerate eutrophication of aquatic ecosystems and can be unfavourable for some species of fish (Smoron 1998). The investigation indicated a lack of relationship between $N-NH_4$ concentration and sites and condition within the sites.

Though peat soil degradation is caused mainly by groundwater lowering and accompanying dewatering, the process also depends on others factors. According to the present state of knowledge (Gotkiewicz 1996), the development of agriculture ecosystems and forests on peatlands is known to accelerate moorsh-forming processes. Therefore, birch and alder forests may exert a negative impact on water quality in peatlands. Drying up of peat soils under alder and birch forests is caused by the intensive evapotranspiration of mature forests. Retention of considerable amounts of water from summer rainfalls by the tree canopy also has a negative effect. The deep root system of trees causes considerable moisture deficit in deeper soil layers and transformation of soil in these layers (Gotkiewicz and Szuniewicz 1984; Okruszko, Szuniewicz and Szymanowski 1987). As a result of the uptake of large amounts of water by trees and excessive aeration, peat soils overgrown by trees are more transformed than those in grasslands or other open ground.

The study carried out on the area of the BNP and its buffer zone confirmed the influence of ground cover and land management on $N-NO_3$ concentration in groundwater. However, low mean concentrations of $N-NO_3$ and high standard deviations (Table 2) indicated a high variability of $N-NO_3$ release depending on site conditions, which in turn depended on peatland cover. In Kuwasy some areas are left as alder and birch forests. In samples of groundwater taken from forest areas, the concentration of $N-NO_3$ was more then five times higher than from grasslands (Table 3). Observed differences were confirmed by statistical analysis (Table 5).

The problem of bush and tree ecological succession has recently appeared on areas of the BNP. Initially, the problem was considered as an aspect of saving the natural landscape and flora and fauna characteristics of this area. The results of the present study show that ecological succession of trees and bushes has significant influence on groundwater quality. In Grobla Honczarowska, $N-NO_3$ concentration in groundwater was about seven times higher than in samples taken from open ground (Table 3).

Although the sites managed as forest in Kuwasy and the sites covered by trees on the area of the BNP should be recognized as sources of $N-NO_3$, the pollution is insignificant, as shown by low concentration of $N-NO_3$ in the surface water. Active ecological succession may have a negative influence on the quality of ground and surface waters in the future. Currently, an excessive concentration of $N-NH_4$ in ground and surface water may cause some concern.

Drinking water taken from dug wells was in most cases polluted by $N-NO_3$ and $N-NH_4$. Drinking water from deep wells, apart from in one case, conformed to Polish standards of drinking water quality (Table 6). Based on the results, it is impossible to confirm the impact of mineral nitrogen release on drinking water quality because the investigated water intakes are open to point sources of pollutants (Sapek and Urbaniak 2001).

Our studies did not confirm the relationship between groundwater level variability and mineral nitrogen concentration.

CONCLUSIONS

1. High groundwater levels and periodical flooding protect peat soils from degradation and prevent the release of nitrate nitrogen to ground and surface water resources.
2. Active ecological succession of bushes and trees may be one of the reasons for increasing nitrate nitrogen concentration in groundwater.
3. Peatland areas managed as forests and areas covered by trees as a result of ecological succession should be recognized as sources of pollution.
4. Grassland management and high groundwater level have a positive effect protecting water resources from nitrate nitrogen contamination and preventing peat soil degradation.
5. Excessive concentrations of ammonium nitrogen in ground and surface water were observed on study areas.
6. The fate of nitrogen from mineralized peat material on studied peatlands is unknown. Observed concentrations of $N-NO_3$ (except for groundwater from under forest) did not confirm advanced peat soils degradation on agriculturally managed areas.

The research was carried out within the European Union international project Prowater – Contract no EVK1-CT-1999-00036 PROWATER 'Program for prevention of diffuse pollution with phosphorus from degraded and re-wetted peat soils'.

REFERENCES

Gotkiewicz, J. 1996: The release and transformation of mineral nitrogen in the hydrogenic soils (orig. title: Uwalnianie i przemieszczanie azotu mineralnego w glebach hydrogenicznych). *Zeszyty Problemowe Postępów Nauk Rolniczych* z. 440 Azotany w ekosystemach rolniczych, Olsztyn, pp. 121–129.
Gotkiewicz, J. and Szuniewicz, J. 1984: Fen transformation under birch forests. Proceedings of the 7th International Peat Congress, Dublin, June 18–23 1984, Volume 3, pp. 286–297.
Krom, M. 1980: Spectrophotometric determination of ammonia; a study of modified Berthelot reaction using salicylate and dichloroisocyanurate. *The Analyst 105*: 305–316.
Navone, R. 1964: Proposed method for nitrate in potable waters. *Journal American Water Works Association. 56*: 781.
Okruszko, H. 1993: Transformation of peat soils under the impact of daring. *Zeszyty Problemowe Postępów Nauk Rolniczych. 406*: 3–73.
Okruszko, H., Szuniewicz, J. and Szymanowski, M. 1987: Differentiation of physico-hydrological properties of soils developed from fen-peat under the effect of their different utilisation.

Conference proceedings – Internationales Symposium Zum Thema Bodentwicklung Auf Neidernoor Und Konsequenzen Fur Die Landwirtschaftliche Nutzung 25 bis 29 Mai, Eberswalde, DDR.

Sapek, A., Sapek, B. and Gawlik, J. 1990: Nitrogen mineralisation rate in peat soils within the depressions funnel of the Bełchatów strip mine (orig. title, Rozpoznanie nasilenia mineralizacji azotu w glebach torfowych w zasięgu leja depresyjnego kopalni Bełchatów). *Wiadomości IMUZ* T. XVI zeszyt 3 Warszawa, pp. 79–86.

Sapek, B. and Urbaniak, M. 2001: Ocena zanieczyszczeń gleby z terenu zagrody i jej otoczenia składnikami nawozowymi w gospodarstwach demonstracyjnych projektu BAAP II. *Wiadomości melioracyjne i łąkarskie 1*: 32–36.

Searle, P.L. 1984: The Berthelot or indophenol reaction and its use in the analysis of the chemistry of nitrogen. *The Analyst 109*: 549–565.

Smoroń, S. 1998: Eutrophication of surface water as an effect of biogenic compounds penetration from agricultural sources to the environment (orig. title, Przenikanie substancji biogennych ze źródeł rolniczych do środowiska – czynnik eutrofizacji wód powierzchniowych). *Zeszyty Edukacyjne*, Falenty *5*: 57–70.

Stanisz, A. 1998: Basic course of statistics (orig. title, Podstawowy kurs statystyki). StatSoft Polska, Cracow pp. 263, 269.

CHAPTER 8

Estimating microbial denitrification parameters from *in situ* bioreactor operation under groundwater conditions

U. Naumann, P. Maloszewski, I. Ghergut, K.-P. Seiler and W. Stichler
GSF Institute of Hydrology, D-85764 Munich-Neuherberg, Germany

ABSTRACT: In order to interpret the results of microbial denitrification experiments carried out under groundwater conditions with rock samples of the Southern Franconian Alb (a karstic area in Germany, with high exposure to nitrate contamination), a nutrient transport–consumption model is formulated and its predictions compared to observed denitrification levels, measured enzyme activities and short- or medium-term changes in bacterial numbers as recorded in five bioreactors operating *in situ*. The model presented relies on a number of simplifying assumptions, of which the most important is that of a meagre biofilm development, for a single (bulk) bacterial population in up to three phases, free of predation/competition and feeding on a two-component substrate. Non-linear kinetics is assumed for nutrient consumption and biomass growth, first-order exchange is assumed between similar species in different phases, coupled with a double-porosity model for nutrient and biomass flow and transport.

AIMS

The experiments described are part of a broader investigation programme launched by Seiler in the 1980s (Seiler, Maloszewski and Behrens 1989; Seiler and Behrens 1992; Seiler 1999) in the Southern Franconian Alb, a karstified Upper Jurassic (Malm) carbonate formation in Southern Germany (north of Ingolstadt, with the main investigation area lying between Danube and the river Altmuehl), whose character, determined by two consecutive sedimentation cycles, is dominated by two contrasting facies types: the bedded limestones with a fissure/solution channel porosity amounting to 2% but with syngenetic porosities under 2%, and the dolomitised reefs with matrix porosities reaching 5–10%. The present investigation is specifically concerned with the reasons behind the lower denitrification activity in the bedded facies, as compared to the reef facies. With variable, but generally low DOC supply, and average 'bedded'/'reef' concentration ratios differing from species to species: chloride – 1.5, nitrate – 2.2, sulphate – 1.2, sodium – 1.6, potassium – 1.6, atrazine – 1.9 (values reproduced after Seiler and Hartmann 1997), it is clear that the porosity contrast alone cannot account for these values, and that several (possibly coupled) chemical transformations are involved alongside with transport in this karst system. A number of papers have already been devoted to biodenitrification in the Southern Franconian Alb (Hausner, Lawrence, Wolfaardt *et al.* 2000; Müller, Assmus, Hartmann *et al.* 2000; Naumann, Vomberg, Hartmann *et al.* 2001 etc.) and the issue of (organic) carbon availability has been strongly emphasised, but there has been no attempt, so far, to look at the experimental findings through the eyes of a kinetic model for nutrient consumption and biomass growth. Here, a qualitative interpretation for the observed differences in nitrate levels is first sought in terms of carbon availability and/or limiting nutrient residence times. Second, one would like to obtain quantitative estimates for the relevant denitrification parameters as a basis for predicting future contamination levels reliably (as diffuse nitrate input also evolves with time under the influence of changing policies). Despite some ambiguity in parameter estimation, the model helps identify the factors determining the fate of nitrate in different facies types of the Southern Franconian formation. The question of what is actually used as an electron donor in biodenitrification, additionally or alternatively to DOC, has yet to be answered by different methods.

EXPERIMENTAL SETTING AND MATHEMATICAL MODELLING

The *in situ* operating bioreactors consist of columns homogeneously filled with porous/ rock material, submersed in artesian spring water and maintained under dark, anaerobic conditions (using a Helium atmosphere to exclude oxygen), in a constant or stepwise-constant flow (with the aid of a pump). The column filling is mechanically tailored in a way that is supposed to enhance plankton attachment (whether this also favours biofilm growth could not be ascertained), and sterilized prior to the experiments, during which it is to be colonized anew under specified nitrate supply (whereas carbon supply may contain labile components escaping direct observation). The hydrodynamic parameters of the column under different flow regimes (Peclet number, bulk or mobile-phase residence times) can be determined with the aid of non-reactive tracers (tritiated water). The selected artesian spring which served as a drinking water supply for many years before (Klingenbachquelle) is physico-chemically representative of the region's groundwaters, and it can be assumed that its typical DOC and nitrate loads, when applied under dark, anaerobic conditions, favour a bacterial community structure and physiology representative for the groundwaters of the deeper Franconian Alb formations. The basic transformation performed by these 'bioreactors' can be summarized as: $10[H] + 2H^+ + 2NO_3^- \rightarrow N_2 + 6H_2O$ in which [H] usually stands for some organic carbon compound acting as a hydrogen donor and substrate for biomass growth; whereby only the evolution of *primary* electron-donor/acceptor concentrations are of relevance to the model on a more than very short term.

The processes to consider within a bioreactor representative elementary volume (REV) are summarized in Figure 1: any REV contains a mobile water phase (occupying a fraction θ_{mob} of the REV), an immobile water phase (occupying a fraction θ_{imm} of the REV water) and a rock matrix phase (occupying a fraction $(1 - n)$ of the REV); in both water phases there are dissolved nutrients (nitrate concentrations $N_{mob,imm}$ and biologically available carbon compound, or electron donor concentrations $C_{mob,imm}$) and loose or 'planktonic' denitrifiers (concentrations alive or active $W_{mob,imm}$ and dead or inactive $\Omega_{mob,imm}$), whereas in the rock matrix phase there are no sorbed nutrients but only sessile denitrifiers ('biofilm' with concentrations X for the active and Ξ for the inactive), nutrient supply from the immobile water phase being shared by loose immobile plankton and the sessile phase. For the sake of simplicity we assume that only the active ('living') become inactive ('dead') with a first-order probability rate λ (analogous to radioactive decay) and there is no reverse process (the inactive never become active

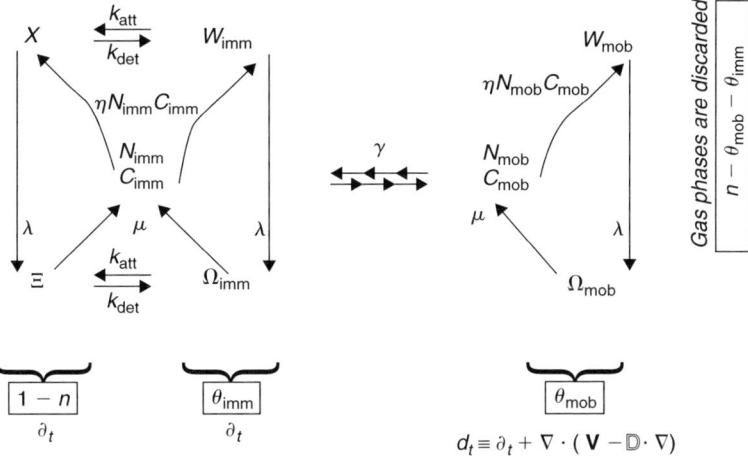

Figure 1. Schematic representation of REV processes described by the model.

again) but the dead biomass may release carbon (electron donor) in a form available for denitrification again, at a rate κ which may depend upon the time τ elapsed since 'cell death' (the dead-biomass 'age'); in particular, this can be a constant, first-order probability rate μ. Exchange between mobile and immobile phases is governed by first-order rate γ for all species, biomass exchange between sessile and loose phases is assumed to observe first-order kinetics (with rates $k_{att,det}$ for attachment and detachment, respectively); conversion of nutrients to biomass takes place with a reaction rate efficiency η which is generally concentration-dependent and in this particular model represents the sole source of non-linearity, however of a rather weak type (see below). All mobile species are transported by advection and hydrodynamic dispersion described by the same operator (velocity **v** and dispersion coefficient D), which in all our experiments can be treated as one-dimensional.

Nitrate sorption to carbonate-rock material (sometimes reported in the literature) did not appear to play a role in our experiments (also confirmed by Vomberg 2002, for similar rock samples), so it was not included in the model. We also discard gas phases, since the slight N_2O and N_2 production in the course of denitrification does not influence flow and transport on a macroscopically relevant scale. With reaction efficiencies η factorable into independent Monod-type kinetics:

$$\eta_{mob} \equiv r_{max} \frac{N_{mob}}{N_{mob} + N_{HS}} \frac{C_{mob}}{C_{mob} + C_{HS}}, \; \eta_{imm} \equiv r_{max} \frac{N_{imm}}{N_{imm} + N_{HS}} \frac{C_{imm}}{C_{imm} + C_{HS}},$$

balance equations for the ⌐model observables⌐ read:

$$\boxed{\begin{array}{c} \text{dissolved Nitrate,} \\ \text{immobile water phase} \end{array}} \; \partial_t N_{imm} = -\eta_{imm}\left(W_{imm} + \frac{\alpha}{\beta S}X\right) + \frac{\gamma}{\beta}(N_{mob} - N_{imm})$$

$$\boxed{\begin{array}{c} \text{dissolved available (organic) carbon,} \\ \text{immobile water phase} \end{array}} \; \partial_t C_{imm} = -y_{C/N}\eta_{imm}\left(W_{imm} + \frac{\alpha}{\beta S}X\right) + \frac{\gamma}{\beta}(C_{mob} - C_{imm})$$

$$+ \frac{y_{C/N}}{y_{B/N}} \int_0^\infty \kappa(\tau)\left[\omega_{imm}(\tau) + \frac{\alpha}{\beta S}\xi(\tau)\right]d\tau$$

| denitrifying plankton, immobile water phase | $\partial_t W_{\mathrm{imm}}$ | $=$ | $y_{\mathrm{B/N}}\eta_{\mathrm{imm}}W_{\mathrm{imm}} + \dfrac{\gamma}{\beta}\left(W_{\mathrm{mob}} - W_{\mathrm{imm}}\right)$ |

$$- k_{\mathrm{att}}W_{\mathrm{imm}} + k_{\mathrm{det}}\frac{\alpha}{\beta S}X - \lambda W_{\mathrm{imm}}$$

| dead-plankton age density, immobile water phase | $\omega_{\mathrm{imm}}\left(\tau = 0; \mathbf{r}, t\right) = \lambda W_{\mathrm{imm}}\left(\mathbf{r}, t\right)$ |
| | $\left(\partial_\tau + \partial_t\right)\omega_{\mathrm{imm}} = -\kappa\left(\tau\right)\omega_{\mathrm{imm}} + \dfrac{\gamma}{\beta}\left(\omega_{\mathrm{mob}} - \omega_{\mathrm{imm}}\right)$ |

$$- k_{\mathrm{att}}\omega_{\mathrm{imm}} + k_{\mathrm{det}}\frac{\alpha}{\beta S}\xi$$

| sessile denitrifiers (biofilm), rock matrix phase | $\partial_t X$ | $=$ | $y_{\mathrm{B/N}}\eta_{\mathrm{imm}}X + k_{\mathrm{att}}\dfrac{\beta S}{\alpha}W_{\mathrm{imm}} - k_{\mathrm{det}}X - \lambda X$ |

| dead-plankton age density, rock matrix phase | $\xi\left(\tau = 0; \mathbf{r}, t\right) = \lambda X\left(\mathbf{r}, t\right)$ |
| | $\left(\partial_\tau + \partial_t\right)\xi = -\kappa\left(\tau\right)\xi + k_{\mathrm{att}}\dfrac{\beta S}{\alpha}\omega_{\mathrm{imm}} - k_{\mathrm{det}}\xi$ |

| dissolved Nitrate, mobile water phase | $\mathrm{d}_t N_{\mathrm{mob}}$ | $=$ | $-\eta_{\mathrm{mob}}W_{\mathrm{mob}} + \gamma\left(N_{\mathrm{imm}} - N_{\mathrm{mob}}\right) + \left(\dfrac{\delta N}{\delta t}\right)_{ext}$ |

| dissolved available (organic) carbon, mobile water phase | $\mathrm{d}_t C_{\mathrm{mob}}$ | $=$ | $-y_{\mathrm{C/N}}\eta_{\mathrm{mob}}W_{\mathrm{mob}} + \gamma\left(C_{\mathrm{imm}} - C_{\mathrm{mob}}\right)$ |

$$+ \frac{y_{\mathrm{C/N}}}{y_{\mathrm{B/N}}}\int_0^\infty \kappa\left(\tau\right)\omega_{\mathrm{mob}}\left(\tau\right)d\tau + \left(\frac{\delta C}{\delta t}\right)_{ext}$$

| denitrifying plankton mobile water phase | $\mathrm{d}_t W_{\mathrm{mob}}$ | $=$ | $y_{\mathrm{B/N}}\eta_{\mathrm{mob}}W_{\mathrm{mob}} + \gamma\left(W_{\mathrm{imm}} - W_{\mathrm{mob}}\right)$ |

$$- \lambda W_{\mathrm{mob}} + \left(\frac{\delta W}{\delta t}\right)_{ext}$$

| dead-plankton age density, mobile water phase | $\omega_{\mathrm{mob}}\left(\tau = 0; \mathbf{r}, t\right) = \lambda W_{\mathrm{mob}}\left(\mathbf{r}, t\right)$ |
| | $\left(\partial_\tau + \mathrm{d}_t\right)\omega_{\mathrm{mob}} = -\kappa\left(\tau\right)\omega_{\mathrm{mob}} + \gamma\left(\omega_{\mathrm{imm}} - \omega_{\mathrm{mob}}\right)$ |

$$+ \left(\frac{\delta\omega(\tau;\mathbf{r},t)}{\delta t}\right)_{ext}$$

Here $\mathrm{d}_t \equiv \partial_t + \nabla\cdot\left(\mathbf{v} - \mathbb{D}\cdot\nabla\right)$ just an abbreviation (no 'material derivative'!)

Model parameters are:

$\mathtt{MWRT, MMWRT} =$ bulk or mobile water residence time in the bioreactor

$\mathrm{Pe} =$ Peclet number (defined for the mobile water phase)

$$\alpha \equiv \frac{1 - n}{n} = \frac{\text{rock matrix volume}}{\text{void (pore) volume}}, \quad \beta \equiv \frac{\theta_{\mathrm{imm}}}{\theta_{\mathrm{mob}}} = \frac{\text{immobile water volume}}{\text{mobile water volume}}$$

$\gamma =$ transfer coefficient mobile $\leftrightarrows$ immobile water phase

$r_{\max} =$ maximum utilization rate of dissolved Nitrate

$N_{\mathrm{HS}}, C_{\mathrm{HS}} =$ half-saturation nutrient (nitrate, carbon) concentrations (for each *factor* in the reaction rate independently)

$y_{\mathrm{B/N}}, y_{\mathrm{C/N}} =$ yield and need coefficients for biomass and carbon, respectively (measured against nitrate)

$\lambda =$ intrinsic cell loss rate (e.g. due to inability to divide)

$\kappa(\tau) =$ carbon release rate from dead biomass $\begin{matrix}\text{IF AGE-}\\ \equiv\\ \text{INDEPENDENT}\end{matrix}$ μ

$k_{\mathrm{att}}, k_{\mathrm{det}} =$ attachment/detachment coefficients plankton $\leftrightarrows$ biofilm plus parameters describing initial and boundary conditions. For our bioreactor operation modi, only initial (dead or living) biofilm residues after imperfect rock sterilization, as well as dead biomass and labile carbon levels in the time-dependent Dirichlet boundary condition at column top may represent or contain unknown values. For saturated flow conditions, the saturation degree S equals 1. For model formulation it is not necessary to specify measuring units at this stage, but in any application the units of model parameters will be consistent with those of model observables, for example, if W and X are expressed as 'biomass' per volume rather than 'cell numbers' per volume, and the electron acceptor concentration is expressed as Nitrate-N molarity, then the yield coefficient $y_{B/N}$ will be expressed as biomass gain per Nitrate-N mol, etc. Given the meagre biofilm development (1−2 cell layers were reported under similar conditions by Paulsen, Oppen and Bakke 1997; Hausner *et al.* 2000; Müller *et al.* 2000; Naumann 1999, 2000; Vomberg 2002, 2003), as expectable in groundwaters 'normally' contaminated from diffuse input sources, a modelling of the *spatial* biofilm development as an essentially 3-D process, with biofilm growth itself representable as a concentration-dependent diffusion process, under diffusion-limited nutrient availability, becomes dispensable here.

PARAMETER ESTIMATION AND DISCUSSION

So far, a mechanism of effective reuse, within the denitrification reactions performed by a given species, of labile carbon released from accumulating dead biomass of its own production could not be identified experimentally. This hypothesis should − before extending the model with too many parameters − first be checked upon its biochemical likelihood; the use of this alternative carbon source cannot be ascertained unambiguously relying on model fitting alone. A complementary test, with the possible role of 'own' recycled carbon quantified within the limits of a first-order model for its release rate (proportional to dead biomass, by factor μ) is suggested in Figure 2: if external nitrate is supplied in stoichiometric excess (such that there is, say, a 50% C deficit), bacterial population sizes, after exhaustion of external carbon, relate *monotonously* to the release rate μ for electron-donor of dead biomass origin.

In the experiment shown in Figure 3, poor denitrification was seen during the first $\sim$200 days of bioreactor operation under the natural flow conditions of the installed site. After a reduction of flow rates (q) from $4\,\mathrm{ml\,h}^{-1}$ to $0.3\,\mathrm{ml\,h}^{-1}$ (bulk MWRT in the column increased from $10\,\mathrm{d}$ to $136\,\mathrm{d}$) and a change from continuous to pulse-wise flow, a notable reduction of nitrate was seen. Thus, denitrification 'failure' during the first phase of *in situ* bioreactor operation cannot be attributed entirely to a lack of biochemically available carbon. The observed increase in denitrification performance can be explained in terms of the two-phase flow model formulated in the previous section (with an 'immobile' and a mobile water phase), taking the externally applied flow rate q as an explicit parameter instead of the former (scaling) parameter MWRT. For nutrient-depleted conditions ($\eta/q \ll 1$ as both substrates are very dilute, but one cannot say in advance which of them, nitrate or carbon? is in excess or deficit, respectively), the CSTR version of the balance equations

$$\frac{d}{dt}N_{\mathrm{mob}} = q(N_{\mathrm{INPUT}} - N_{\mathrm{mob}}) - \eta_{\mathrm{mob}}W_{\mathrm{mob}} + \gamma(N_{\mathrm{imm}} - N_{\mathrm{mob}})$$

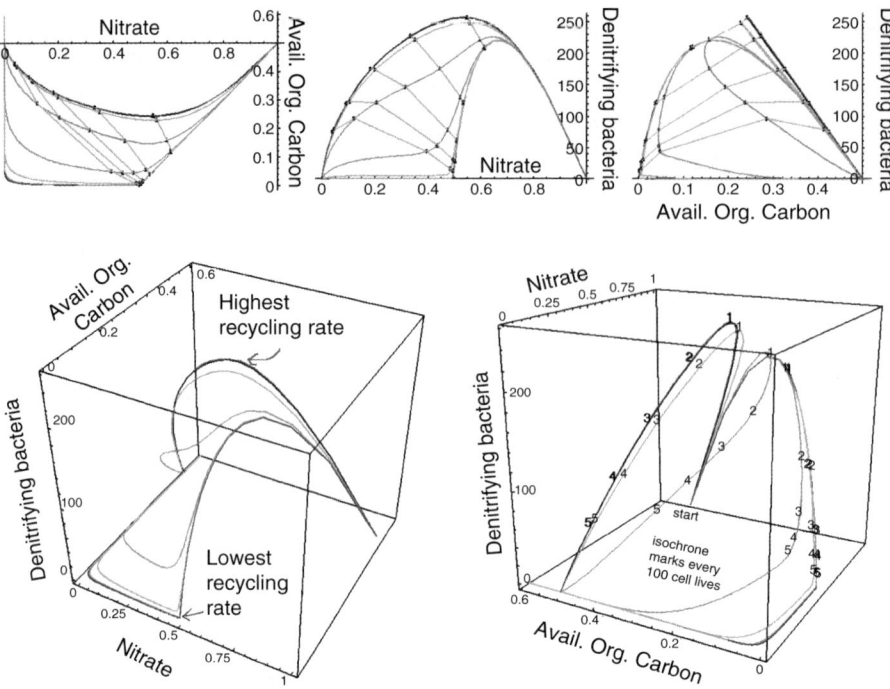

Figure 2. Carbon recycling scenario.

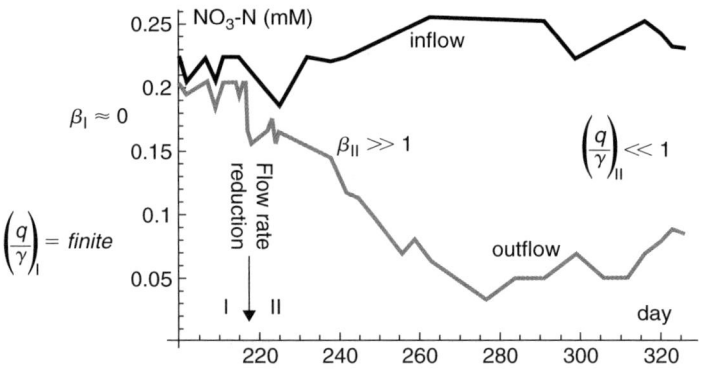

Figure 3. *In situ* bioreactor operation with an abrupt change in flow conditions.

$$\frac{d}{dt}C_{\text{mob}} = q(C_{\text{INPUT}} - C_{\text{mob}}) - y_{\text{C/N}}\eta_{\text{mob}}W_{\text{mob}} + \gamma(C_{\text{imm}} - C_{\text{mob}}) + \frac{y_{\text{C/N}}}{y_{\text{B/N}}}\mu\Omega_{\text{mob}}$$

$$\frac{d}{dt}W_{\text{mob}} = q(W_{\text{INPUT}} - W_{\text{mob}}) + (y_{\text{B/N}}\eta_{\text{mob}} - \lambda)W_{\text{mob}} + \gamma(W_{\text{imm}} - W_{\text{mob}})$$

$$\frac{d}{dt}\Omega_{\text{mob}} = q(\Omega_{\text{INPUT}} - \Omega_{\text{mob}}) + \lambda W_{\text{mob}} + \gamma(\Omega_{\text{imm}} - \Omega_{\text{mob}}) - \mu\Omega_{\text{mob}}$$

$$\frac{d}{dt}N_{\text{imm}} = -\eta_{\text{imm}}\left(W_{\text{imm}} + \frac{\alpha}{\beta S}X\right) + \frac{\gamma}{\beta}(N_{\text{mob}} - N_{\text{imm}})$$

$$\frac{d}{dt}C_{\text{imm}} = -y_{\text{C/N}}\eta_{\text{imm}}\left(W_{\text{imm}} + \frac{\alpha}{\beta S}X\right) + \frac{\gamma}{\beta}(C_{\text{mob}} - C_{\text{imm}}) + \frac{y_{\text{C/N}}}{y_{\text{B/N}}}\mu\left(\Omega_{\text{imm}} + \frac{\alpha}{\beta S}\Xi\right)$$

$$\frac{d}{dt}W_{\text{imm}} = (y_{\text{B/N}}\eta_{\text{imm}} - \lambda - k_{\text{att}})W_{\text{imm}} + \frac{\gamma}{\beta}(W_{\text{mob}} - W_{\text{imm}}) + k_{\text{det}}\frac{\alpha}{\beta S}X$$

$$\frac{d}{dt}\Omega_{\text{imm}} = \lambda W_{\text{imm}} + \frac{\gamma}{\beta}(\Omega_{\text{mob}} - \Omega_{\text{imm}}) - (\mu + k_{\text{att}})\Omega_{\text{imm}} + k_{\text{det}}\frac{\alpha}{\beta S}\Xi$$

$$\frac{d}{dt}X = (y_{\text{B/N}}\eta_{\text{imm}} - \lambda - k_{\text{det}})X + k_{\text{att}}\frac{\beta S}{\alpha}W_{\text{imm}}$$

$$\frac{d}{dt}\Xi = \lambda X + k_{\text{att}}\frac{\beta S}{\alpha}\Omega_{\text{imm}} - (\mu + k_{\text{det}})\Xi$$

yields the following estimate for nitrate reduction:

$$|\Delta N| = N_{\text{INPUT}} - N_{\text{OUT}} = \left[\beta\left(\frac{k_{\text{a}}}{k_{\text{d}}} + 1\right)\left(1 + \frac{q}{\gamma}\right) + 1\right]\frac{\eta}{q}W_{\text{INPUT}} + O\left[\left(\frac{\eta}{q}\right)^2\right]$$

Missing bacterial numbers (for the active denitrifiers specifically) prevents an unambiguous parameter reconstruction from the nitrate breakthrough curve alone. However, the asymptotic approximation of relative nitrate reduction enables estimating the ratio of bacterial attachment/detachment coefficients:

$$\frac{|\Delta N|_{\text{II}}}{|\Delta N|_{\text{I}}} \approx \frac{N_{\text{OUT,II}}}{N_{\text{OUT,I}}}\frac{q_{\text{I}}}{q_{\text{II}}}\beta_{\text{II}}\frac{k_{\text{a}}}{k_{\text{d}}} \implies \frac{k_{\text{a}}}{k_{\text{d}}} \approx \frac{1}{\beta_{\text{II}}}\frac{q_{\text{II}}}{q_{\text{I}}}\frac{|\Delta N|_{\text{II}}/N_{\text{OUT,II}}}{|\Delta N|_{\text{I}}/N_{\text{OUT,I}}} \approx 0.9$$

This indicates that biofilm development as attained after more than 300 days *in situ* bioreactor operation is still in a rather 'young' phase (mature stages are more likely to be characterized by a stronger $k_{\text{a}}/k_{\text{d}}$ imbalance).

Some of the mechanisms/components included in the above model are asymptotically excluded in a short-term experiment; exclusion of dead biomass influence, alongside with exclusion of immobile-phase storage by the choice of rapid flow, reduces the model dimension from '10' (actually, 7 + three times a τ-continuum) to 4, and facilitates parameter reconstruction (Table 1 and Figure 4, upper part). On the other hand, population sizes attained in this 'accelerated' bioreactor operation for about 50 days (starting from virtually sterile initial conditions) may not be representative for the population sizes attained after the far longer times at which aquifer-scale processes are normally evaluated (Figure 4, lower part). The biokinetic parameter estimations given in Table 1 confirm that in the short-term experiment the groundwater 'bioreactor' was operated at a reaction rate far under its theoretical maximum, so (in combination with a rapid flow) the asymptotic approximation is valid.

PROVISIONAL CONCLUSIONS AND DISCUSSION

Alongside with the interpretation of available environmental isotope data, a solid package of artificial tracer methods has been established for investigating solute transport and particle migration processes in karst systems under saturated flow conditions (cf. Behrens, Drost, Wolf *et al.* 1997) and, following a decade of research (Seiler, Maloszewski and Behrens 1989; Seiler, Behrens and Wolf 1996; Seiler and Behrens 1992; Seiler 1999),

Table 1. Parameter estimations from a short-term bioreactor operation (cf. Figure 4).

Parameter	Probable variation limits
Maximum N utilization rate	0.15–0.25 µg NO_3-N/cell/d
Half-saturation conc. for Nitrate	2.5–3.5 g NO_3-N/l
Half-saturation conc. for Carbon	The stoichiom. equiv. of (i.e., $y_{C/N}$ times) 1–2 g NO_3-N/l
Presumable 'Available Organic Carbon' in inflow	The stoichiom. equiv. of (i.e., $y_{C/N}$ times) 3–4 mg NO_3-N/l
Biomass specific yield $y_{B/N}$	$0.5–1.5 \times 10^9$ cell/mg NO_3-N
Cell attachment rate	$0.25\,d^{-1}$ (down to 1/10 of it)
Cell detachment rate	$0.025\,d^{-1}$ (down to 1/20 of it)
Intrinsic population 'half-life'	>200 d
Initial biofilm traces	75–125 cell/g rock

flow and transport processes in Southern Franconian Alb groundwaters are satisfactorily understood. However, the understanding of 'natural attenuation' processes in the Southern Franconian Alb is still at its beginning, especially as regards the role of vadose zone processes (Einsiedl, Maloszewski and Stichler 2003). Earlier biodenitrification experiments carried out for this system (Hausner *et al.* 2000; Müller *et al.* 2000; Naumann *et al.* 2001) in batch mode have been interpreted in terms of a lack of 'organically available carbon'. Our experiments involving macroscopic flow and transport indicated that nutrient residence times are no less decisive a limiting factor for biodenitrification under natural flow conditions. Under saturated flow conditions, residence times are mainly controlled by porosity values in the dual rock structure; when flow becomes unsaturated the dynamics of the immobile/mobile water ratio can no longer be predicted from porosity values alone (cf. Ghergut, Maloszewski, Naumann *et al.* 2003), but in any flow situation the reef facies will offer an advantage over the bedded facies, as had been expected by Seiler and Hartmann, 1997. On the other hand it appears that more e^--donor was available for biodenitrification in at least part of our experiments, than what was measured as 'DOC' (Dissolved Organic Carbon) by common methods, its origin being yet unclear. Once experimental contaminations have been excluded, three mechanisms could be suggested: dead biomass recycling within the given bacterial population, slow exhaustion of sedimentary deposits in the rock material, or a direct use of DIC (Dissolved Inorganic Carbon) from the carbonate-abundant rock (autolithotrophy), any such mechanism, however, presupposing the existence of an adequate energy source; and, in this respect, none of the facies types (reef or bedded) appears as privileged over the other. The availability and origin of the additional electron-donor source cannot be clarified by merely fitting a transport–consumption model to the nitrate reduction data. Nutrient fate in a representative, laboratory-scale denitrifying system is currently being investigated in detail with the aid of ^{13}C- and ^{15}N-labelled nutrients by Wolf and Einsiedl (publication details not available as yet) following hypotheses formulated in Einsiedl, Maloszewski and Stichler 2003 with respect to the transport of humic substances through the vadose (epikarst) zone. Also, the identification and characterization, by gene-molecular techniques, of denitrifying bacteria occurring in Southern Franconian Alb groundwaters is in progress (Vomberg 2002, 2003).

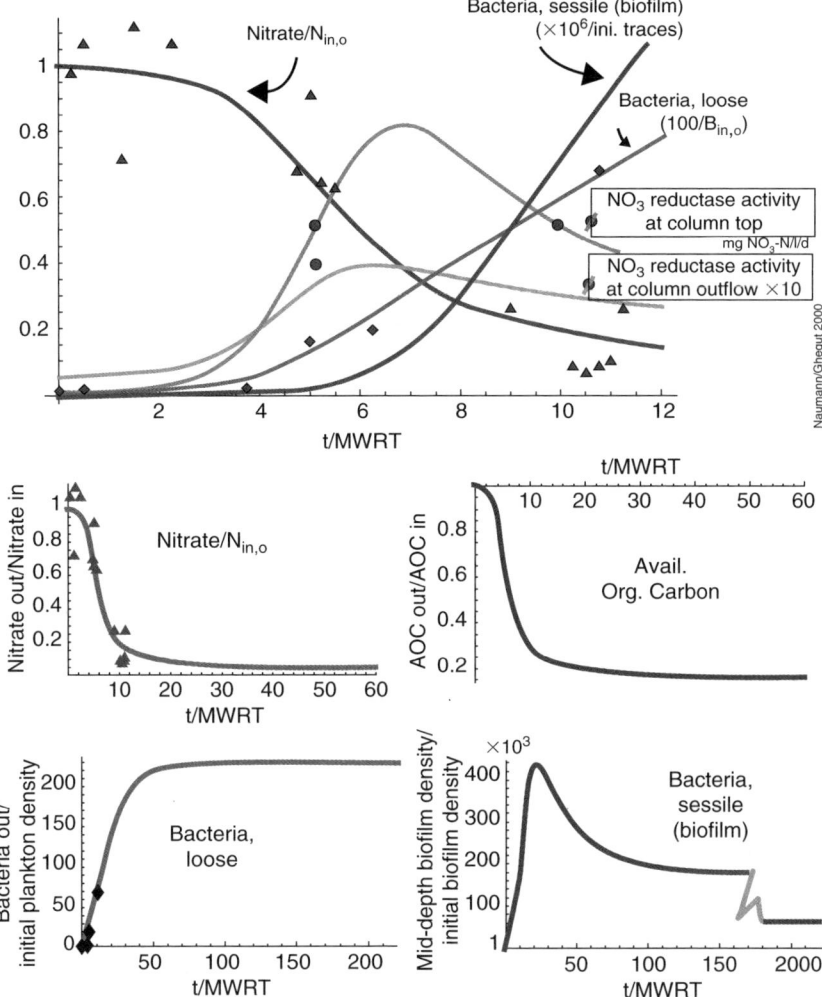

Figure 4. Comparison between model predictions and measured bioreactor performance.

ACKNOWLEDGEMENTS

One of the authors expresses thanks to the German Academic Exchange Office (DAAD) for a research scholarship, to the GSF Research Centre for the permission to use computer facilities of its Hydrological Institute, and to professors Behrens, Sauter and Walcher for useful discussions.

REFERENCES

Behrens, H., Drost, W., Wolf, M., Orth, J.P. and Merkl, G. 1997: *Groundwater exploration and contaminant migration testing in a confined karst aquifer of the Swabian Jurassic* (Germany). In: A. Kranjc (ed.): Tracer Hydrology 1997. Balkema, Rotterdam, pp. 305−312.

Einsiedl, F., Maloszewski, P. and Stichler, W. 2003: *'Natural Attenuation' Prozesse in der vadosen Zone eines Karstsystems.* In: W. Walther, F. Reinstorf, E. Worch and S. Wohnlich (eds): Diffuse input of chemicals into soil and groundwater – assessment and management, Proc. IGW Workshop. Technical University, Dresden, pp. 49–58.

Georgescu, A. 1990: *Aproximatii asimptotice (in Romanian language).* Editura Tehnica, Bucharest, pp. 52–66.

Ghergut, I., Maloszewski, P., Naumann, U., Trimborn, P., Stichler, W. and Seiler, K.-P. 2003: *Understanding bioreactor behaviour upon changing flow conditions.* In: W. Walther, F. Reinstorf, E. Worch and S. Wohnlich (eds): *Diffuse input of chemicals into soil and groundwater – assessment and management,* Proc. IGW Workshop. Technical University, Dresden, pp. 59–64.

Hausner, M., Lawrence, J.R., Wolfaardt, G.M., Schloter, M., Seiler, K.-P. and Hartmann, A. 2000: *The use of immunological techniques and confocal LSM for the characterization of* Agrobacterium tumefaciens *and* Pseudomonas fluorescens *atrazine-utilizing biofilms.* In: H.-C. Flemming, U. Szewzyk and T. Griebe (eds): Biofilms, investigative methods and applications. Technomic, Lancaster, pp. 143–153.

Logan, J.D. 2001: *Transport modeling in hydrogeochemical systems.* Springer, Heidelberg.

Müller, E., Assmus, B., Hartmann, A. and Seiler, K.-P. 2000: *The* in situ *detection of a microbial biofilm community on karst rock coupons in a groundwater habitat.* In: H.-C. Flemming, U. Szewzyk and T. Griebe (eds): Biofilms, Investigative methods and application. Technomic, Lancaster, pp. 155–163.

Naumann, U. 1999: *Der Karst im Labor – Untersuchungen zur Mikrobiologie im Grundwasserleiter.* In: GSF Institute of Hydrology Annual Report 1998 (in German language). GSF Research Centre, Neuherberg, pp. 182–191.

Naumann, U. 2000: *Bakterielle Lebensgemeinschaften im Karstgrundwasser – Einsatz molekularbiologischer Methoden.* In: GSF Institute of Hydrology Annual Report 1999 (in German language). GSF Research Centre, Neuherberg, pp. 69–78.

Naumann, U., Vomberg, I., Hartmann, A. and Seiler, K.-P. 2001: *Microbial denitrification in a karst aquifer.* In: K.-P. Seiler and S. Wohnlich (eds): New approaches characterizing groundwater flow, Vol. 1. Balkema, Rotterdam, pp. 607–610.

Paulsen, J.E., Oppen, E. and Bakke, R. 1997: Biofilm morphology in porous media, a study with microscopic and image techniques. *Water Science and Technology 36*: 1–9.

Segel, L.A. (ed.) 1991: *Biological kinetics.* University Press, Cambridge.

Seiler, K.-P. (ed.) 1999: *Grundwasserschutz im Karst der südlichen Frankenalb* (in German language). GSF Special Report 4/99. GSF Research Centre, Neuherberg.

Seiler, K.-P. and Behrens, H. 1992: *Groundwater in carbonate rocks of the Upper Jurassic in the Franconian Alb and its susceptibility to contaminants.* In: H. Hötzl and A. Werner (eds): Tracer Hydrology 1992. Balkema, Rotterdam, pp. 259–266.

Seiler, K.-P. and Hartmann, A. 1997: *Microbiologic activities in karst aquifers with matrix porosity and consequences for groundwater protection in the Franconian Alb, Germany.* In: A. Kranjc (ed.): Tracer Hydrology 1997. Balkema, Rotterdam, pp. 339–345.

Seiler, K.-P., Behrens, H. and Wolf, M. 1996: *Use of artificial and environmental tracers to study storage and drainage of groundwater in the Franconian Alb, Germany, and the consequences for groundwater protection.* In: Proc. on Isotopes in Water Resources Management. IAEA, Vienna, pp. 135–145.

Seiler, K.-P., Maloszewski, P. and Behrens, H. 1989: Hydrodynamic dispersion in karstified limestones and dolomites in the Upper Jurassic of the Franconian Alb. *Journal of Hydrology 108*: 235–247.

Starr, R.C. and Gillham, R.W. 1993: Denitrification and organic carbon availability in two aquifers. *Ground Water 31* (6): 934–947.

Vomberg, I. 2002: *Denitrifikationspotential von Biofilmen unter Grundwasserbedingungen.* In: GSF Institute of Hydrology Annual Report 2001 (in German language). GSF Research Centre, Neuherberg, pp. 118–125.

Vomberg, I. 2003: *Analyse von Biofilmen aus Quellen der Fränkischen Alb.* In: GSF Institute of Hydrology Annual Report 2002 (in German language). GSF Research Centre, Neuherberg, pp. 117–120.

CHAPTER 9

Denitrification in a karst aquifer with matrix porosity

K.-P. Seiler and I. Vomberg
GSF-National Research Centre, D-85758 Neuherberg, Germany

ABSTRACT: In the Franconian Alb, Germany, bedded carbonates and reef dolomites of the Upper Jurassic each represent a facies with syngenetic porosities; both facies are fissured and karstified, occur site by site, receive the same groundwater recharge and are subject to the same mode of land use. A comparative artificial and environmental tracer study evidenced that in reef dolomites matrix flow is in the range of metres per year, and in fissures of the bedded carbonates in the range of kilometres per day; additionally non-reactive tracer dilution is much stronger in reef dolomites (dispersivities >50 m) than in bedded carbonates (dispersivities <10 m). As a consequence, reef dolomites have higher storage capacities (porosities 6–10 vol.%) than the bedded carbonates (porosities <2.5 vol.%), which may lead to a long-term accumulation of contaminants in reef groundwater as far as it does not undergo microbial disintegration. Collecting micro-organisms by traditional water sampling and by incubation of sterilized rock coupons proved that microbial numbers and functional groups in the reef facies are more abundant under agro- than beneath forest lands, and very few microbial communities occur in old groundwater (>1,000 years). Obviously, the abundance of colonies and functional groups reflect the prevailing nutrient and energy supply of micro-organisms through the land use. Laboratory experiments with groundwater samples from the study area without and with incubated rock coupons resulted in low respectively high disintegration activities, which are attributed to the bio-films upon rocks. Since denitrification in water needs anaerobic conditions, while the redox potential of groundwater in the karst is in the range of +400 mV, it was supposed that under field conditions the bio-films provide a chemical environment, which may significantly differ from flowing water. This hypothesis has been proofed by field investigations through an enrichment of ^{34}S and ^{18}O in sulphates, both in the groundwater, and will be extended to ^{15}N and ^{18}O in nitrates.

BACKGROUND AND GEOLOGIC SETTING OF THE STUDY AREA

Mesozoic sandstones, chalks, compacted clays and reef carbonates are fissured and mostly have significant matrix porosities. Both fissure and matrix porosities contrast greatly in their hydraulic conductivities. As a result, these consolidated rocks dispose of an important drainage and storage capacity for water and contaminants as well, which often

- leads to short and long-term problems in the groundwater quality, issued from diffuse pollution sources (Seiler et al. 1992),
- is difficult to monitor in real time, and
- implies serious groundwater protection measures.

This statement holds if physical, chemical and microbial elimination processes do not sufficiently contribute to natural attenuation; among these attenuation processes

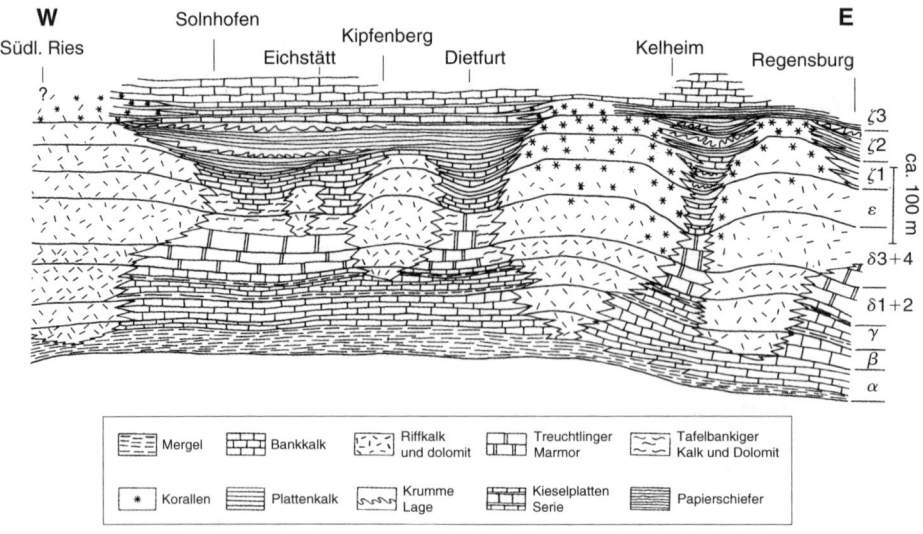

Figure 1. General stratigraphic section of the carbonates of the Franconian Alb, starting from the Dogger/Malm interface (scanned from Meyer and Schmidt-Kaler 1989).

microbial disintegration in karst aquifers has not yet been studied. Therefore this investigation aims to contribute in describing such a bi-porous media using traditional and artificial as well as environmental tracer techniques, to study boundary conditions for microbial incubation of rock surfaces, microbial disintegration processes on the laboratory scale and to extrapolate the laboratory results to the field scale.

Such results are of general interest for most Mesozoic rocks. Although, the results found for the karst of the Frankonian Alb have still a local character; further research may contribute to this general question.

The carbonates of the Franconian Alb, Germany, belong to two consecutive cycles of sedimentation, each starting with a marl facies, which changes gradually into bedded carbonates. In the upper cycle, however, dolomite reefs replace bedded carbonates over significant stratigraphic intervals (Figure 1) and therefore allow a comparative study of the contaminant behaviour in rocks with and without matrix porosities. This comparative study was further facilitated by the facts that

• both facies occur site by site (Figure 1) with sharp vertical interfaces,
• groundwater recharge is the same ($300 \, \mathrm{mm \, a^{-1}}$) all over the study area and
• agriculture, forest and urban land use is practised independently of the facies.

As compared to bedded lime-stones, reef dolomites commonly have porosities of syngenetic and dolomitisation origin (>6 vol.%) (Weiss 1987). Since hydraulic conductivities in the matrix pores ($<10^{-8} \, \mathrm{m \, s^{-1}}$) are more than five orders of magnitude lower than in fissures (Michel 2001; Weis 1987), the spectra of flow velocities and dispersivities are expected to be different in both these facies; as a consequence, the susceptibility of the reef and the bedded facies for contaminants had to be assessed in different ways.

Matrix porosities of both facies have been measured using the mercury penetration method. The result of these measurements (Figure 2) is that in the reef facies matrix

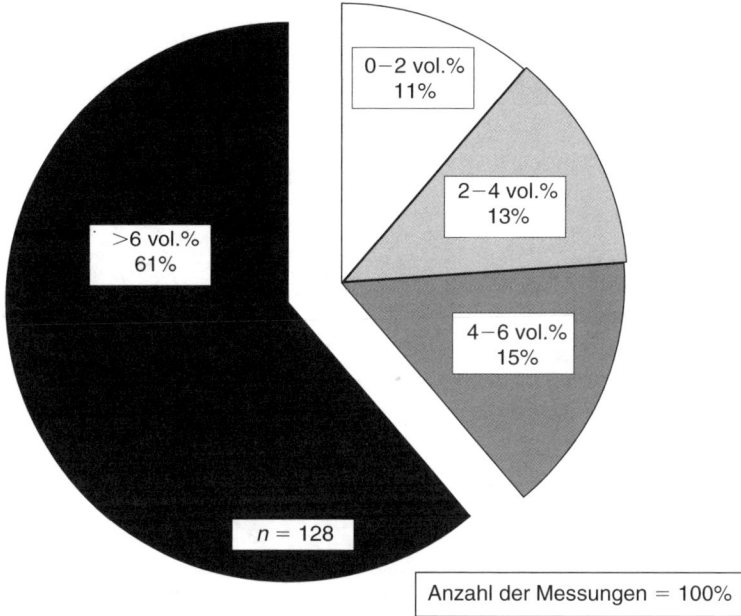

Figure 2. Matrix porosities of carbonates of the Franconian Alb. Porosities lower than 2 vol.% correspond to the bedded carbonates, higher than 2 vol.% to the reef facies. Average reef porosities are between 6 and 10 vol.%; maximum was 25 vol.%. Number of investigated cores: 127 = 100%.

porosities exceed 6 vol.%, whereas in the bedded carbonates they are lower than 2 vol.% and mostly close to none. In between 2 and 6 vol.% of porosities (Figure 2) the bedded dolomite facies groups, which is not very abundant in the study area.

Pfaff (1987) studied the role of fissures in the research area and found that fissure porosity is at maximum 2.5 vol.% and that an average of 40 vol.% of the groundwater recharge discharges immediately through fissures and 60 vol.% through the matrix pores. Similar geologic and hydrogeologic settings are known from the Bunter in the Saar region (Seiler 1969).

RESULTS OF THE ARTIFICIAL AND ENVIRONMENTAL TRACER INVESTIGATIONS

The research area covers $1,000 \, km^2$; more than 150 dye tracer tests have been executed applying Fluoresceine and Eosin, which both behave conservatively with respect to subsurface flow (Behrens 1971). All tracer experiments were executed during late spring and early summer, the season with small groundwater recharge.

Parallel to these tracer experiments, at the end of the dry weather period (November–December) of every year, water samples from reef springs have been systematically collected for Tritium analysis; during this period of the year fissures drain the water from the porous rock matrix. A respective systematic sampling of groundwater from the bedded limestones was not undertaken.

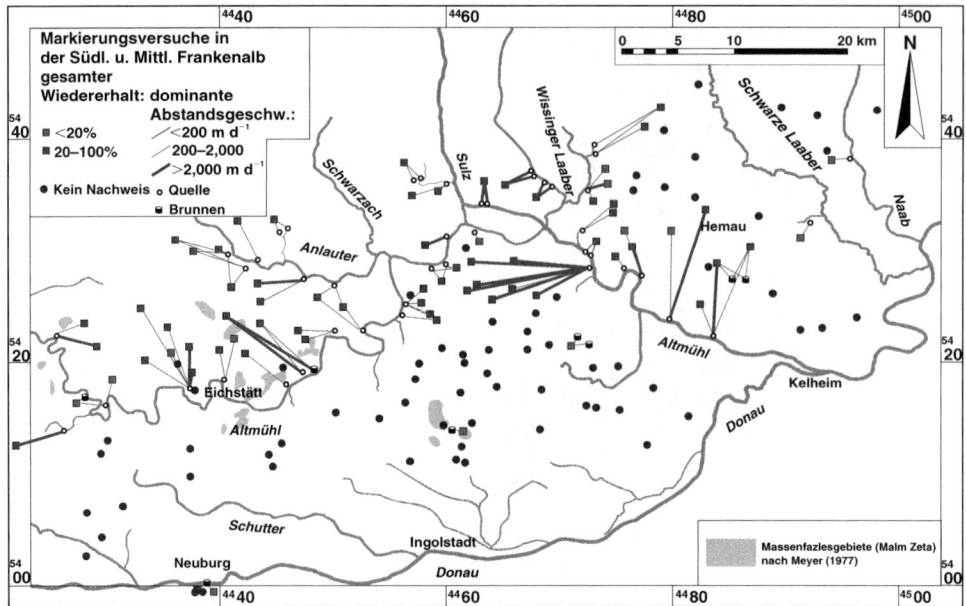

Figure 3. Results of tracer experiments with (■—○) and without (●) recovery in the Karst of the Franconian Alb (Glaser 1998). In the area between Altmühl and Donau reef dolomites prevail, in the other areas bedded carbonates.

Results of artificial tracer experiments

All tracer experiments have been executed with 5 kg of Fluoresceine or Eosin; the detection limit for these tracers is 2, respectively 20 μg l^{-1}.

Sampling for tracer detection was performed at the beginning of the tracer experiment four times a day and switched continuously till the end of a 7-year sampling period to once every three months. The shortest tracer-break-through lasted one day and occurred in the bedded carbonates on a distances of several kilometres, the longest tracer-break-through exceeded 7 years in the reef facies at a distance of 1 km from the injection point.

About half of the tracer experiments (Figure 3) have been performed in the bedded carbonates and yielded high tracer recoveries (>20%), and flow velocities ranging from 200 to >2,000 m d^{-1} (Figure 3). The other half of the tracer experiments was performed in the reef facies and always ended at distances exceeding 1.5 to 2 km without measurable tracer recovery; this, however, is not linked to any tracer sorption, but to tracer dilution (Seiler et al. 1989) which gets enhanced in the reef facies by diffusive tracer exchanges from the fissure to matrix water pathway and by dead-end-fissures.

As referred to an input of 5 kg of a tracer, the break-through curves in the study area can be subdivided into three categories (Figure 4):
1) A high concentration maximum and a narrow geometry (Figure 4, bedded facies) refers to flow velocities exceeding 1,500 m d^{-1} and dispersivities lower than 10 m (Seiler *et al.* 1989); both stand for some kind of piston flow in solution channels.
2) A concentration maximum close to the detection limit (Figure 4, reef facies), a very wide geometry of the tracer-break-through and a recovery of <5% are attributed to enhanced hydrodynamic dispersivities (>50 m) supported by molecular diffusion

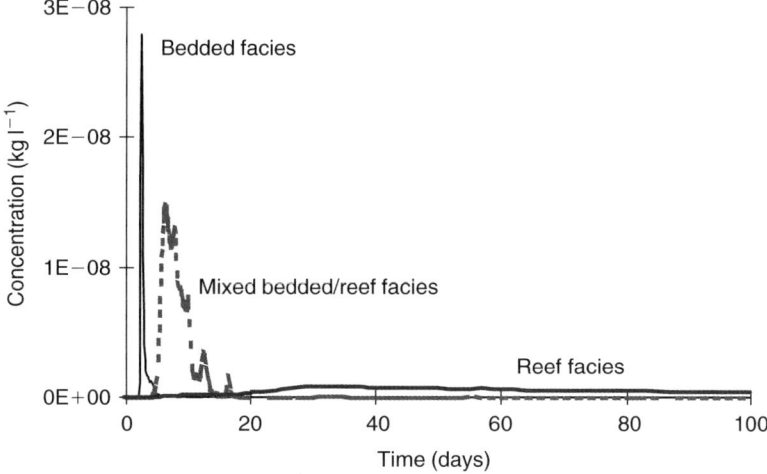

Figure 4. Break-through curves for characteristic facies units.

(Sudicky and Friend 1981). Such tracer break-through-curves have only been observed in the reef facies on distances shorter than 1.5 km to 2 km; over longer distances no tracer recovery was ever observed although wells and springs have been sampled over a seven-year period of time. As shown by laboratory experiments, these observations are not linked to tracer sorption or microbial disintegration of the organic tracer; it is due to hydrodynamic dispersion enhanced by a molecular tracer exchange between fissure and matrix path ways and favoured by dead-end-fissures. Respective numerical modelling (Maloszewski and Seiler, unpublished) proves the correctness of this interpretation.

3) A concentration maximum and tracer-break-through geometry intermediate to the two mentioned before (Figure 4, mixed bedded/reef facies) is attributed to flow in fissures with a wide array of fissure openings.

Tritium in groundwater out of the reef facies

Tracer tests provide instantaneous, preferential and, due to the weak transverse hydrodynamic dispersion (Seiler *et al.* 1989), sectorial information about groundwater flow in an aquifer. In contrast, environmental isotopes result in catchment integrated and radioactive environmental tracers, such as ^3H, time integrated information.

In the study area, ^3H concentrations of groundwater (Pfaff 1987) express during the wet season mostly the contribution of actual infiltration and throughout the dry season an origin from the storage system of the aquifer. From repeated sampling in the dry season during investigation time (Seiler et al. 1995) results show (Figure 5) that in groundwater out of

•   the bedded carbonates ^3H is close to the ^3H concentration in precipitation of the same or the few preceding years,

•   the reef facies ^3H concentrations are systematically low and stand for long mean residence times.

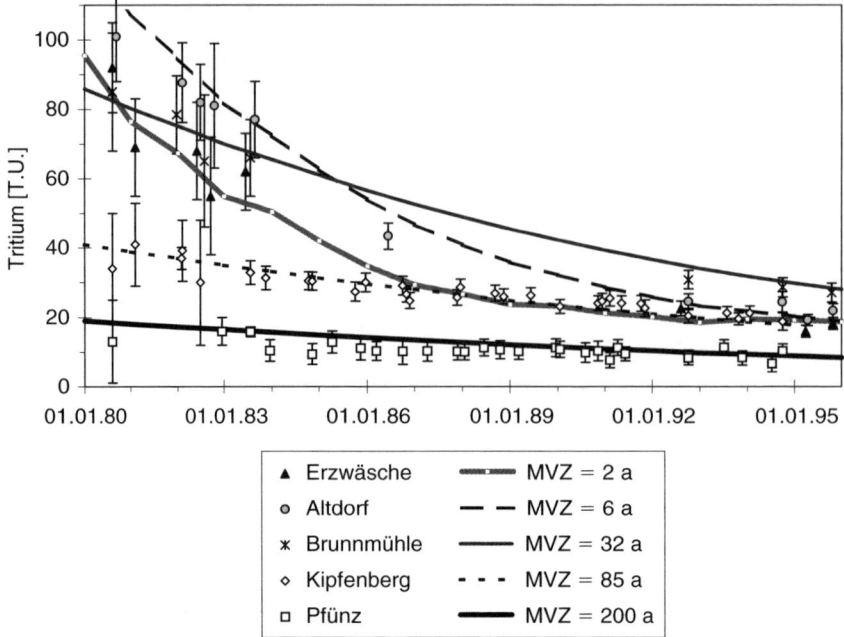

Figure 5. ^{3}H concentrations in springs out of the bedded (Erzwäsche, Altdorf) reef.

Long-term Tritium records (Figure 5) have been evaluated supposing an exponential age distribution in the subsurface water; this evaluation results in mean residence times of the matrix groundwater of 50 to 200 years as compared to residence times of days and weeks in fissures of the bedded carbonates.

With respect of the fate of contaminants in the study area, the consequences of very short residence times in the bedded carbonates have to be assessed in contrast with residence times of many decades in the reef facies (Kipfenberg, Pfünz).

THE FATE OF NITRATES AND MICROBIAL ACTIVITIES IN THE REEF FACIES

The study area is almost equally used by agriculture and forestry; cities exceeding the size of a village are only located in valleys and close to springs.

In the study area nitrogen excess from agriculture amounts to 50–40 kg N/(ha year) and groundwater recharge is about 300 mm/year. From these numbers a concentration of 17 to 10 mg N/l or 66 to 40 mg NO_3^-/l plus the natural input (see below) is expected in the groundwater. Indeed, measured concentrations in groundwater out of the bedded carbonates range in agro-lands from 60 to 80 mg l^{-1}. Under similar conditions, in the reef facies the nitrate concentrations are mostly below 20 mg l^{-1} (Figure 6), which is close to the natural nitrogen input of 20 to 25 mg NO_3^-/l.

Apparently, the groundwater out of the bedded carbonates approximates the current pollution situation quite correctly and the groundwater out of the reef facies was not yet

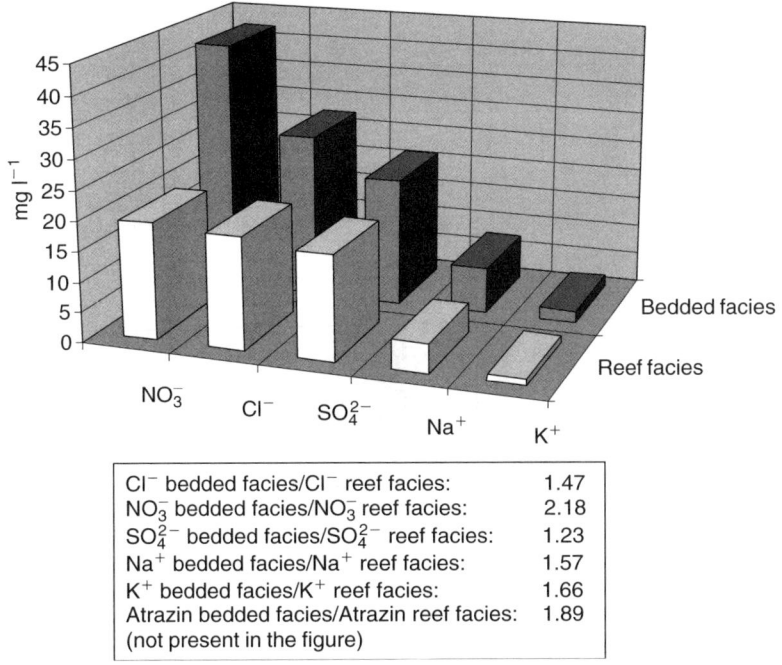

Cl⁻ bedded facies/Cl⁻ reef facies:	1.47
NO₃⁻ bedded facies/NO₃⁻ reef facies:	2.18
SO₄²⁻ bedded facies/SO₄²⁻ reef facies:	1.23
Na⁺ bedded facies/Na⁺ reef facies:	1.57
K⁺ bedded facies/K⁺ reef facies:	1.66
Atrazin bedded facies/Atrazin reef facies:	1.89
(not present in the figure)	

Figure 6. Average concentrations of some agrochemicals in the groundwater out of the bedded carbonates without and in the reef facies with matrix porosity in the Franconian Alb, Germany (Glaser 1998).

fully charged with pollutants, either because of
• the high dilution volume and the long residence times of water in the matrix, or
• microbial processes occur in the reef facies and decrease the pollutant concentration.

Taking into account the ratio of the concentrations of agrochemicals in the bedded and the reef facies (Figure 6), this ratio is similar for K^+, Na^+ and the non-reactive Cl^-, high for nitrates and Atrazin and low for sulphates. Atrazin, which has been included in this consideration, however, can not yet really be assessed, because it has not been applied in an as long period of time as the before mentioned agro-chemicals. The deficit of nitrate and the surplus of sulphate, however, could be attributed to a coupled microbial oxidation and reduction process of both agrochemicals, the deficit of Atrazin possibly to a microbial metabolisation.

Bacteria in groundwater out of the reef facies

If microbial activities play any role in the reef facies, this should be reflected in differences in the number and composition of microbial communities (Figure 7) beneath forest (A3–A4) as compared to agro-lands (A1–A2), which both differ from the input of contaminants. Repeated sampling of groundwater was performed in representative reef areas to determine the total bacterial numbers (DAPI-stain) and the colony forming units (cfu) on R_2A agar (aerobic incubation, 22°C) and on a denitrification agar (R_2A agar + 0.5%

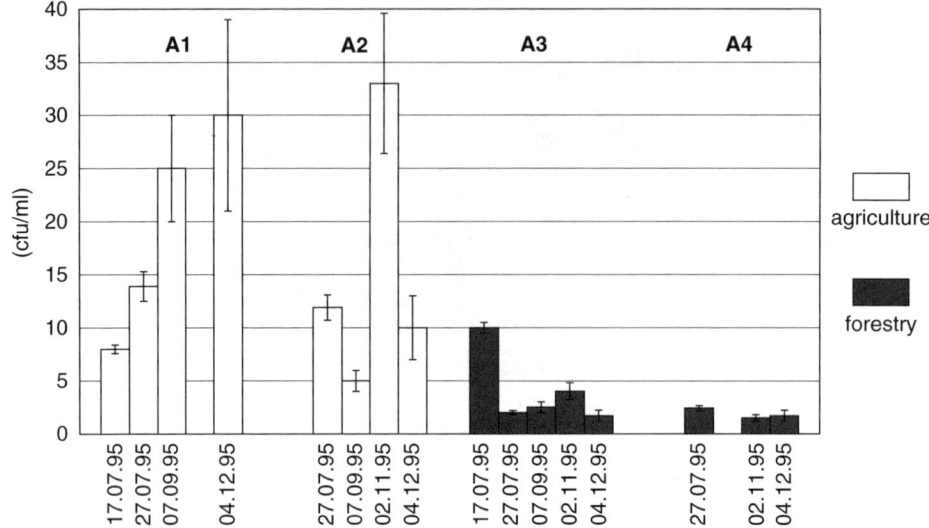

Figure 7. Colony forming units (cfu) on denitrification agar (R_2A agar $+ 0.5\%$ KNO_3) under anaerobic conditions.

KNO_3, anaerobic incubation, $22°C$). As compared to the total counts, less than only 1% could be cultivated on R_2A agar. The cfus did not show significant seasonal differences during July to December; however, the water samples out of agro-land always had higher numbers of total and viable counts than those of forest areas. This is interpreted as an adaptation of microbial numbers and functional groups to the available energy and nutrient sources (C, N, P); any incubation time to adapt to changing input conditions could not be considered although it is known to exist (Seiler and Alvarado-Rivas 2002).

Laboratory denitrification experiments

The natural DOC content of the groundwater, used for denitrification experiments, was $2\,mg\,C/l$ and the original nitrate concentration was $10\,mg\,l^{-1}$; all samples have been completed to $50\,mg\,NO_3^-/l$.

Since it is important not only to study the microbial activity of cultivated bacteria from the groundwater, but also of bio-films, the disintegration experiments have been performed both with

- only groundwater (300 ml),
- groundwater and a microbial incubated rock coupon (incubation time in springs and wells of 11 months), and
- with adding further organics to the before mentioned experiments.

All experiments were incubated for 3 weeks under anaerobic conditions and each experiment was executed in five replicates.

The NO_3^--, NO_2^-- and N_2O-concentrations in the reactor were repeatedly measured (Figure 8) and the water volume was always kept at the same level. No N-balance could be performed for all these experiments, because of the N-abundance in the air. From all these

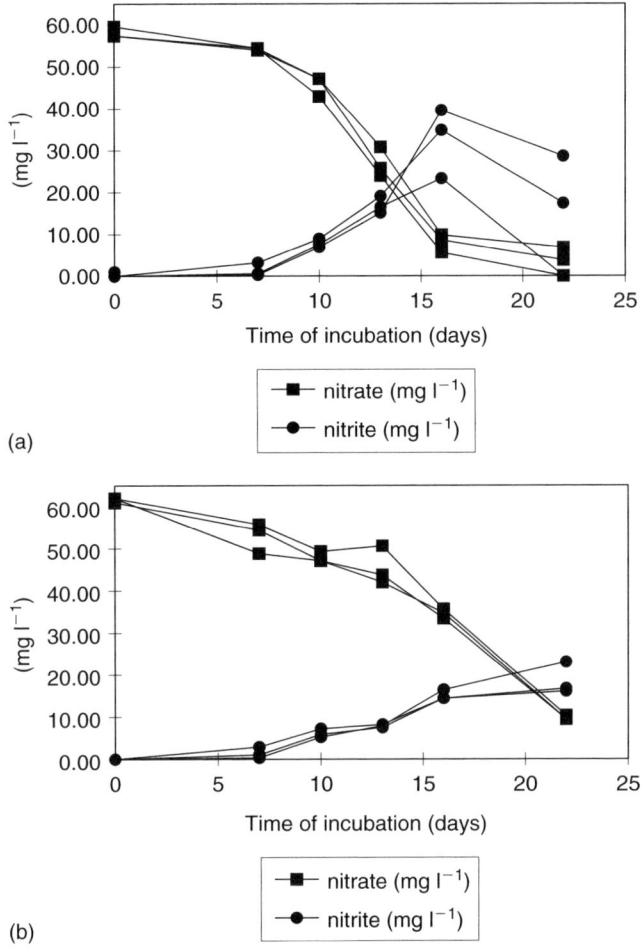

Figure 8. Concentration changes of nitrate and nitrite in anaerobically incubated water samples from agriculture (a) and forest areas (b); nitrate was filled up to $50\,mg\,l^{-1}$ and no additional carbon source was added.

experiments it becomes evident that denitrification occurred, however,

- it was less intensive under forest than agro-land (Figure 8);
- it was slowest in the case of groundwater without any additives;
- it was accelerated when adding organics;
- it sped up significantly when adding incubated rock coupons (rock pieces with bio-films); and
- it did not change in the presence of bio-films, when adding organics.

Obviously, under anaerobic conditions denitrification is most significant in the presence of bio-films. This bio-film may not only be considered as a catalyst but also as a special nutrient store for the microbial community. This raises many questions on the interaction of bio-films with both water and the solid phase, which are actually under research by the lady co-author.

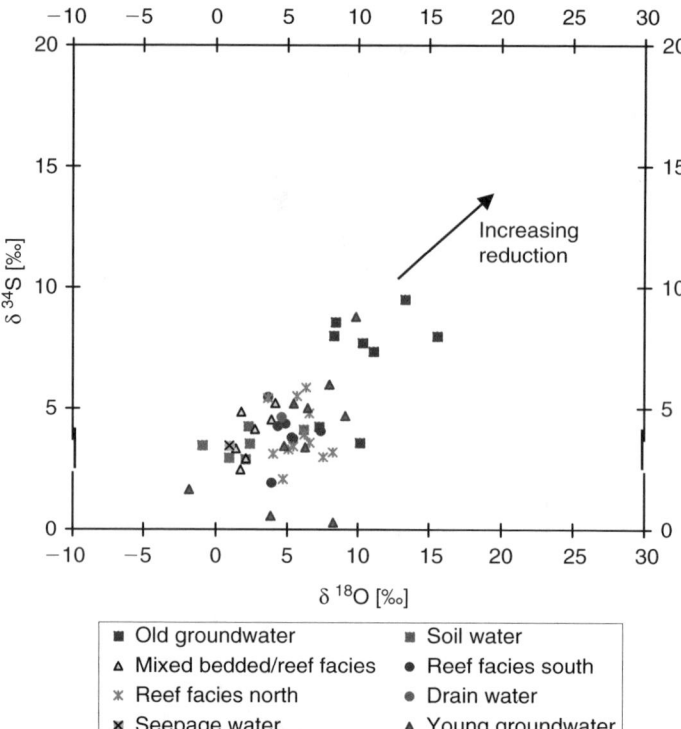

Figure 9. ^{34}S and ^{18}O concentrations in groundwater with high redox potential (+400 mV). Measuring accuracy for ^{18}O: ±0.15‰ and for ^{34}S: ±0.3‰.

Since the redox potential of groundwater in the Frankonian Alb is in the range of +400 mV and denitrification needs anaerobic conditions, the following working hypothesis was set up:

• In groundwater, bio-films also create a chemical environment, which differs from that in the flowing groundwater, and
• The denitrification process in the study area is focused on bio-films, which are more abundant in the matrix pores than on fissure surfaces.

If the bio-films produce a reducing chemical environment, which differs from the flowing groundwater, the reduction process should express in an isotope enrichment of both ^{34}S and ^{18}O in sulphates and ^{15}N and ^{18}O in nitrates. This, indeed, can be observed (Figure 9) comparing, for example, ^{34}S and ^{18}O of sulphates in seepage (light squares), young (triangles) and old groundwater (dark squares). Without doubt this enrichment expressed much stronger if dilution of sulphates, released from bio-films to flowing water, was not so strong.

It is supposed that denitrification predominantly takes place in the pores of the matrix. Two arguments are in favour of this working hypothesis:

• Till now, no denitrification was recognized in the bedded carbonates, not even during dry weather discharge conditions; since fissure flow is rather quick (>200 m d^{-1}), shear forces would destabilize the bio-films on fissure walls and monthly observation of rock coupons showed that bio-films grow very slowly and discontinuously.

- In the matrix pores slow flow certainly will not destabilize bio-films and measurement of the bottleneck widths of these pores (Michel 2001) showed that a significant portion exceeds the body size of bacteria (0.5–5 μm) by at least one order of magnitude. The final proof of this interpretation could be delivered by sterile cores, which are difficult to acquire due to technical reasons.

CONCLUSIONS

In biporous aquifers like the reefs of the Franconian Alb there exists a discontinuous distribution of flow velocities; mean residence times of metres per year are co-existing with flow velocities of kilometres per day. This leads to a significant long-term storage of pollutants, which is difficult to monitor in real times. In most Mesozoic consolidated rocks, groundwater recharge and non-reactive agrochemicals undergo both to an average of 60% long-term storage, whereas 40% is instantaneously flushed out from soils to springs by infiltration events.

Stored pollutants may undergo microbial decomposition if the energy and nutrient supply is in favour of it. Such a denitrification capacity exists under the hydrogeological and petrographic conditions in the Frankonian Alb, Germany. Here the respective microbial communities are higher in number and functional groups beneath agro- than forestlands.

Bio-films focus mostly microbial activities; they act as a kind of catalyst and create a special reducing chemical environment for the microbial community; therefore denitrification can also be observed in groundwater with an oxidizing character: this has been stated using ^{34}S and ^{18}O in groundwater sulphates.

In bi-porous media bio-films grow and endure best in the porous matrix as far as pore sizes allow it and get destabilized on fissure walls by shear stress exerted by the fast flowing water.

A very open question is to what extend the bio-films need both the interaction with flowing water and with the solid rock surface in order to provide additional nutrients and energy sources for the microbial disintegration of pollutants.

REFERENCES

Behrens, H. 1971: Untersuchungen zum quantitativen Nachweis von Fluoreszenzfarbstoffen bei ihrer Anwendung als hydrologische Markierungsstoffe. *Geologica Bavarica* (München) *64*: 120–131.

Glaser, St. 1998: Der Grundwasserhaushalt in verschiedenen Faziesbereichen des Malms der Südlichen und Mittleren Frankenalb. *GSF-Bericht* (Neuherberg) 2 (98): 135.

Meyer, R.K.F. 1977: Startigraphie und Fazies des Frankendolomits and der Massenkalke (Malm), 3. Teil: Südliche Frankenalb. *Erlanger Geologische Abhandlungen 104*: 40.

Meyer, R.K.F. and Schmidt-Kaler, H. 1989: Paläogeographischer Atlas des Süddeutschen Oberjuras. *Geol Jb. A 115*: 77.

Michel, U. 2001: *Petrophysikalische Eigenschaftzen der dolomitischen Massenfazies (Kimmeridge) der Südlichen Frankenalb in Abhängigkeit von der faziellen und diagenetischen Entwicklung nebst ihrer Bedeutung für die Verdünnung und den Abbau von Schadstoffen im Karstgrundwasser.* Unpublished PhD-thesis, Univeirtät Erlangen-Nürnberg, Erlangen, p. 127.

Pfaff, Th. 1987: Grundwasserumsatzräume im Karst der südlichen Frankenalb. *GSF-Bericht* (Neuherberg) *3* (87): 187.

Seiler, K.-P. 1969: Kluft- und Porenwasser im Mittleren Buntsandstein des südlichen Saarlandes. *Geologische Mitteilungen* (Aachen) *9*: 75–96.

Seiler, K.-P. and Alvarado-Rivas, J. 2002: *Tools of groundwater protection below the city of Caracas, Venezuela*. In: E. Bocanegra, D. Martínez and H. Massone (eds): Groundwater and human development, CD-Proceedings, pp. 1804–1810.

Seiler, K.-P., Behrens, H. and Hartmann, H.-W. 1992: Das Grundwasser im Malm der Südlichen Frankenalb und Aspekte seiner Gefährdung durch anthropogene Einflüsse. *Deutsche Gewässerkundliche Mitteilungen* (Koblenz) *35*: 171–179.

Seiler, K.-P., Behrens, H. and Wolf, M. 1995: *Use of artificial and environmental tracers to study storage and drainage of groundwater in the Franconian Alb, Germany, and the consequences for groundwater protection*. Proc. Isotopes in Water Resources Management Vol. 2. IAEA, Vienna, pp. 135–146.

Seiler, K.-P., Maloszewski, P. and Behrens, H. 1989: Hydrodynamic dispersion in karstified limestones and dolomites in the Upper Jurassic of the Franconian Alb. *Journal of Hydrology 108*: 235–247.

Sudicky, E.A. and Frind, E.O. 1981: Carbon-14 dating of groundwater in confined aquifers: Implication of aquitard diffusion. *Water Resources Research 17*: 1060–1064.

Weiss, E.G. 1987: *Porositäten, Hydraulic conductivitäten und Verkarstungserscheinungen im Mittleren und Oberen Malm der Südlichen Frankenalb*. Unpublished PhD-thesis, University of Erlangen-Nürnberg, Erlangen, p. 211.

Section 3

Modelling of nitrate transport and chemistry

O NEILL GROUND WATER ENGINEERING
7 SOUTH MAIN STREET
NAAS, CO. KILDARE
TEL: 353-45-895668
FAX: 353-45-881705
Email: info@groundwatereng.ie

CHAPTER 10

Development of a groundwater abstraction modelling environment for drinking water supply

L. Hubrechts and J. Feyen
Laboratory for Soil and Water, Katholieke Universiteit Leuven (K.U. Leuven),
Vital Decosterstraat 102, B-3000 Leuven, Belgium

J. Patyn and J. Bronders
Flemish Institute for Technological Research (VITO), Boeretang 200, B-2400 Mol, Belgium

A. Refsgaard, L. Basberg and O. Larsen
DHI, Water and Environment, Agern Allé 11, DK-2970 Hoersholm, Denmark

C. Vlieghe
PIDPA, Desguinlei 246, B-2018 Antwerp, Belgium

M. Buysse
Centre for Water Research and Co-operation of Flanders (SVW), Mechelsesteenweg 64,
B-2018 Antwerp, Belgium

ABSTRACT: Intensive agriculture in Flanders (Belgium) as elsewhere threatens the quality of groundwater in infiltration areas of abstraction wells. Drinking water companies and the Flemish authorities in particular are very concerned about the short- and long-term quality status of groundwater, particularly in those areas where groundwater is extracted for drinking purposes. To assess the evolution of the water quality in infiltration areas and the effect of N-leaching out of the root zone on the infiltrating water, a research project had been set up. The aim of the project was the development and testing of a methodological approach assessing the fate of N in the zone of influence of groundwater abstraction well fields, based on geohydrological and hydrogeochemical field data and modelling approaches.
   The methodological approach was applied to the PIDPA groundwater abstraction well field of Olen (Province of Antwerp, Belgium). The paper presents the general methodological approach, the data monitoring scheme and the simulation results.

INTRODUCTION

Project background

Nitrate pollution of surface and groundwater is an increasing problem in most European and North American countries (Postma, Boesen, Kristiansen *et al.* 1991). One of the major threats is the pollution of drinking water supplies from groundwater. Since high concentrations are believed to cause potential health problems, EU directives fix the maximum admissible NO_3-concentration in drinking water at $50\,mg\,l^{-1}$, corresponding to $11\,mg\,l^{-1}$ NO_3-N.

Extensive application of N-fertilisers and manure for agricultural purposes are assumed to be the main cause of the increasing NO_3-concentrations in soil and groundwater. In Flanders, groundwater abstraction sites for drinking water production are often situated in areas where agriculture, in combination with intensive cattle breeding, is dominantly present. In order to evaluate quantitatively the effects of changing pumping regimes and/or measures taken to protect the water quality from NO_3-pollution, good knowledge of the transport and chemical behaviour of N-species in the unsaturated soil and in the aquifer system of well field infiltration areas is an essential prerequisite. Application of this knowledge to the infiltration areas of individual abstraction wells or well fields can be supported by calibrated and validated mathematical models.

Project objectives

The research project aimed to develop and to test a methodology to set up an integrated and dynamic hydrologic and hydrogeochemical modelling environment, describing quantitatively the fate of N in infiltration areas of individual groundwater abstraction sites for drinking water production. The methodology accounts for local site characteristics, for example, climate, topography, soils, geology, mineralogy, land use and agricultural management practices.

These objectives include: i) the definition study, including the formulation of the problems in the particular well field infiltration area, directly or indirectly related to the leaching of nitrate from the crop root zone to the aquifer system, ii) the development, calibration and validation of a local hydrogeochemical model describing the fate of N in the unsaturated zone and in the aquifer system, and iii) the collection of existing and new data on the case site by field observations and controlled laboratory experiments to hydrogeochemically characterize the physical system and to provide for model calibration and validation data.

The DHI software tools (DAISY, MIKE-SHE and MIKE11) were selected to develop the modelling environment.

THE METHODOLOGY

Figure 1a shows a schematic overview of the developed methodology.

As can be inferred from Figure 1b, the definition study emphasizes the collection and screening of existing data and shows that attention has to be paid to the characterization of the geohydrological system. In this phase, it is also of the utmost importance to depict the relevant hydrochemical processes and to define clearly the problem concerned.

The research strategy includes: i) the delineation of the study area; ii) an inventory of needed data; iii) an overview of data to collect (including a time schedule); iv) the strategy to assess the geohydrological conditions in the study area (including an overview of the modelling tools to be used); and v) the strategy to assess the hydrogeochemical conditions in the study area related to N and its direct and indirect effect on the abstracted groundwater quality (including an overview of the modelling tools to be used).

Scheme of the developed methodology

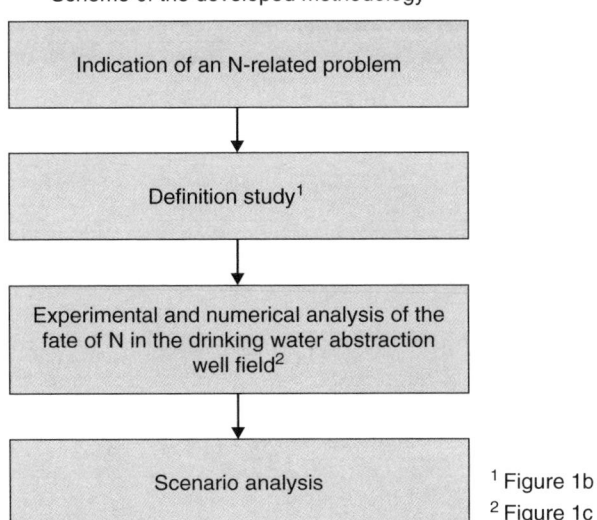

Figure 1a. Methodological approach assessing the fate of N in the infiltration area of water abstraction well fields (general scheme).

Definition study

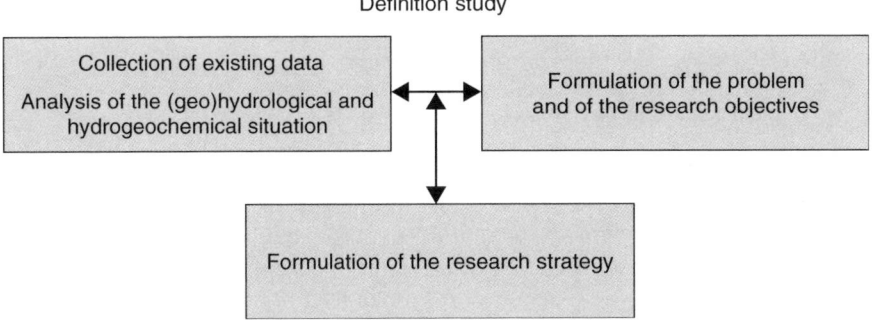

Figure 1b. Schematic overview of the 'Definition study'.

The implementation of the research strategy (Figure 1c) will be illustrated by means of the case of the 'drinking water abstraction well field of PIDPA in Olen'. Section 3 of this paper describes the characterization of the physical system, whereas Section 4 discusses the development, testing and application of the aimed modelling environment and the main results of the methodological approach for the case site.

CHARACTERIZATION OF THE INFILTRATION AREA OF THE PIDPA GROUNDWATER ABSTRACTION SITE OF OLEN (ANTWERP, BELGIUM)

Site description

The case site is situated in the immediate neighbourhood of a sluice complex on the Albert canal in Olen in the Nete catchment area (Figure 2). The topographic level is about 20 m

Implementation of the research strategy

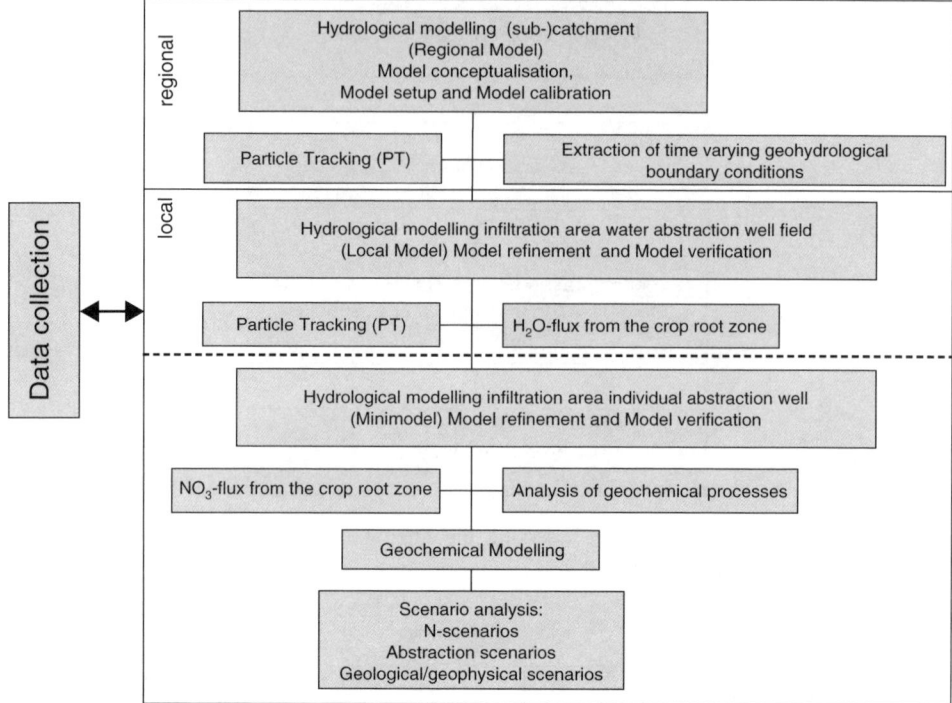

Figure 1c. Implementation of the research strategy.

above sea level. The well field is abstracting groundwater from the geological Formation of Berchem and the Formation of Diest. Daily discharges, from a variable number of the 11 wells with filter depths between 40 and 80 m, ranges between 5,000 and 7,000 m^3.

The regional geological composition of the aquifer system of the Nete-catchment can be schematized as a four-layer system with the top of the Formation of Boom as its bottom boundary.

East of the site, Quaternary sands cover the Formation of Kasterlee. West of the site the Formation of Kasterlee is absent. At the site, the Formation of Kasterlee was found to be a discontinuous layer, covering the sandy Formation of Diest and Berchem. The basis of the Formation of Kasterlee and the top of the Formation of Diest are clayey, dividing the aquifer system in a semi-confined aquifer in the Formations of Diest and Berchem, and in a phreatic aquifer in the Formation of Kasterlee and in the Quaternary (loamy) sands.

The overall groundwater flow direction in the 75 m thick semi-confined aquifer of the Formations of Diest and Berchem at the site is SE-NW. Its transmissivity of 1,200 m^2 d^{-1} was estimated from well tests. The phreatic aquifer in the Formation of Kasterlee and in the overlaying Quaternary sediments east of the site is draining towards the Grote Nete and the Kleine Nete through the small drains and river system.

Sandy to loamy sandy, internally well-drained, soils are present with an average thickness of nearly 2 m developed within the Quaternary sediments. Soils under agriculture show well developed anthropogenic top horizons (thickness of about 0.5 m) with high organic matter contents. Pre-podzolic or podzolic soil profile development is

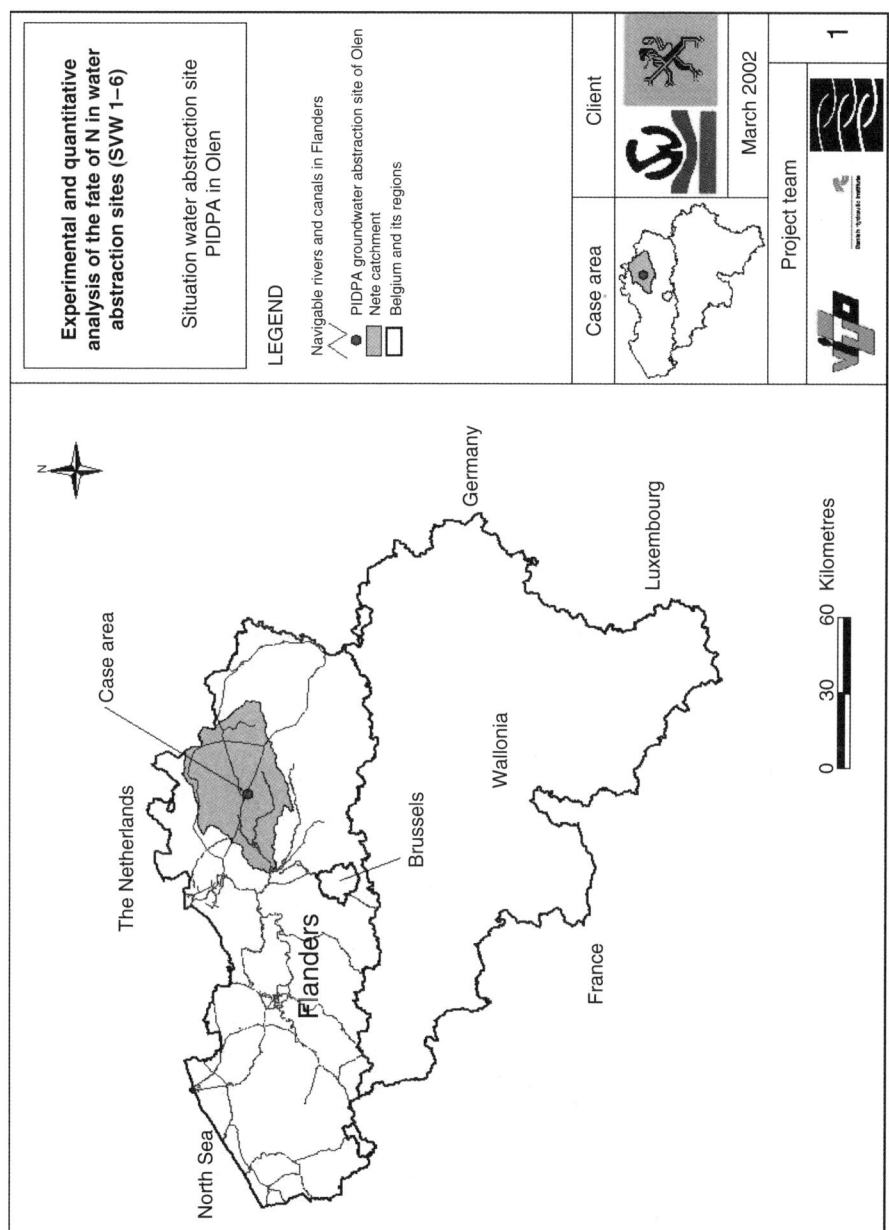

Figure 2. Location of the PIDPA-drinking water abstraction site in Olen (Prov. Antwerp, Belgium).

found mainly under forest. Riverbanks and valleys are covered by poorly developed sandy to loamy alluvial soils.

The area is characterized by a moderate climate, an average yearly rainfall of about 750 mm, equally distributed over the year, and an average yearly evapotranspiration of about 500 mm. Rainfall surplus is observed from October till March, whereas the period between April and September is characterized by an overall precipitation deficit.

Agriculture, leaf and pine forests, wetlands, and urban and industrial areas are the main land use types in the area. Meadow, grassland and maize under intensive N-management cover about 95% of the local agricultural land, occupying 56% of the total land use.

Field monitoring and laboratory experiments

The unsaturated zone and the phreatic aquifer

In the root zone of most soils (the upper 0.5 to 1.0 m) N appears under different forms as a result of a number of complex and interactive processes, such as mineralization, nitrification, denitrification, volatilization, adsorption and desorption, immobilization, plant uptake and transport, among others, resulting in changing NO_3-concentrations and in NO_3-leaching from the bottom of the crop root zone towards deeper layers and finally towards the aquifer system.

Based on differences in well field influence on the local geohydrology, differences in soil type, land use and N-management, three experimental field plots within the influence area of the well field were selected for a more detailed characterization of the physical and chemical behaviour of N in the unsaturated zone. The three plots are covered by maize, grassland and poplar forest respectively. The test fields were equipped for measuring the top and bottom boundary conditions of the water and the N-balance, and for assessing time series of water and N-state variables.

The soils of the three experimental plots were described in detail. Individual pedo-genetic horizons were sampled and chemically analysed on pH_{H_2O}, pH_{KCl}, total N- and C-content on a total destruction sample, and on NH_4-, NO_2-, NO_3- and SO_4-concentrations of a 1/5 KCl-extract, among others.

On each experimental field one Trime access tube ($L = 2.5$ m) for monitoring the soil moisture profile, one observation well in the shallow phreatic aquifer (filter depth from 2.0 to 3.0 m), two drainage pans (respectively at the bottom of the A- and E-horizon) and six suction cups (at a depth of 0.20, 0.60, 1.20, 1.80, 2.80 and 3.80 m below field surface) were installed. Cumulative pluviometers were installed at the grassland and the poplar forest experimental field.

Water balance data were collected weekly during the growing season (April till September) and on a two-weekly basis for the rest of the year. Site-specific daily rainfall data were obtained from processing the collected cumulative rainfall data jointly with the daily rainfall data from a meteorological station measured by the Royal Meteorological Institute of Belgium and situated in Geel at a distance of 8 km from the experimental fields. Data from this station were also used to calculate daily reference evapotranspiration (ETo) using the Penman-Monteith equation.

Water samples from pluviometers, drainage pan collectors, and shallow observation wells and suction cups in the unsaturated zone and in the phreatic aquifer are collected on a monthly basis during the growing season. Analyses on parameters reflecting the N-dynamics were carried out.

Data on the N-management practices on the grassland and maize parcel were obtained from the respective farmers.

A number of undisturbed soil columns ($L = 1.0$ m, $\varnothing = 0.8$ m or 0.3 m) has been taken in the topsoil of the three experimental sites, transported to the laboratory and equipped with an irrigation system, a drainage water collection system, and with a number of thermistors and TDR-probes for measuring soil temperature, soil moisture and solute breakthrough. A number of steady state breakthrough curves were measured, applying a step-input flux of a $CaCl_2$-solution at the top of the samples. Calibrated small, undisturbed soil samples (100 cm^3) were taken from the columns for assessing gravimetrically the soil moisture content profile and for determining the moisture retention and hydraulic conductivity characteristics of the different pedogenetic horizons. Breakthrough curves were fitted through the experimental data using the CDE (convection dispersion equation), the MIM (mobile–immobile) model and the CLT (stochastic convective log-normal transfer function) transport model, aiming to quantitatively characterize the nitrate transport processes in the topsoil.

The semi-confined aquifer

Since NO_3^- does not form insoluble minerals, neither precipitates or is adsorbed, the only way of removing it from aquifers is by reduction. The concentration and availability of O_2, organic matter and/or pyrite (FeS_2) mainly determine the latter process.

To refine the existing hydrogeological and hydrogeochemical characterization of the abstraction site and of its area of influence, and to provide for model calibration and validation data, a number of new observation wells with filter depths between 5 and 6 m in the phreatic aquifer, and between 9 and 10 m, 13.5 and 14.5 m, 19.5 and 20.5 m and 63.5 and 64.5 m in the semi-confined aquifer were installed. One metre long undisturbed samples were taken from a deep drilling (65 m) for depth ranges between 4 and 14 m and between 60 and 65 m. The textural and mineralogical compositions of those samples were analysed.

The local piezometry and the hydro-geochemical conditions of the aquifer system were monitored on a regular base from already present and from recently installed observation wells.

A detailed description and sub-sampling of an undisturbed 7.5 m long drilled sample and analyses of the sub-samples on pyrite (FeS), CEC, exchangeable Ca, Na, K, Mg, NH$_4$ and Fe and of the soil solution of respective samples on the concentration of Cl, SO_4^{2-}, NO_3^-, NO_2^- and NH_4^+ were carried out.

Field monitoring results

The unsaturated zone and the phreatic aquifer

The phreatic water table depth ranges between ±0.5 m and ±2.5 m. NO_3^--concentration in the soil solution and in the phreatic groundwater show much higher concentrations (up to 150 mg l^{-1}) for intensively managed agricultural fields (grassland and maize field), than is the case for the poplar forest. The increase of NO_3-concentration with depth in the phreatic groundwater of the experimental sites, especially in the case of the grassland and of the maize field, and the overall low NO_2^--concentrations, below the detection limit, exclude important denitrification processes within the unsaturated zone and the phreatic aquifer.

The semi-confined aquifer

Complementary drilling in the well field area reveals the presence of a semi-pervious layer at the bottom of the Formation of Kasterlee at depths ranging from 4 to 6 m. Upstream of the abstraction site and outside its zone of influence, the second aquifer within the Formation of Diest and Berchem is confined. Within the zone of influence, the two consecutive phreatic aquifers reflect the effect of the abstraction on the second (main) aquifer piezometry.

No NO_3^- could be observed till present in the abstracted groundwater. Analyses of groundwater samples from existing observation wells, with filters in the Formation of Diest, revealed concentrations of NO_3^- mostly lower than the detection limits.

Qualitative analyses of the hydrogeochemical data revealed i) significant groundwater flow between the phreatic and the semi-confined aquifer through the semi-confining clay layer, ii) the complete removal of NO_3^- from the infiltrating phreatic groundwater in the clay layer due to pyrite oxidation, and iii) the increase of the concentration of Fe(II) and SO_4 in the confining clay layer and the Fe(II)- and SO_4^{2-}-leaching towards the semi-confined aquifer.

DEVELOPMENT OF THE MODELLING ENVIRONMENT

A local hydrological model was set up, based on assumptions concerning general piezo-metry and geohydrological boundary conditions that could be obtained from a large set of available groundwater head measurements over the area. The infiltration area of one single abstraction well was delineated within the infiltration area of the PIDPA water abstraction well field, applying the PT-module on the results of the hydrological modelling approach for the local model. The delineation of the infiltration area within the water abstraction well field infiltration area can be seen from Figure 3.

The hydrology of the single abstraction well infiltration area was modelled. The hydrological modelling accounted for 19 calculation layers over the aquifer system. In the horizontal plane a grid with 25 m square grid cells was applied.

Evaluation of the DAISY-code, modelling the crop growth, the soil water balance and the fate of N and C within the crop root zone, was based on comparison of measured and calculated nitrate concentration time series for the monitored experimental fields. Spatial analyses for the local model area were made on the time varying water- and NO_3-fluxes at the bottom of the crop root zone (1 m below the field surface) using the DAISY-GIS modelling tool and accounting for the location-specific soil profile development and for the hydrodynamic properties of the composing soil horizons, obtained from lysimeter breakthrough experiments as well as from extended soil databases, accounting for the land use pattern and for general agricultural N-management practices.

Quantitative analyses of the hydrogeochemical conditions were based on the results of PHREEQC-modelling and revealed that i) NO_3-removal by pyrite oxidation within the confining clay layer is the dominant process; ii) NO_3-removal by oxidation of organic carbon within the particular layer can be neglected; and iii) N_2, SO_4 and Fe(II) are the dominant reaction products resulting from the NO_3-reduction.

Qualitative and quantitative analyses results on the fate of NO_3 within the saturated zone were formulated in a number of stoichiometric reaction equations and implemented in a 3D hydrogeochemical modelling environment, using the MIKE-SHE AD

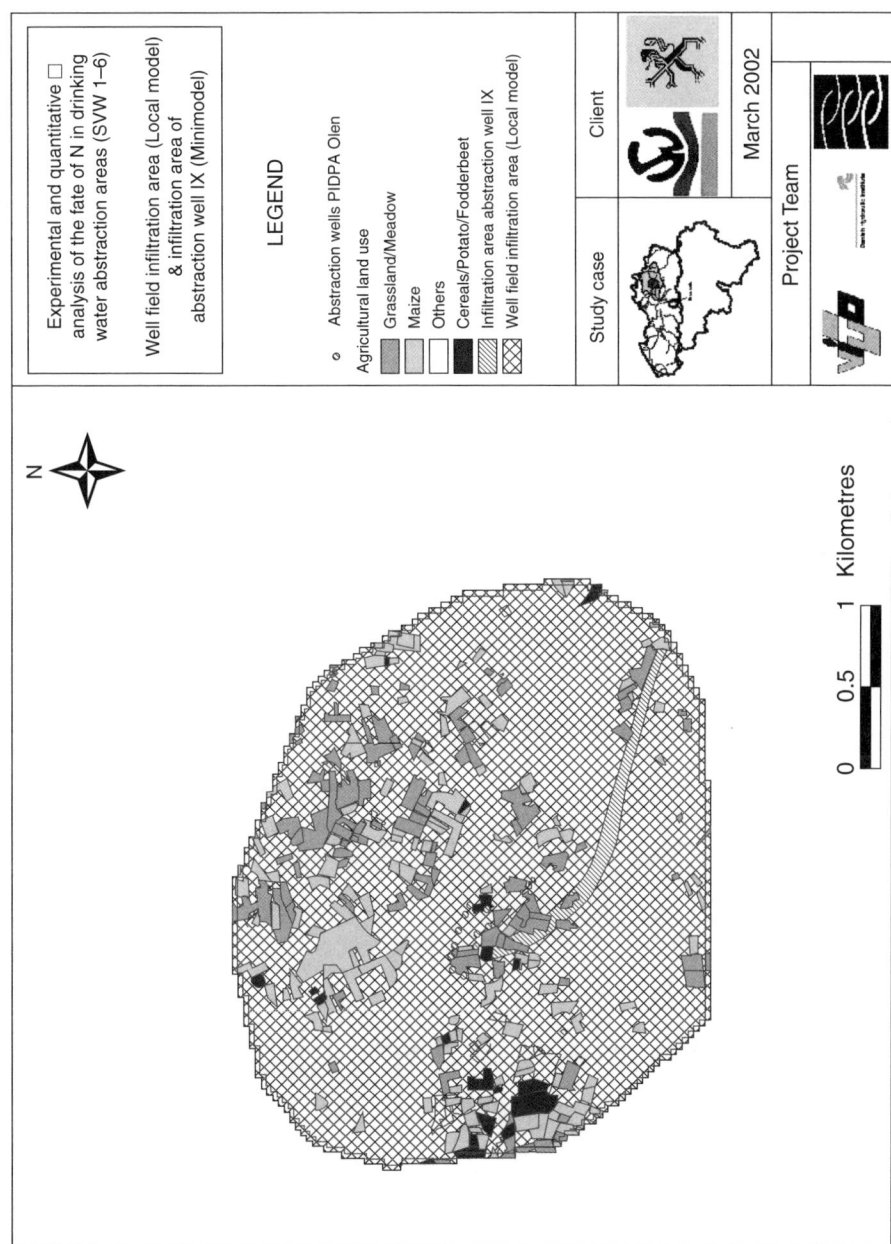

Figure 3. Infiltration area of abstraction well IX within the PIDPA drinking water abstraction well field in Olen.

(Advection–Dispersion) and BM (Biological Module) code. Spatially distributed time series for water and NO_3-fluxes, resulting from running the DAISY-GIS modelling set up, were input as hydrological and hydrochemical top boundary conditions respectively.

Modelling approaches for the actual situation (Figure 4) revealed the effective chemical buffering capacity of the pyrite-containing clay layer. It could be concluded that (actual) pyrite oxidation due to NO_3-reduction did not affect the abstracted groundwater quality in terms of increases in the concentration of Fe(II) nor of SO_4. However, important increases

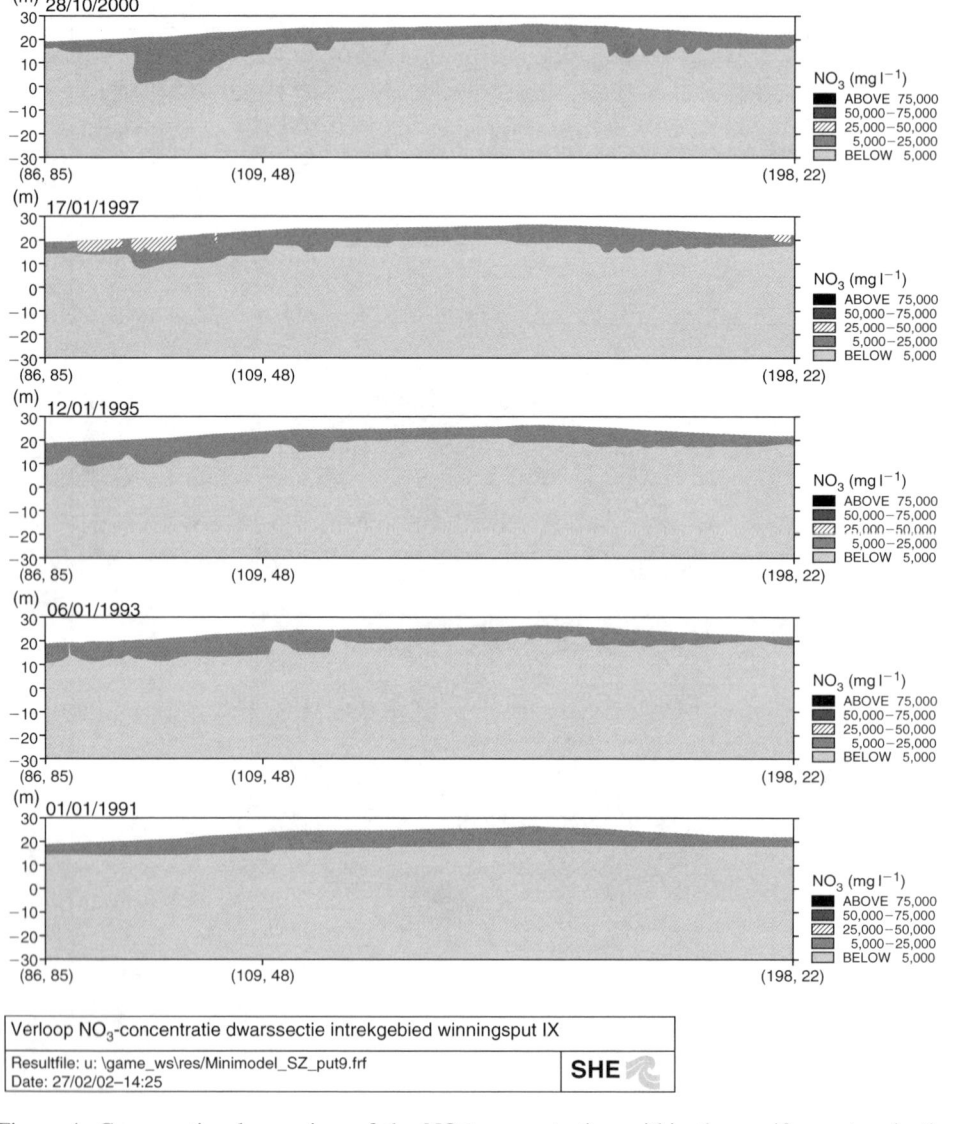

Figure 4. Cross-sectional overview of the NO_3^--concentration within the aquifer system in the infiltration area of abstraction well n° IX.

in surface water NO_3^-- concentration through subsurface drainage of the phreatic groundwater might be expected.

CONCLUSIONS

A methodology for assessing the fate of N in the infiltration area of abstraction well fields was set up. The methodology is based on the experimental and numerical analysis of the geohydrological and hydrogeochemical conditions of the infiltration area supported by *in situ* monitoring and modelling approaches. The methodology was developed as a generally applicable framework and was tested for the groundwater abstraction well field of PIDPA in Olen.

An extensive monitoring scheme was set up in the selected study area with the objective to collect model parameters and data for the calibration and validation of the modelling used to analyse the transport and fate of nitrogen in the unsaturated and saturated zone of the infiltration area.

This paper describes the methodological framework as well as the characteristics of the PIDPA water abstraction site, the field monitoring and laboratory experiments. Information is provided on the qualitative and quantitative analysis of the monitoring data and on the modelling approach chosen (DAISY, MIKE-SHE, and MIKE11), on the evaluation of the consecutive modelling phases and on their results.

The modelling approach was also applied for assessing the fate of N within the infiltration area of one single abstraction well. The modelling for actual conditions of abstraction discharges and N-application conditions revealed no significant impact on the actual nor future N-related quality of the abstracted raw groundwater from the semi-confined aquifer.

REFERENCES

Abbott, M.B. and Refsgaard, J.C. (eds) 1996: *Distributed hydrological modelling*. Kluwer Academic Publishers, Dordrecht.
Basberg, L., Dagestad, A. and Engesgaard, P. 1998: Geochemical modelling of natural geo-hydrochemical stratification dominated by pyrite oxidation and calcite dissolution in a glaciofluvial Quaternary deposit, Gardermoen, Norway. *NGU-BULL* 434: 35−44.
Christiaens, K., Feyen, L., Abu El-Nasr, A., Vasquez, R., Van Hoorick, M. and Feyen, J. 1998: Data requirements, data sources and data flow for the distributed physically based hydrological MIKE-SHE model, with application to the Gete basin. Institute for land and water management, Katholieke Universiteit Leuven, Belgium, Internal report No. 50, p. 68.
Hubrechts, L., Patyn, J., Bronders, J., Basberg, L., Thorsen, M., Vlieghe, C., Buysse, M. and Feyen, J. 2001: Development of a groundwater abstraction modelling environment for water supply. Paper # 098 at the 4th DHI Software Conference, 6−8 June 2001, Helsingør, Denmark.
Pedersen, J.K. 1992: Nitratreduktion I jord og grundvang (in Danish). Ph.D. thesis, Department of Environmental Engineering, Technical University of Denmark.
Postma, D. 1990: Kinetics of nitrate reduction by detrital Fe(II)-silicates. *Geochim. Cosmochim. Acta 54*: 903−908.
Postma, D., Boesen, C., Kristiansen, H. and Larsen, F. 1991: Nitrate reduction in an unconfined sandy aquifer: water chemistry, reduction processes, and geochemical modeling. *Water Resources Research 27*: 2027−2045.
Refsgaard, A. and Kristensen, M. 2000: Groundwater protection based on an integrated particle tracking model. In: P.L. Bjerg, P. Engesgaard and T.D. Krom (eds): Groundwater Research. Balkema, Rotterdam pp. 509−510.

CHAPTER 11

Groundwater-borne nitrate intakes into the river Elbe (German part)

R. Kunkel and F. Wendland
Research Centre Jülich, Systems Analysis and Technology Evaluation, 52425 Jülich, Germany

M. Bach
University of Gießen, Department of Natural Resources Management, Heinrich-Buff-Ring 26–32, 35392 Gießen, Germany

H. Behrendt
Institute of Water Ecology and Inland Fisheries (IGB), Müggelseedamm 310, 12587 Berlin, Germany

ABSTRACT: An integrated model approach to quantify the nitrogen loads entering the surface waters via the groundwater path is applied to the German part of the river Elbe basin. For this purpose, the results of a nitrogen balance model (Bach, Behrendt and Frede 2000), which considers the most important N-inputs to and N-removals from the soil and the groundwater residence time/denitrification model WEKU (Kunkel and Wendland 1997; Wendland, Kunkel, Grimvall, *et al.* 2001) are combined. The modelled groundwater-borne nitrogen inputs into surface waters were validated using results of the MONERIS model (Behrendt, Huber, Kornmilch *et al.* 2000) concerning riverine nitrogen retention, nitrogen inputs from point sources as well as nitrogen inputs through direct runoff (for example, drainage).

In the vicinity of surface waters, and in solid rock areas, the groundwater-borne nitrogen inputs into surface waters are considerably higher compared to the inputs into the aquifer due to predominantly unfavourable denitrification conditions and short residence times of groundwater. In the North German lowlands, however, the groundwater-borne nitrate inputs into surface waters are considerably lower compared to the inputs into the aquifer. There, the residence time of groundwater in the aquifer is high and the groundwater is predominantly oxygen free and contains pyrite and/or organic carbon compounds, allowing a halving of the nitrate loads in the groundwater within a period of 1–4 years (see Wendland and Kunkel 1999).

INTRODUCTION

At present, the river Elbe, having a catchment size of about 150,000 km^2, transports about 105,000 tons of nitrogen each year to the North Sea (Behrendt *et al.* 2000). By this nitrogen load the river Elbe still contributes considerably to the total nitrogen pollution of the North Sea. Agricultural-borne emissions from excess fertilized soils have been identified as the main source of the nitrogen. In the framework of the International Conference for the Protection of the River Elbe, the German government assumed the obligation to reduce the riverine nutrient inputs into the North Sea by 50%. The realization of this obligation makes it necessary to develop reduction measures, which take into consideration the different agricultural usages as well as the different site conditions in the Elbe basin (for example, Wendland, Albert, Bach *et al.* 1994; Wendland, Bach and Kunkel 1998).

In order to quantify nitrate loads which enter the receiving waters via the groundwater path, the results of a diffuse nitrogen balance model (Bach *et al.* 2000) are combined with a model concerning the quantification of relevant runoff components, nitrogen residence time in the soil and groundwater and the natural nitrate degradation ability of the soil and aquifer. Through combination with information on riverine nitrogen retention and nitrogen inputs from point sources (Behrendt *et al.* 2000), all nitrogen inputs into subsurface and surface waters can be quantified. The results presented here for the German part of the Elbe basin (ca. 100,000 km^2) were obtained within the framework of Germany's federal environmental agency project International Harmonization of the Quantification of Nutrient Input into the German Surface Waters from Diffuse and Point Sources.

METHOD

The approach used for the quantification of groundwater-borne nitrate inputs into surface waters is designed for area-differentiated modelling on a supraregional scale. According to the model's applicability to the entire catchment area, the hydrological, pedological and hydrogeological input parameters needed for modelling were taken from thematic maps. The scale of these maps, ranging from 1:50,000 to 1:200,000, determine the degree of detail of the model input values and define in connection with the suitable model approaches the validity range of the model results. The nitrogen inputs into groundwater are calculated from the nitrogen surpluses reduced by the denitrification losses in the soil related to the groundwater recharge/total runoff ratio. These inputs are transported with the groundwater to the surface waters. On the way nitrate degradation may occur. Thus, a calculation of the groundwater-borne nitrate inputs into surface waters requires knowledge of the groundwater flow paths, the total residence time of the nitrate and the denitrification kinetics in the upper aquifer. These processes were considered by different models.

Diffuse nitrogen surpluses are quantified by a nitrogen balance model (Bach *et al.* 2000). Agricultural statistics with data on crop yields, livestock farming, land use etcetera served to balance the actual nitrogen supplies and extractions for the agricultural acreages. The long-term nitrogen balance averaged over several vegetation periods is calculated considering the organic nitrogen fertilization, the mineral nitrogen fertilization, the symbiotic N fixation, the atmospheric N inputs and the N extractions with the crop substance. As a rule, the sum of nitrogen supplies, primarily by mineral fertilizers and farm manure, and nitrogen extractions, primarily by field crops, leads to a positive balance.

The GROWA98 model (Kunkel and Wendland 2002) was used to carry out an area-differentiated water balance analysis for the hydrological period 1961–1990. The mean long-term total runoff was modelled as a function of the regional interaction of the climate, soil, geology, topography and land use conditions. The total runoff was separated into the direct runoff (interflow and surface runoff) and groundwater runoff (groundwater recharge) using base-flow indices, depending on area characteristics (for example, geology, depth of groundwater). The total runoff/groundwater recharge ratio was taken as a measure for the extent diffuse nitrogen surpluses are displaced from soil to groundwater.

Nitrate degradation in soils was calculated using a Michaelis-Menten kinetics using the approach of Köhne and Wendland (1992). Denitrification losses occur mainly in the effective root zone of the soils, and can be described as a function of the nitrogen surpluses, the average field capacity and the site-specific denitrification conditions.

Reactive nitrate transport in groundwater is modelled with the WEKU-model (Kunkel and Wendland 1997; 1999). In the first step groundwater velocities are calculated according to Darcy's law from hydraulic conductivity, effective yield of pore space of the aquifer and the slope of groundwater surface (hydraulic gradient). The calculation of the residence times of the groundwater runoff is performed in a second step. Based on groundwater contour maps, a digital relief model of the groundwater surface is generated. This is analysed paying attention to information on the water network as well as the groundwater discharge or transfer areas with respect to lateral flow dynamics and groundwater-effective recipients. The residence times of the groundwater runoff are then obtained for each initial grid by summation over the individual residence times in the grids resulting from the groundwater velocities and individual flow distances along the flow path until they enter the surface water.

The WEKU model was extended by a module for the quantification of nitrate degradation in groundwater. According to extensive field studies by Böttcher, Strebel and Duynisveld (1989) in a catchment area in the North German Lowlands and van Beek (1987) for a site in the Netherlands, a first order denitrification kinetics has been assumed with a reaction constant in the range of 0.17 to 0.56 a^{-1}. This corresponds to a halving of the nitrogen leached to the groundwater after a residence time of between 1.2 and 4 years. Rather simple indicators, such as the presence of Fe(II), Mn(II) and the absence of O_2 and NO_3 can be used to decide whether a groundwater province has hydrogeochemical conditions in which denitrification is possible or such transformation of nitrogen can be neglected (Wendland and Kunkel 1999).

Validation of the groundwater-borne nitrate inputs into rivers was done based on results of the MONERIS model (Behrendt *et al.* 2000; Behrendt, Dannowski, Deumlich *et al.* 2002). The model distinguishes between point source emissions from waste water treatment plants and direct industrial discharges and six diffuse pathways, including the inputs via groundwater. According to Behrendt *et al.* (2002) it was assumed that the observed nitrogen concentrations in rivers under base flow conditions correspond to groundwater-borne nitrate inputs. Thus the modelled groundwater borne nitrogen inputs into surface waters are compared to the corresponding numbers from the MONERIS model. In case the overall agreement between the values given by the MONERIS model and the calculated values given by the WEKU model is satisfying we conclude that the procedure gives reliable estimations for the groundwater-borne nitrate input into the aquifers.

RESULTS AND DISCUSSION

Nitrogen

The nitrogen surpluses as a result of balancing the inputs by mineral fertilizers and farm manure, and nitrogen extractions, primarily by field crops, are shown in Figure 1. The

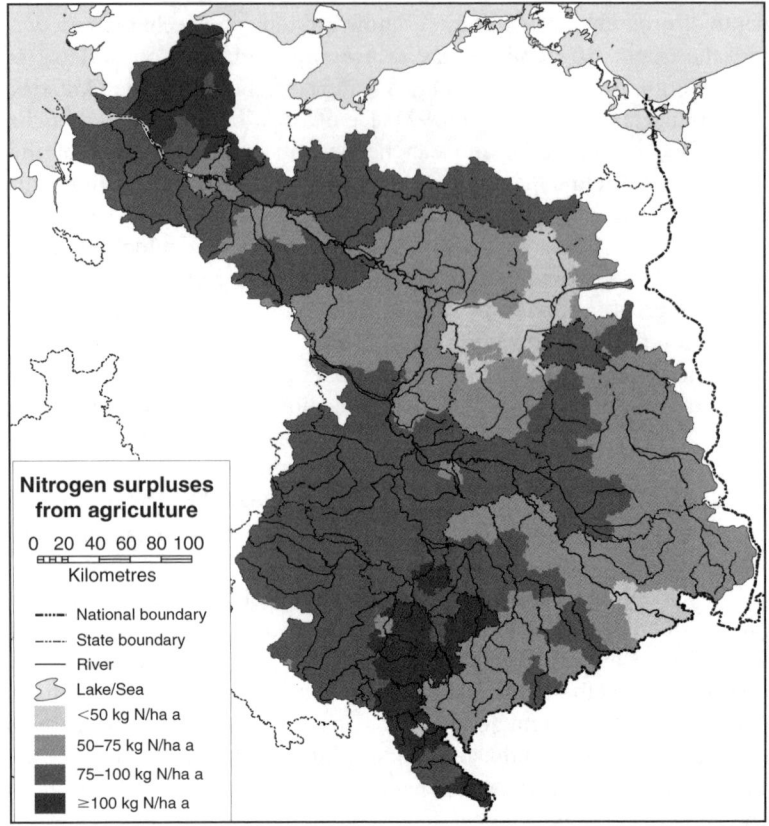

Figure 1. Annual nitrogen surpluses from agriculture.

surpluses were calculated on the basis of agrarian statistics available on community level for the period 1990–1995. On average for the whole catchment area, about 78 kg nitrogen per hectare and year are applied in excess of extraction to the agricultural-used land (Bach *et al.* 2000). Particularly in regions with area-independent animal processing (intensive animal production) large annual nitrogen surpluses up to more than 100 kg N/ha result from the animal excretions. This form of land use management occurs mainly in the north-western part of the catchment. In addition, all regions with commercial and specialty crops predominantly cultivated on mostly fertile loamy soils and favourable climatic conditions display large nitrogen surpluses. This includes, for example, the fertile loessian plains north of the midlands, such as the Thuringian Basin. Low nitrogen surpluses are calculated for regions with mostly forage crops production, which is the typical situation for the middle and eastern parts of the basin.

Ratio of groundwater recharge to total runoff

The ratio of groundwater recharge to total runoff is determined on the basis of annual average precipitation levels from 1961 to 1990 using the GROWA model (Kunkel and

Wendland 2002). This long-term annual rate is used to determine from the nitrogen surpluses the part of nitrate leached to the groundwater. In some areas groundwater runoff is not more than 20–40% of the total runoff. This situation is typical for regions close to the water table, for example in the marshy regions in the north western part of the catchment area. The same situation applies for areas where groundwater runoff is bound at paleozoic and crystalline rocks, mainly in the southern part of the Elbe basin. Hence, in these regions direct runoff is the main runoff component. As a consequence, the intake of excess fertilizers from land to surface waters takes place a short time after fertilizer application.

In areas where unconsolidated rocks predominate and which show deep groundwater tables, groundwater runoff is to a large extent equal to the total runoff. Consequently in these regions aquifer systems are the main pathway for nitrate into rivers. Because of the low flow velocity of groundwater in aquifer systems, the unconsolidated rock areas show a long-term pollution risk. Actual displacement of nitrate from soil to groundwater may be detectable in surface waters only after decades. Consequently, remediation measures will be effective in the same time range. For a classification with respect to a nitrate emission risk assessment however, it has to be taken into account urgently that in a number of aquifers in especially unconsolidated rock areas hydrogeochemical conditions occur, which may enhance natural nitrate degradation.

Nitrate inputs into the aquifers

As already discussed, the summarized N-balance results in nitrogen surpluses in a range between ca. 30 kg N/ha a and 120 kg N/ha a. However, the nitrogen surplus is in general not equal to the nitrogen leached from the root zone to the groundwater systems. On one hand a certain amount of nitrate may be degraded in the root zone due to denitrification processes. With the help of the model approach developed by Köhne and Wendland (1992) denitrification in the root zone is quantified as a function of soil properties and crop types. On the other hand it has to be taken into account, that a certain fraction of the nitrogen leached from the root zone is coupled to the direct runoff components. In order to separate the fraction of nitrogen leached to groundwater from the total nitrogen leached from the root zone, the total nitrogen leaching from the root zone is weighted by the base flow ratio calculated according to GROWA model for the whole area of Germany. The results are shown in Figure 2. High to very high nitrogen leaching to the groundwater of more than 50 kg N/ha a were calculated for all regions with a high groundwater runoff fraction and an intensive agricultural use on fertile luvisol soils. Regions with relatively low nitrogen intakes to the groundwater are obtained for areas where either less fertile soils and/or high direct runoff fractions predominate. On average for the whole catchment area the remaining nitrogen leaching from the root zone is quantified to be about 15 kg N/ha a.

Groundwater residence times in the upper aquifer

Denitrification in the upper aquifer depends on the groundwater residence times (Böttcher *et al.* 1989). Thus, considering groundwater residence times in the upper aquifer enables the calculation of nitrate concentrations in the aquifer at the outflow into the surface water. Moreover, groundwater residence times specify the time scales, until nitrate reduction strategies to remediate polluted groundwater resources may lead to a substantial groundwater quality improvement.

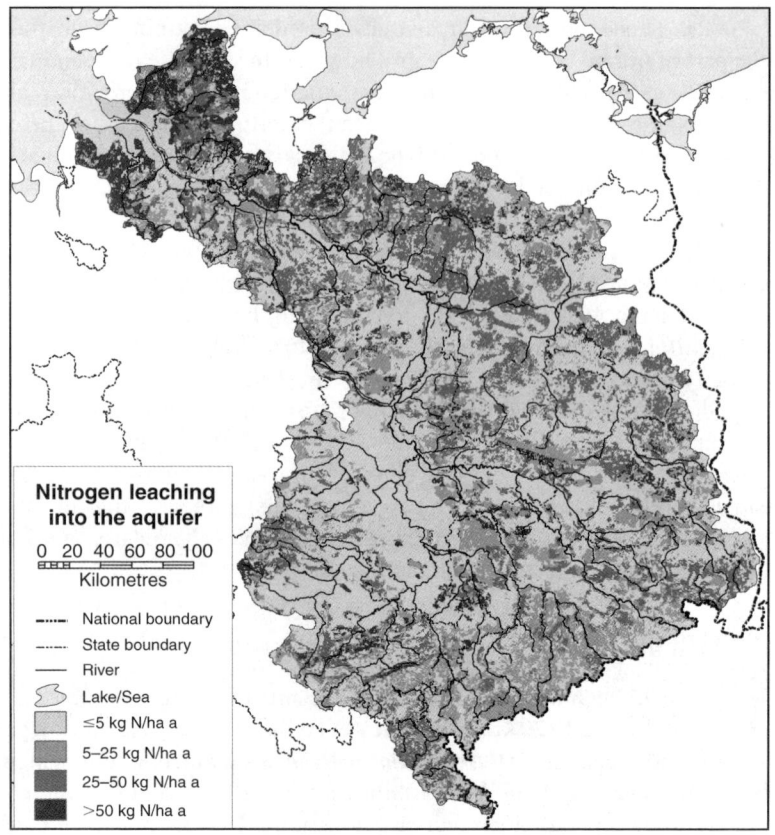

Figure 2. Annual nitrogen inputs into the aquifer.

The mean values of the calculated groundwater residence times for Germany vary between less than 1 year to more than 150 years. Long residence times may result both from small groundwater velocities as well as from long flow paths up to the recipient, pointing at the long time periods, after which nitrate inputs into the aquifer can contribute to the pollution of surface waters in some regions. Short residence times generally result for areas in the vicinity of rivers and/or regions with high groundwater velocities.

Denitrification conditions in the aquifer

The analysis of the denitrification conditions in the groundwater was done separately for the different groundwater-bearing formations occurring in Germany. In total about 8,800 groundwater samples were evaluated and classified with respect to their nitrate-degrading capacity, primarily on the basis of groundwater quality data on Iron(II), Manganese(II), Oxygen and Nitrate.

In case more than 75% of the sampling points in the hydrogeological units display oxygen concentrations below $2\,mg\ O_2/l$, the hydrogeological lithologic units was classified as belonging to the predominantly reduced aquifers, in which denitrification may occur. It was also discovered that Iron(II) concentrations above $0.2\,mg\ Fe(II)/l$ and

Manganese(II) concentrations above 0.05 mg Mn(II)/l occurred for roughly 90% of the sampling points. This is a further indication that the aquifers display predominantly reduced groundwater and have a significant nitrate degradation capacity. At least, more than 80% of the observed nitrate concentrations in the groundwater was below 1 mg NO_3/l. In connection with the high iron(II) and manganese(II) and low oxygen contents, this is an indication that nitrate compounds leached out of the soil can be degraded in the groundwater systems. The groundwater-bearing formations of the North German lowlands (glacio-fluviatile sands, moraine deposits) were classified as nitrate degrading aquifers. In most areas, where aquifers are composed of consolidated rocks (for example, sandstones, schists, limestone), usually non-nitrate degrading conditions occur. Groundwater-bearing formations, which could not be uniformly classified as nitrate degrading or non-nitratedegrading, were attached to an intermediate denitrification type, for example, the groundwater-bearing formations 'glacial outwash' and 'greywacke'.

Groundwater-borne nitrate input into the rivers

As a result, Figure 3 shows the remaining nitrogen outputs to surface waters from groundwater after its residence time in the aquifer. The outputs are shown in the initial cell for which the inputs into the soil have been calculated. As can be seen, nitrogen intakes in the vicinity of surface waters and high nitrogen leaching levels contribute considerably to the groundwater-borne nitrate inputs input into the rivers. Even with good conditions for a complete degradation of nitrate in the aquifer, the brief residence times in the aquiferous layers are not sufficient for an adequate degradation of high nitrate inputs. There is, furthermore, a hazard potential in many regions where high nitrate inputs are associated with relatively short residence times of the groundwater, as well as restricted and/or insignificant degradation conditions in the aquifer. These regions include mainly some parts of the Thuringian Basin as well as some solid-rock areas in the southern part of the catchment. In contrast, even with high nitrate concentrations in the recharged groundwater only very slight nitrate concentrations result for the loose rock aquifers of the North German Plain (northern part of the catchment) upon entry into the main receiving waters. There, nitrate input via groundwater pathways was quantified to about 2 kg N/ha a. The comparison of this value with the average nitrogen load leached from the root zone (Figure 2) indicates that through denitrification in the groundwater more than 90% of the nitrogen inputs into groundwater are retained in the aquifer systems. The, as a rule, long residence times of the groundwater and good hydrogeochemical conditions cause unrestricted nitrate degradation, with the result that the groundwater of the North German Plain is almost nitrate-free after maximum residence time when it enters the rivers.

The modelled groundwater-borne nitrogen inputs into surface waters have been compared with observed nitrogen concentrations in rivers at low flow conditions and low temperature for about 100 monitoring stations mainly in the Elbe basin. The method for the calculation of these indicators is described in detail by Behrendt *et al.* (2002). The values calculated with this model approach show only relatively small differences to the observed values (about 10–20%), as indicated in Figure 1.

Nevertheless, this fact should not be used as an argument to abandon a general reduction of the nitrate inputs into the groundwater or to delay such measures. Nitrate degradation is associated with the consumption of iron sulphide compounds and/or organic carbon.

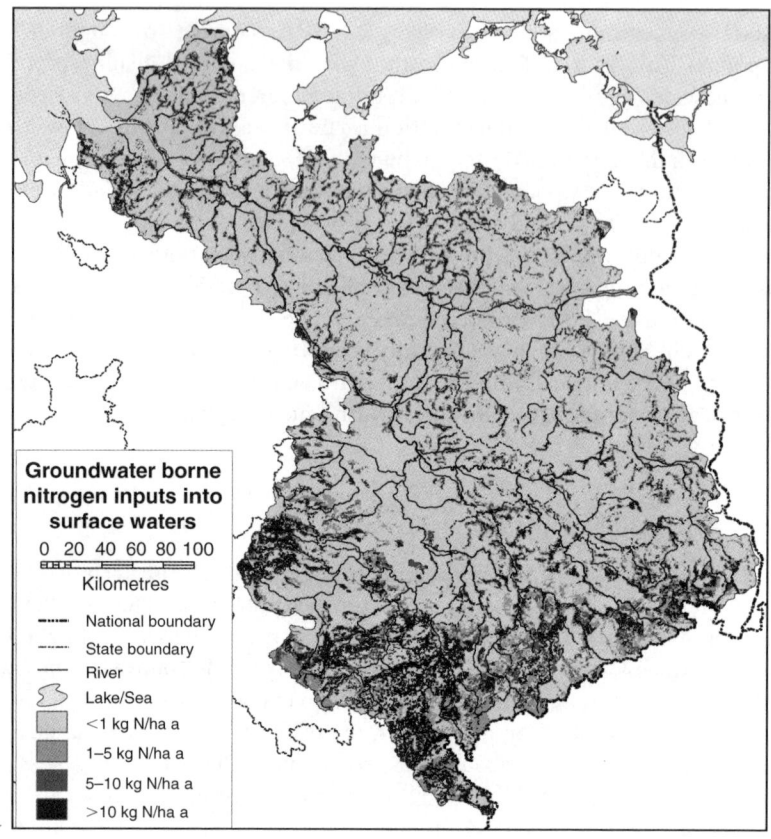

Figure 3. Groundwater-borne nitrogen inputs into surface waters.

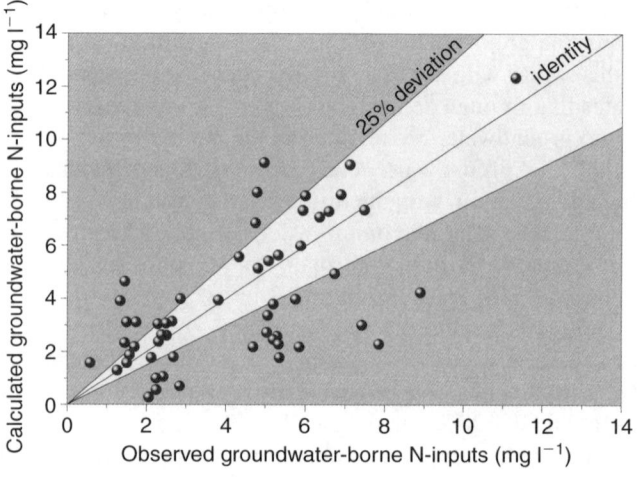

Figure 4. Comparison between measured and calculated groundwater-borne nitrogen inputs into surface waters.

Once the natural content of these components in an aquifer is exhausted, a rapid increase of the nitrate concentrations of the groundwater results, the so-called nitrate breakthrough. With regard to the idea of a sustainable management of groundwater resources in Germany natural denitrification capacities have to be protected against avoidable nitrate pollutions.

CONCLUSIONS

Studies of groundwater-borne nitrate input into surface waters of the river Elbe basin showed that both the groundwater residence time and the amount of nitrogen that is denitrified can vary strongly with the site in which the nitrogen enters the saturated zone. For the loose rock areas of the North German Plain in the northern part of the basin about 90% of the diffuse nitrogen input into the groundwater are degraded in this way. For the consolidated rock regions in the southern part of the basin it could be shown that only 30% of the diffuse nitrogen input into the groundwater are degraded. The comparison with observed groundwater-borne riverine nitrogen adds proof of the reliability of the modelled values.

With regard to the European perspective of the nitrogen loading of rivers via groundwater path, we think that it is possible to generalize and transfer the results. The northern part of the Elbe basin is part of the supraregional geological structure of the European Pleistocene Lowland, which ranges from the Netherlands in the west and Denmark in the north to the Baltic States and the Ukraine in the east. As could be shown, the key factors controlling the subsurface nitrate retention (residence times, denitrification capacity) are correlated to specific hydrogeological site conditions, occurring in the North German Plain. Parallel studies carried out in the Uecker basin (Germany) and the Gjern basin (Denmark) showed an equal distribution of groundwater residence times as well as the same hydrogeochemical characteristics (Wendland *et al.* 2001). We would assume on the basis of these results that 'long' residence times and 'good' denitrification conditions should in general be valid for the entire European Pleistocene Lowland. Due to the results obtained for all regions in Germany, which, from a geological point of view, are dominated by consolidated rock areas, it becomes obvious that there, no significant denitrifying aquifer conditions can be expected. Also it is well known (see Kunkel and Wendland 2002) that direct runoff, rather than groundwater runoff, is the dominant runoff component in solid rock areas. Generalizing this observation with regard to the European perspective, we would assume that nitrogen retention in groundwater would be of minor significance only in large parts of Southern Europe and Scandinavia.

Consequently we think that the transfer and application of the chosen procedure to other areas in this supraregional geological structure should be possible without major problems. The advantage of such a broader application of the model concept would lie in the supraregional quantification of the dimension subsurface nitrogen retention may play in Europe. In the same way, it would be possible to evaluate sensitive regions, that is, areas with high nitrogen surpluses and low groundwater residence times and hence incomplete subsurface nitrogen retention. However, it has to be taken into account that so far this generalization is not more than a rough estimate on the basis of the results from our study and on knowledge about the general geological structure in Europe. Statements about the applicability of the combined model for these areas can only be proven in the course of further selective studies.

REFERENCES

Bach, M., Behrendt, H. and Frede, H.-G. 2000: *Agricultural nutrient balances.* UBA/Texte 30/2000: 25–30.

Beek, C.G.E.M. van (ed.) 1987: *Landbouw en Drinkwatervoorziening, orientierend Onderzoek naar de Beinvloeding can de Grondwaterkwaliteit door Bemesting en het Gebruik van Bestrijdingsmiddelen*; Onderzoek 1982–1987. Report Meded. 99, Keuringsinstituut voor Waterleidingsartikelen KIWA N.V., 99, Nieuwegein, The Netherlands.

Behrendt, H., Dannowski, R., Deumlich, D., Dolezal, F., Kajewski, I., Kornmilch, M., Korol, R., Mioduszewski, W., Opitz, D., Steidl, J. and Stronska, M. 2002: *Investigation on the quantity of diffuse entries in the rivers of the catchment area of the Odra and the Pomeranian Bay to develop decision facilities for an integrated approach on waters protection (Phase III).* Final report, German Federal Environmental Agency, Berlin, p. 271 S.

Behrendt, H., Huber, P., Kornmilch, M., Opitz, D., Schmoll, O., Scholz, G. and Uebe, R. 2000: *Nutrient balances of German river basins.* UBA-Texte 23/2000: 261.

Böttcher, P., Strebel, O. and Duynisveld, W.H.M. 1989: Kinetik und Modellierung gekoppelter Stoffumsetzungen im Grundwasser eines Lockergesteinsaquifer. *Geologisches Jahrbuch 51*: 3–40.

Köhne, Ch. and Wendland, F. 1992: *Modellgestützte Berechnung des mikrobiellen Nitratabbaus im Boden.* Internal Report, KFA-STE-IB-1/92, Jülich, Germany.

Kunkel, R. and Wendland, F. 1997: WEKU-a GIS supported stochastic model of groundwater residence times in upper aquifers for the supraregional groundwater management. *Environmental Geology 30*(1/2): 1–9.

Kunkel, R. and Wendland, F. 1999: *Das Weg-/Zeitverhalten des grundwasserbürtigen Abflusses im Elbeeinzugsgebiet.* Schriften des FZ Jülich, Reihe Umwelt/Environment, 19, Jülich, Germany.

Kunkel, R. and Wendland, F. 2002: The GROWA98 model for water balance analysis in large river basins – the river Elbe case study. *Journal of Hydrology 259*: 152–162.

Kunkel, R., Wendland, F. and Albert, H. 1999: Zum Nitratabbau in den grundwasserführenden Gesteinsschichten des Elbeeinzugsgebietes. *Wasser und Boden 51* (9): 16–19.

Wendland, F. and Kunkel, R. 1999: *Das Nitratabbauvermögen im Grundwasser des Elbeeinzugsgebietes.* Schriften des FZ Jülich, Reihe Umwelt/Environment, 13, Jülich, Germany.

Wendland, F., Albert, H., Bach, M. and Schmidt, R. 1994: Potential nitrate pollution of groundwater in Germany: a supraregional differentiated model. *Environmental Geology 24*: 1–6.

Wendland, F., Bach, M. and Kunkel, R. 1998: The influence of nitrate reduction strategies on the temporal development of the nitrate pollution of soil and groundwater throughout Germany – a regionally differentiated case study. *Nutrient Cycling in Agroecosystems 50*: 167–179.

Wendland, F., Kunkel, R., Grimvall, A., Kronvang, B. and Müller-Wohlfeil, D.I. 2001: *Model system for the management of nitrogen leaching at the scale of river basins and regions.* In: C.A. Brebbia (ed.): Water Pollution VI – Modelling, Measuring and Prediction, WIT Press, Southampton, Boston.

CHAPTER 12

Development of tools needed for an impact analysis for groundwater quality due to changing of agricultural soil use

W. Mioduszewski, M. Fic, A. Ślesicka and A. Zdanowicz
Institute for Land Reclamation and Grassland Farming at Falenty, Poland

W. Walther, M. Paetsch, F. Reinstorf and D. Weller
Institute for Groundwater Management, Technical University, Dresden, Germany

S. Diankov, G. Velovsky, S. Radoslavov, W. Marinov and O. Nicheva
Institute for Water Problems of the Bulgarian Academy of Science, Sofia, Bulgaria

E.P. Querner and J. Roelsma
ALTERA Wageningen, The Netherlands

ABSTRACT: The international joint research project was carried by four research institutions in the framework of EC INCO-COPERNICUS programme. The intention of the research is to develop a software tools to enable the user to simulate changes in agricultural soil use and their effect on groundwater for small catchment areas, smaller than $10 \, km^2$. In order to do this the following research has been done:
- Studies on water and substances (mainly nitrogen) circulation in the field experimental plot (quality of soil and groundwater, meteorology, land use, moisture content etc. have been monitored). The experimental plots have been created in Poland, Germany and Bulgaria.
- Selection of the numerical models for calculation of the transport of nitrogen in unsaturated and saturated zones. Verification and calibration of these models.
- Calculation for different scenarios representative for changing agricultural soil use.

INTRODUCTION

The international joint research project was carried by four research institutions in the framework of the EC INCO-COPERNICUS programme.

The aim of the study was to develop tools on the basis of available software instruments to enable the user to simulate scenarios of variations in agricultural soil use and their effect on groundwater for small catchment areas, smaller than $10 \, km^2$. As a result of the planned reconstruction of Polish and Bulgarian agriculture, an increase in the intensity of agricultural soil use due to the conditions of the market is expected.

On the other hand, in Germany and the Netherlands the decrease of fertilizers is expected. It is necessary to predict the influence of intensification or extensification of agriculture on groundwater quality. The developed tools will enable the user to make scenario analyses and predict the reactions of the system before the agricultural soil use starts to change. This general objective is divided into three tasks:

- Exchange of expertise between partners and harmonization of suitable measurement equipment for saturated and unsaturated zones, choosing common methods for chemical analysis and organization of the experimental field plot.
- Process studies on investigation fields, recording of nutrient fluxes through the root and vadose zones and the zone of saturation.
- Calibration and verification of flow and transport models on the basis of the results from experimental plot, scenario calculations.

The Institute for Groundwater Management (IGW), Technical University in Dresden was responsible for overall coordination of the programme, especially management and harmonization of measurement, model technique and installation of monitoring networks. Institute for Land Reclamation and Grassland Farming (IMUZ) was responsible for the scientific coordination of the project and for problems of process studies on investigation plots. The Institute of Water Problems (IWP) has performed research connected with testing and calibrating of selected simulation models under concrete land use conditions. Also, in collaboration with the other partners, it has performed work on problems related with model studies, model comparison, model linking and development of coupling–transforming programmes. The main objective of ALTERRA contribution to the project was the assistance of the Polish partner in selecting and using the appropriate models for water and substances, both on a local and on a regional scale.

DESCRIPTION OF THE RESEARCH ACTIVITIES

The project was run over a period of 3 years (1/01/1999–31/12/2001). The activities in the project are divided into three work packages.

Harmonisation of measurement and modelling techniques, installation of monitoring networks

A common methodology of investigation was selected. In particular, harmonisation of techniques for measuring parameters of water and various substances in the saturated and unsaturated zone, and the modelling techniques which may be applied to a selected drainage water basin.

Each partner had chosen a model combination for his test field. From the German group (IGW) Visual MODFLOW (Andersen 1993) in combination with MT3D was chosen for saturated zone and HERMES for unsaturated zone. For the Polish group (IMUZ) the SWAP and ANIMO (Groenedijk and Kroes 1997) models were chosen for unsaturated zone and FLOTRANS (Guiguer, Molson, Franz *et al.* 1993) for saturated zone. This part of the project was conducted with the close cooperation and help of the Netherlands group (ALTERRA). The Bulgarian team (IWP) worked with the WAVE model (unsaturated zone) and MODFLOW and MT3D for saturated zone.

All partners (except the Netherlands group) organized the field experimental plots with two multilevel piezometers. Measurement equipment for unsaturated zone on each experimental plot was installed near the multilevel well. For estimation of soil moisture measurement of matrix potential, tensiometers were chosen by the Bulgarian and German teams. For monitoring soil moisture, the reflectometer method (TDR) for Falenty was chosen (Malicki 1999). For analyses of substances, the suction lysimeters were installed by all groups.

Process studies of water and substances, parameter
estimation (*in situ*/in laboratory)

The meteorological parameters were measured or collected from nearby stations. A detailed survey and analysis of land use was made for each field experimental plot. Data obtained helped to estimate the amount of nutrients.

Measurements of soil properties involved: determination of grain-size distribution, content of organic matter, hydraulic conductivity, the content of sulphur compounds and the pF curve (Hötzl and Witthüser 1999). Contents of organic nitrogen and organic carbon were investigated by the German and Bulgarian groups as well.

Continuous measurements were made of groundwater level, soil moisture and chemical properties of water (saturated and unsaturated zones). The parameters NH_4^+, NO_3^-, pH, O_2 and Ec were determined in samples every one to three months; every four months HCO_3^-, Ca^{2+}, Mg^{2+}, K and Na^+ were measured. Additionally the Polish group made groundwater level measurements at a piezometer set up at the meteorological station (twice a week). Micropiezometers were installed to take 'point water samples' at 12 points.

Estimation of groundwater age by isotopes (Zuber 1986; Małoszewski and Zuber 1996) was made for the water extracted at the experimental plots in Poland and Bulgaria. The German group estimated the groundwater age by determination of CFC in groundwater (Matthess 1994). The analysis of nitrogen turnover was made on the basis of field measurements and groundwater age measurement. On a laboratory scale, nitrate decomposition was measured using the so called 'batch test' (Fic and Isenbeck-Schröter 1989). The nitrogen turnover was analysed on the basis of field experiments as well (Walther 1995).

Model calibration, model linking

Taking into account plant growth, modelling vertical water flow and nitrogen transport in the soil was based on SWAP/ANIMO, WAVE and HERMES models, which are one-dimensional models used for calculating groundwater flow in the unsaturated zone with some elements of the saturated zones. The models were calibrated using moisture content, groundwater level and nitrate content measured in the field plots.

Modelling groundwater flow was based on the MODFLOW, MT3D and the FLOTRANS models which simulate two or three-dimensional flow and predict advective–dispersive contaminant transport. The models were verified using data (groundwater level and nitrate contents in the saturated zone) from the field plot (time dependent measurements) and from a single measurement made in the catchment.

A methodology has been worked out for linking the models for the unsaturated and saturated zone. Application guidelines have been developed for the practical use of the results obtained during the work on the project. Scenario calculation has been made to show the influence of agricultural soil use changes on groundwater quality.

ANALYSIS OF RESULTS

The main result of the study was elaboration of tools (methods) for assessing the impact of changing agricultural land use on groundwater quality. It has been documented that the assessment of the load transported out from the unsaturated zone and forecasting groundwater quality may be made using the following combinations: (SWAP/ANIMO and FLOTRANS), (HERMES and MODFLOW), (WAVE and MODFLOW/MT3D).

The results of the field measurements showed that installed multilevel wells give best opportunities for sampling of groundwater in precise depths. Methods applied for measuring soil moisture for Falenty (Poland) field (TDR technique) have demonstrated high usefulness of this type of studies. A major advantage of this method is the possibility for conducting continued measurements, including during the winter season (tensiometers used previously proved highly unreliable). An original achievement was using micro-piezometers to extract water from the saturated zone as well. It was demonstrated that, in periods of drought, using ceramic cups for extracting water from the unsaturated zone made it very difficult to get the necessary volume of water. Due to the extended time of storage of water in the cup measurement results are not entirely trustworthy. The calibration of the model for the aeration zone required the most complete data set for the moisture regime in this zone.

Laboratory and field study of solute transport gave interesting results. From batch tests in the laboratory, heterotrophic denitrification was encountered. After a period of about 70 days an autotrophic denitrification could be recorded in some cases. But many more experiments are necessary in order to be able to find a more reliable statement. With regard to the nitrate turnover in the aquifer Thuelsfeld (Germany) a 1st order kinetic was accepted:

$$-\frac{dc_{NO_3}}{dt} - k \times c_o$$

The decay-coefficient (k) was determined using the relationship of water ages and nitrate concentrations in different depths. Investigation of the oxygen concentrations revealed different zones in the aquifer: a zone without oxygen at a depth of about 21 m below the surface of the oxygen-containing zone above. Because of the important discrepancy of the turnover, different values of k have to be determined for two zones. The k-values are 0.2 and 0.4 l/year.

Some results showed that at the depth of 0.3 m nitrate content is low due to the high uptake of nutrient by the plants (Falenty field). Substantial differences in nitrate concentrations were registered between measurements made at 0.9 and 1.2 m below soil surface. At this depth, nitrogen uptake by plants is slight. Consequently it is thought that at this location, there is a substantial decomposition of the compound, observed also during the winter season. Problems in taking water samples from the unsaturated zone made it difficult to explain the observed phenomenon.

It was established that groundwater quality in a given location may be dependent not only on the agricultural use of the area directly overlying the investigated point but mainly on the quality of water inflowing from a greater distance (the recharge area). At the Falenty site, the presence of three distinct aquifer zones have been registered: surface zone with young but markedly polluted water, the lower zone with pure water inflowing from a greater distance and an intermediate zone where the mixing of these waters takes place. The greatest impact on groundwater quality in the experimental plot is exerted by water inflowing from a more distant part of the catchment.

The results from dating of water samples from the three aquifer layers show that the water in the upper layers is young, while the age of the tested deeper water is higher, for example greater than 50 years. Certain considerations give the possibility of determining the velocity of water movement in the deeper water permeating layers. These considerations have contributed significantly to composing the regional model of the groundwater flow in the investigated area.

Modelling of water and substances transport in saturated zones was performed using different method. SWAP/ANIMO describe in detail the movement of water and nutrients and give good results for the cases with detailed input. For water and nitrogen movement the WAVE model is similar. Both need a lot of very specific data. Those models cannot be used by farmers, but are a very good tool for sophisticated study of the influence of agricultural land use on water quality. Both models are physically based for determining water flux and conceptual models on nutrient turnover and flux.

HERMES was developed for advising farmers on fertilization. It is a conceptual model for water flux, and nutrient turnover and flux. It needs a relatively low amount of data for parametrization. But HERMES is not able to react on fast changes of water content and nitrate transport in soil zones. It considers leaching of nitrate only to a depth of 2 m below the surface.

All three models for the unsaturated zone show sufficient adaptation for soil water content modelling. Results of all models and test fields for the nitrogen cycle (concentrations and fluxes) are generally worse than for water content. The reason is the relatively small amount of observed data. Generally all models are applicable for different soil types, agricultural practices and under various hydrological conditions. Each model is a suitable tool for studying influence of agricultural land use on nitrogen leaching from soils towards deeper zones. But each model has advantages and disadvantages regarding the amount of input data and concerning the flexibility of describing internal processes and precision of results.

Modelling of water and substances transport in saturated zones was performed using different methods. For the saturated zones the model MODFLOW with MT3D is recommended, while the FLOTRANS can be used for simple situations. MODFLOW is a commercial programme available worldwide, and is in ongoing development. It is suitable for simulations of groundwater flow and substance transport. FLOTRANS is a very simple model, which can be used for the first estimation of nutrient transport in an aquifer.

Linking of single models has been undertaken. The concept developed for linking the models of the unsaturated and the water saturated zone is motivated by the necessity to achieve the simulation of a continuous process. The concept organizes uses output parameters from the unsaturated zone models, which are transformed to input parameters for the regional groundwater models. The linking concept is built taking into account two aspects: 1) a technology for direct linking of models at a local scale, and 2) regional linking of models for different conditions (scenarios) of agricultural land use. Linking of the following three pairs of models have been considered in particular: 1. WAVE with MODFLOW and MT3D; 2. HERMES with MODFLOW and MT3D and 3. SWAP and ANIMO with MODFLOW and MT3D (FLOTRANS).

CONCLUSIONS

The principal benefit following from the project is that a method (tool) was developed for calculating the impact of changes in agricultural land use on groundwater quality. The method may be applied in future at the stage of designing programmes for reconstructing Poland's and Bulgaria's agriculture, contributing to the protection of groundwater quality. In addition, the project generated substantial benefits for science, helping to resolve a number of questions and phenomena associated with the movement of biogenic compounds in the unsaturated zone below the root zone, as well as in the saturated zone (aquifer).

It has been documented that water and substances transport in the unsaturated zone can be calculated using the verified HERMES and WAVE models. These models are suitable for use with a database, which can be developed in direct contact with the farmers, and with a relatively low amount of additional measurements and parameters. ANIMO model needs a lot of very specific data. This model is a very good tool for sophisticated study of influence of agricultural land use on water quality.

The MODFLOW model is a worldwide-available commercial three-dimensional programme, which is suitable for simulations of groundwater flow (saturated zone) and substance transport. The FLOTRANS model, as a very simple one, can be used for the first estimation of the nutrient transport in the aquifer.

The models were calibrated using moisture content, groundwater level and nitrate content measured for saturated and unsaturated zones in the field plots.

REFERENCES

Andersen, P.F. 1993: A manual of instructional problems of the USGS Modflow model. pp. 85.
Groenedijk, P. and Kroes, J.G. 1997: *Modelling the nitrogen and phosphorous leaching to groundwater and surface water.* ANIMO 3.5, Report 144, DLO – Winand Staring Centre, Wageningen.
Guiguer, N., Molson, J., Franz, T. and Frind, E. 1993: *FLOTRANS, User Guide Version 2.1.* Waterloo Centre for Groundwater Research. WHS/WCGR. 79.
Hötzl, H. and Witthüser, K. 1999: *Methoden für die Beschreibung der Groundwasser-beschaffenheit* – Schriften DVWK nr. 125 Bonn s 113.
Kersebaum, K.C. 1995: Application of a simple management model to simulate water and nitrogen dynamics. *Ecological Modelling 81*: 145–156.
Malicki, M.A. 1999: Methodical questions of monitoring of water status in selected biological materials (in Polish). *Acta Agrophysica 19*: 108.
Matthess, G. 1994: *Die Beschaffenteit des Groundwassers* – Lehrbuch der Hydrogeologie Band 2. Gebrüder Borntraeger, Berlin Stuttgart.
Zuber, A. 1986: *Mathematical models for the interpretation of environmental radioisotopes in groundwater systems.* In: P. Fritz and J.C. Fontes (eds): Handbook of Environmental Isotope Geochemistry, pp. 1–55.
Małoszewski, P. and Zuber, A. 1996: *Lumped parameter models for the interpretation of environmental tracer data.* In: Manual on Mathematical Models is Isotope Hydrology. IAEA-TECDOC-910 with Supplement. IAEA, Vienna, pp. 9–58.
Fic, M. and Isenbeck-Schröter, M. 1989: Batch studies for the investigation of the mobility of the heavy metals Cd, Cr, Cu and Zn. Elsevier Science Publishers B.V., Amsterdam. *Journal of Contamination Hydrology 4*: 69–78.
Walther, W. 1995: *Über den Stoffhaushalt der Landschaft und über die diffuse Belastung von Böden. Fließgewässern und Grundwasser, dargestellt an ausgewählten Standorten. Fachbereich für Bauingenieur – und Varmessungswesen,* Techn. Univers. Braunschweig, Habil.

CHAPTER 13

Modelling of nitrate leaching to groundwater under Slovenian conditions

M. Pintar and V. Zupanc
University of Ljubljana, Biotechnical Faculty, Agronomy Department, Centre for Agricultural Land Management and Agrohydrology, Jamnikarjeva 101, SI-1000 Ljubljana, Slovenia

B. Čenčur Curk
Institute for Mining, Geotechnology and Environment, Slovenčeva 93, 1000 Ljubljana, Slovenia

ABSTRACT: Two field experiments were performed to study nitrate leaching towards groundwater. Simulation of fertilization was performed in the year 1999/2000 on a meadow at the first experimental field site in order to study the behaviour of nitrates and their percolation through karstic soil and the underlying unsaturated zone of karstified rock. An irrigation experiment was performed in 2000 at a second field site in an intensive hop garden. GLEAMS could not be applied for soil transport modelling in karstic rock. There is still a lack of commercial models for modelling transport through karstic soil and unsaturated fractured and karstified rock. Most models have been developed for typical agricultural land. The GLEAMS model predicted that with irrigation implementation, nitrate leaching in hop garden decreases.

INTRODUCTION

A large amount of drinking water in Slovenia is located under plains in alluvial aquifers, fields on which intense agricultural activity is ongoing. In addition, the karstic environment is also characteristic of Slovenia. Aquifers of fractured karstic porosity, which are very vulnerable to various types of pollution, are predominant. It can therefore be said that, in general, there are two types of aquifers in Slovenia, representing the majority of the nation's drinking water. Slovenia has rich water resources, but these should be protected from pollution.

In addition to pesticides, nitrates are the most common pollutants found in groundwater underlying agricultural land. They also appear in areas not influenced by human activity, but in these places their concentration usually does not exceed 50 mg NO_3/l, which is the norm for drinking water in Slovenia.

In principle, the fertilization of agricultural areas is the greatest source of nitrates in groundwater. By controlling this input of nitrates into the natural environment, it would be possible to influence the quality of groundwater under intensely cultivated agricultural land. However, the design and implementation of controlling nitrate inputs into the natural environment are possible only when the dynamics of nitrate leaching from the upper soil layer towards groundwater is known. By simulating natural processes of nitrate binding,

Figure 1. Location of nitrate field experiments: Latkova Vas and Sinji Vrh.

transformation and transport, mathematical models enable us to analyse the dynamics of nitrate leaching from the upper soil layer towards groundwater.

In Slovenia, the use of mathematical modelling in predicting groundwater pollution with nitrates and pesticides has not come very far. The GLEAMS mathematical model has been successfully used for the modelling of atrazine leaching (Turk 1995).

Two field experiments were performed to study nitrate leaching towards groundwater. Simulation of fertilization was performed in the year 1999/2000 on a meadow at the experimental field site Sinji Vrh (Figure 1) in order to study the behaviour of nitrates and their percolation through karstic soil and the underlying unsaturated zone of karstified rock (Veselič 2000). Synthetic fertilizer KAN (calcium ammonium nitrate) was used in standard amounts. The second field experiment was in the year 2000 at Latkova Vas (Figure 1) in an intensive hop garden where an irrigation experiment was performed.

MATERIALS AND METHODS

Experimental design of the experimental field site Sinji Vrh

The experimental field site (EFS) Sinji Vrh is located in the western part of Slovenia at the edge of the Trnovski Gozd plateau, which is an overthrust of Jurassic carbonate rock (mostly limestone) over Eocene flysch. The experimental field site at Sinji Vrh is located above the Hubelj spring in the region of the Avče fault, which runs in a NW-SE direction.

The experimental field site Sinji Vrh in the unsaturated zone of fractured and karstified rock presents a 340 m long artificial research tunnel, 5–25 m below the surface. An agrometeorological station has been installed on the surface, where precipitation, evaporation, air temperature, air moisture, and wind speed and direction (both at two levels) are continuously measured.

The surface above the tunnel is covered with grassland and small beech forests, which usually cover outcrops. The soil is a typical karstic soil (calcaric brown soil) with characteristic deeper pockets extending along weak zones like fractures in the underlying rock. In a 0.7 m deep soil profile, the Ah-horizon has a thickness of 15 cm and B-horizon a thickness of 55 cm. In the latter, an upper horizon with more roots and higher organic content can be distinguished. The unsaturated fractured and karstified limestone has

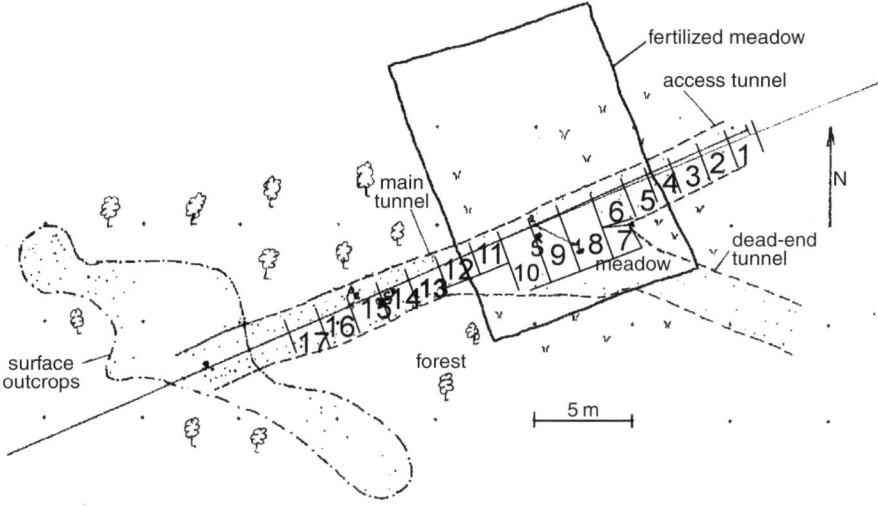

Figure 2. The experimental field site: the position of surface outcrop area, research tunnel with segments numeration and fertilized meadow.

a negligible matrix porosity and very high fracture density with some large conduits (Veselič and Čenčur Curk 2001).

A simulation of fertilization was performed at the north-western part of the research tunnel, where the distance to the surface is $10-15$ m. An area of $150 \, m^2$ of meadow was used for fertilization experiments (Figure 2). The amount of applied fertilizer corresponded to 42 kg N/ha. For this purpose suction cups were installed at two levels (at depths of 20 and 50 cm) in the soil above the research tunnel. The water percolating through the karstic soil and rock was collected by a special construction for collecting water seeping from the ceiling of the tunnel below the fertilized meadow. The construction enables the sampling in segments along the tunnel (Figure 2). Electrical conductivity, temperature and pH were measured *in situ* in water samples during the experiment. Water samples were analysed in the laboratory for nitrate ion and nitrite ion concentration (IRGO) and nitrogen isotope $\delta^{15}N$ (IJS – Institute Josef Stefan).

Experimental design of the experimental field site Latkova Vas

Leaching of nitrates towards groundwater in an intergranular aquifer was monitored in an intensive hop garden in the field experiment at Latkova Vas in the Savinja valley in central Slovenia. Savinja valley is flat, with homogenous land use, surrounded by hills. The area is an agricultural area with very intensive hop production. Hop areas in Slovenia vary between 1,800 and 2,000 ha and represent about 3% of total world hop growing areas.

Three different variants were performed in the experiment: variant 1 = non-irrigated, variant 2 = underground drip irrigation at 30 cm depth and variant 3 = hops irrigated by hose reel system (sprinkler irrigation). Fertilizers were applied in four consecutive ratios to the total amount of 310 kg N/ha. Dates of fertilizer application were 19 April (109th Julian day – j.d.), 8 May (129th j.d.), 20 July (202nd j.d.) and 8 August (221st j.d.). The amount of applied nitrogen was 77, 91.6, 79 and 62 kg respectively. Fertilizer was applied in KAN form, with the exception of NPK 15-15-15 application on the first date.

The ADCON agrometeorological station has been installed on the edge of the hop garden where precipitation and air temperature have been continuously measured. There was 1,089 mm of total precipitation in the year 2000. Most precipitation fell in November (237 mm), September (155 mm), October (151 mm), July (128 mm) and May (116 mm) (Figure 4). The total amount of added water was 57 mm in 15 consecutive applications and 32 mm in one application for the second and the third variant respectively. The soil is 60 cm deep eutric cambisols with a lot of skeleton particles on sand and gravel alluvium as a parent material (Table 1).

The soil data required to run GLEAMS were selected from the GLEAMS manual (Knisel, Davis, Leonard *et al.* 1992) (Table 2). Required data about hop plants were obtained from the Institute for Hop Research and Brewery (Table 3). These kinds of data are presented in the GLEAMS manual for over 70 different crops, but hop plants are not included.

Table 1. Characteristics of soil in the experiment field Latkova Vas (Ap and Bv stand for soil horizons), soil type: eutric cambisols.

Soil prof.	Soil depth (cm)	pH (KCl)	Org. matter (%)	Sand (%)	Silt (%)	Clay (%)	FC* (cm/cm)	WP* (cm/cm)	BS* (%)
Ap	0−27	6.4	3.9	35.5	46.1	18.4	0.36	0.25	71.2
Bv	27−60	5.9	1.3	52.8	18.5	28.7	0.39	0.22	74.2

*FC – field capacity; WP – wilting point; BS – base saturation.

Table 2. Data for eutric cambisols from Latkova Vas selected from GLEAMS manual.

Parameter	Value
Soil evaporation parameter	4
Eff. saturated conductivity of the soil horizon immediately below the root zone $(cm\,h^{-1})$	1.1
SCS curve number	83
Porosity − hor. Ap; hor. Bv	0.42; 0.40
Eff. satur. cond. − hor. Ap; hor. Bv $(cm\,h^{-1})$	0.31; 0.20

Table 3. Data for hop plant required run GLEAMS.

Parameter	Value
Carbon nitrogen ratio for the crop	12
Nitrogen phosphorus ratio for the crop	5
Coefficient C1	4.5
Coefficient C2	−0.31
Potential yield for var. 1 $(kg\,ha^{-1})$	12.000
Potential yield for var. 2 $(kg\,ha^{-1})$	14.600
Potential yield for var. 3 $(kg\,ha^{-1})$	15.000

Four pan lysimeters (Wilson 1995) were installed in the soil layer in each variant, 60 cm deep, in order to sample percolated water. Water samples were analysed by AutoAnalyzer II (ALFA LAVAL, Bran+Luebbe), by AA II Method No. G-016-91. Azo compounds derived from nitrate were determined spectrophotometrically at 520 mm.

The GLEAMS computer model

GLEAMS is used in over 40 North American and European countries. Its use is also recommended by the American Environmental Protection Agency (Knisel, Leonard and Davis 1995). It is mainly used to simulate pesticide leaching through the ground profile. However, since it also has an equivalent module for plant nutrients, it also yields good results when used to simulate nitrate leaching into groundwater. In the nitrogen cycle, it takes into account mineralization, immobilization, denitrification, ammonia volatilization, nitrogen fixation by papilionaceous plants, fertilization, nutrient intake by plants, surface runoff, sedimentation and leaching through the ground profile taking into account the water temperature and moisture content in the soil. Good results for programme validation were obtained only in those cases where local parameters were used, which is as recommended by the programme's authors (Wu, Ward and Workman 1996; Wu, Ward, Workman *et al.* 1997). It is also used to define groundwater protection strategies (Diebel, Taylor, Batie *et al.* 1992; Boisvert, Regmi and Schmit 1997). In the Slovenian environment, GLEAMS has previously been used to monitor atrazine in soils and has yielded satisfactory results (Turk 1995).

RESULTS

Sinji Vrh

The spreading of the synthetic fertilizer KAN (calcium ammonium nitrate) was performed on 6 September. Water quantity from suction cups was often insufficient for analysis. Even so, at 20 cm depth, the highest nitrate concentration (29.2 mg l^{-1}) was reached 22 days (28 September) after the fertilization event. After that nitrate concentrations fell below the detection limit. We assume that nitrate was either used by plants or rinsed to the lower parts of the soil and rock. At 50 cm depth, the nitrate concentration rise appeared in November, after 67 days of fertilization. This rise of the nitrate concentration is more likely to have been caused by natural mineralization in the soil in autumn.

Concentrations of NO_3^- ion in MP2 and MP21 of the research tunnel samples are relatively high also before the fertilization. This could be due to a former fertilization experiment or natural processes. Nevertheless, the later rise of nitrate concentration after fertilization is significant. In the research tunnel samples the nitrate peak appeared 1 week after one major precipitation event (68.8 mm; 21 September), about 22 days after the fertilization (Figure 3). The $\delta^{15}N$ isotope data confirmed that the nitrate source was the mineral manure applied on the meadow (Čenčur Curk, Pintar and Veselič 2000). The nitrate transport along the fast channel at measuring point MP5 was quicker and appeared after 8 days. In some measuring points another rinse of the manure was detected (MP21 and partly also MP10). After a major precipitation the next spring (85.1 mm; 1 March) the nitrate was rinsed again, having been retained in the microfractures of the unsaturated zone.

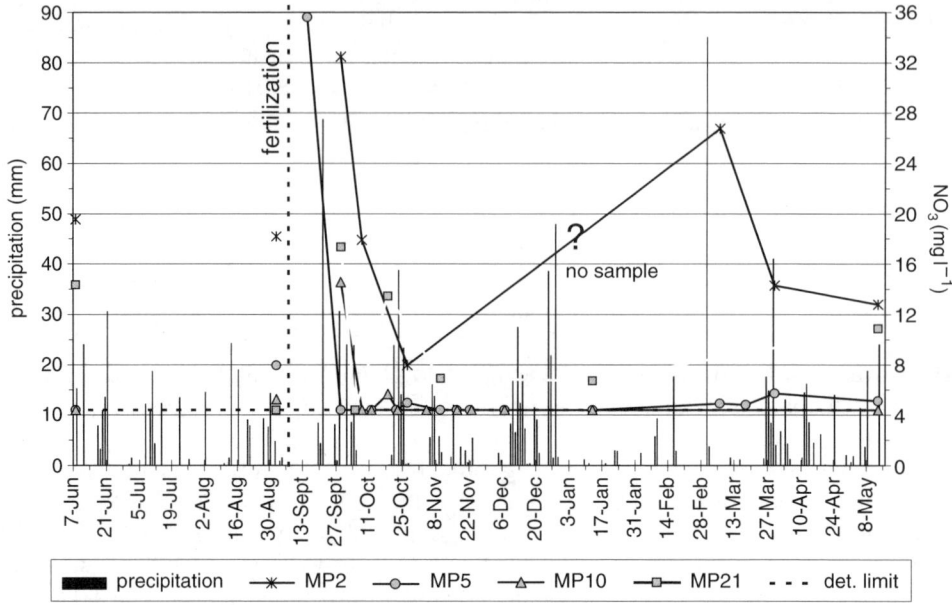

Figure 3. Precipitation events and NO$_3^-$ concentrations (detection limit: 4.4 mg l^{-1}) at sampling points MP2, 5, 10 and 21 for fertilization experiment G2. Note: Before a rain event, due to reduced fracture flow, a problem of sample quantity is often present, so the first sample after rain already contains the highest tracer concentration.

The mathematical model GLEAMS was applied to model nitrate transport through karstic soil. In spite of a long meteorological data set for this area, we could not use GLEAMS for soil transport modelling, since we have only two levels of suction cups with obviously insufficient data. SFDM (single fracture dispersion model) (Maloszewski 1994) was applied to model transport in the rock. The mean transit times for nitrates are in the range of 15 (MP5) to 24 days (MP10), the mean velocity is about 0.5 m d^{-1} and dispersion in the range of 0.5–1 m.

Latkova Vas

Model simulation was conducted for a one-year period representing 1 January (day 001) to 31 December (day 366) 2000. Comparison was made between measured values of N-NO$_3$ concentrations (mg l^{-1}) from the pan lysimeters and the values predicted by GLEAMS. Uncalibrated simulation showed that GLEAMS predicted well N-NO$_3$ concentrations in percolated water in the first half of the year. Values were in the range of the N-NO$_3$ concentrations from the pan lysimeters. But GLEAMS overestimated mean N-NO$_3$ concentrations in leached water for the second half of the year for non-irrigated hop (Var. 1 – Figure 5) and for hop, irrigated with sprinkler irrigation (Var. 2 – Figure 6). There was just one measurement of N-NO$_3$ concentration from pan lysimeters per sampling date in Var. 2 (Figure 6).

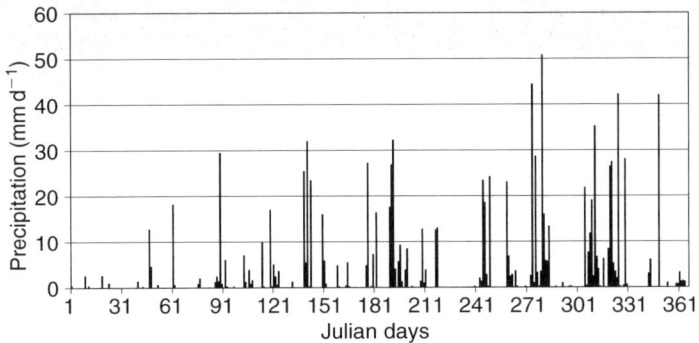

Figure 4. Daily rainfall for the study year 2000.

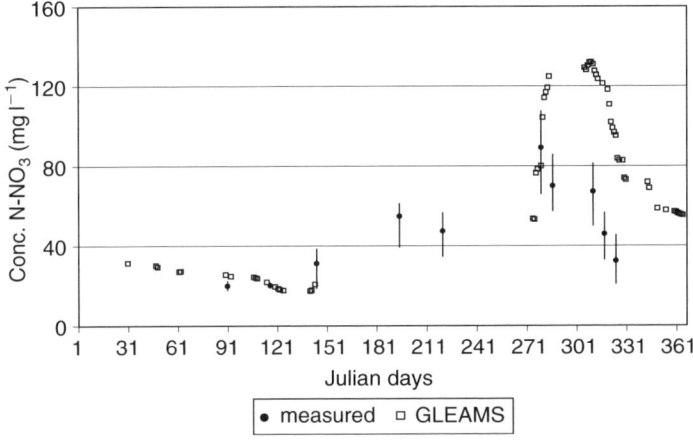

Figure 5. Measured and uncalibrated predicted N-NO$_3$ concentrations in percolated water (mg l^{-1}) in non-irrigated hop garden (Var. 1).

GLEAMS underestimated the mean N-NO$_3$ concentrations in leached water for the second half of the year for drip-irrigated hops (Var. 3 – Figure 7). Predicted and observed N-NO$_3$ concentrations in percolated water were low in the first half of the year and then rose greatly in the second part of the year. However, there were some observed percolation events in the summer months with N-NO$_3$ concentration over 40 mg l^{-1}, whilst GLEAMS has not predicted any percolation.

GLEAMS predicted that in all variants large amounts of N-NO$_3$ would be leached in October and November, which is in accordance with high precipitation and quite high predicted (and observed) N-NO$_3$ concentrations. High N-NO$_3$ concentrations in percolated water in the autumn were a result of high amounts of nitrogen left over from the harvest period (end of August) and additional mineralization of soil organic matter. GLEAMS predicted that there was no nitrogen leached during summer months, which is not in accordance with our measurements. The uncalibrated GLEAMS predicted that drip irrigation would cause the least nitrogen leaching in the hop garden (253 kg N-NO$_3$/ha/year) in comparison with sprinkler irrigation (299 kg N-NO$_3$/ha/year) and non-irrigated control (372 kg N-NO$_3$/ha/year).

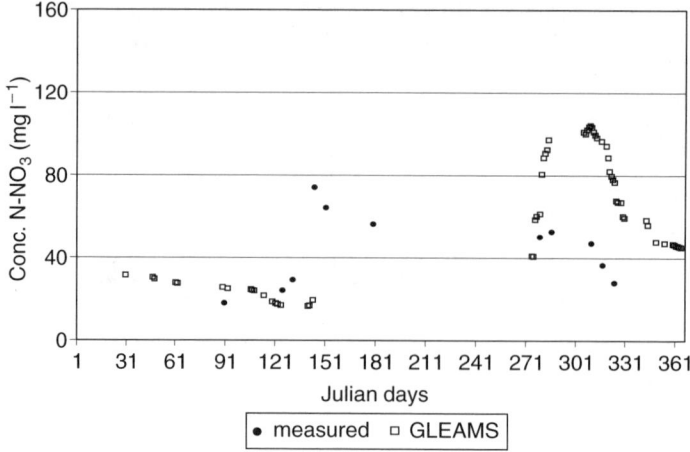

Figure 6. Measured and uncalibrated predicted N-NO$_3$ concentrations in percolated water (mg l^{-1}) in hop garden with hose reel irrigation (Var. 2).

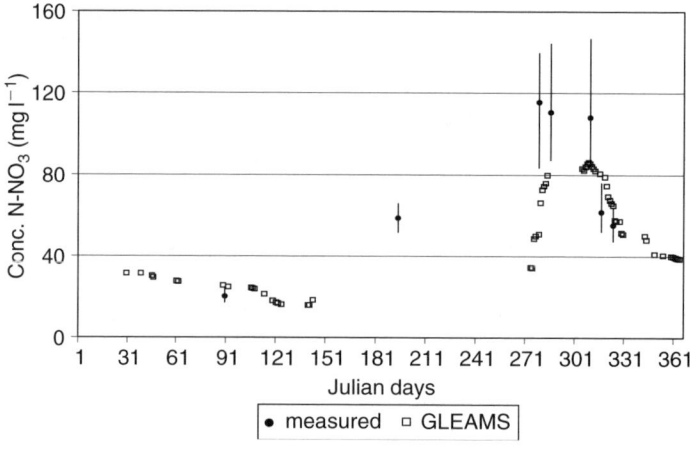

Figure 7. Measured and uncalibrated predicted N-NO$_3$ concentrations in percolated water (mg l^{-1}) in hop garden with drip irrigation (Var. 3).

CONCLUSIONS

GLEAMS could not be applied for soil transport modelling in karstic rock, since there are only two levels of suction cups and limited data on nitrate concentrations. There is still a lack of commercial models for modelling transport through karstic soil and unsaturated fractured and karstified rock. Most models are developed for typical agricultural land (for example, thick fertile soils on gravel) and for the saturated zone of rocks, which do not consider meteorological parameters. A further task of this research is to find appropriate models, which could be coupled together (output of soil model should be input for unsaturated rock model).

The uncalibrated GLEAMS model predicted well N-NO$_3$ concentrations in percolated water in the hop garden for the first part of the observed year 2000. There was less accordance between observed and uncalibrated predicted values in the second part of the year. Uncalibrated GLEAMS predicted that with irrigation implementation nitrogen leaching in the hop garden would decrease. With additional calibration this can be a useful tool for assessing N-NO$_3$ leaching into groundwater from agricultural areas.

ACKNOWLEDGEMENTS

We acknowledge the financial support from the Ministry of Science and Technology of the Republic of Slovenia for research projects under the research grants No. L2-1310-98.

REFERENCES

Boisvert, R.N., Regmi, A. and Schmit, T.M. 1997: Policy implications of ranking distributions of nitrate runoff and leaching from corn production by region and soil productivity. *Journal of Production Agriculture 10* (3): 477–483.

Čenčur Curk, B., Pintar, M. and Veselič, M. 2000: *Nitrate transport through karstic soil and unsaturated karstic rock*. In: G. Günay (ed.): Karst 2000. Hacettepe University, International Research and Application Centre for Karst Water Resources – UKAM, Ankara, Abstracts, pp. 31–32.

Čenčur Curk, B., Pintar, M. and Veselič, M. 2001: *Study of agricultural pollution at in situ rock lysimeter at karstic Trnovo plateau in Slovenia*. In: Bericht [über die] 9. Lysimetertagung zum Thema Gebietsbilanzen bei unterschiedlicher Landnutzung: am 24 und 25 April 2001 an der HBLA Raumberg. Gumpenstein: Bundesanstalt für alpenländische Landwirtschaft, pp. 127–129.

Diebel, P.L., Taylor, D.B., Batie, S.S. and Heatwole, C.D. 1992: Low-input agriculture as a ground water protection strategy. *Water Resources Bulletin 4*: 755–761.

Knisel, W.G., Davis, F.M., Leonard, R.A. and Nicks, A.D. 1992: GLEAMS version 2.1, User manual: 220. Tifton: University of Georgia, Biological and Agricultural Engineering Dept., Tifton, Georgia, USA.

Knisel, W.G., Leonard, R.A. and Davis, F.M. 1995: Representing management practices in GLEAMS. *European Journal of Agronomy 4*: 449–505.

Maloszewski, P. 1994: Mathematical modelling of tracer experiments in fissured aquifer. *Freiburger Schriften zur Hydrogeologie 2*: 1–107.

Turk, I., 1995: *Uporabnost dinamičnega modela GLEAMS za spremljanje migracije atrazina v tleh v slovenskem prostoru: 108*. Ljubljana: Univerza v Ljubljani, Biotehniška fakulteta, oddelek za agronomijo.

Veselič, M. and Čenčur Curk, B. 2001: *Test studies of flow and solute transport in the unsaturated fractured and karstified rock on the experimental field site Sinji Vrh, Slovenia*. In: K.P. Seiler and S. Wohnlich (eds): New approaches characterizing groundwater flow. Proceedings of the XXXI International Association of Hydrogeologists Congress, Munich, Germany, 10–14 September 2001. Munich: Balkema, pp. 211–214.

Veselič, M. 2000: *Karst groundwater protection*. In: I. Vlahović and R. Biondić (eds): Second Croatian Geological Congress, Cavtat – Dubrovnik, 17–20.05.2000. Zagreb: Institute of Geology, pp. 43–45.

Wilson, N. 1995: *Soil water and ground water sampling*. Boca Raton, London; Lewis Publishers, Tokyo, p. 188.

Wu, Q.J., Ward, A.D. and Workman, S.R. 1996: Using GIS in simulation of nitrate leaching from heterogeneous unsaturated soils. *Journal of Environmental Quality 3*: 526–534.

Wu, Q.J., Ward, A.D., Workman, S.R. and Salchow, E.M. 1997: Applying stochastic simulation techniques to a deterministic vadose zone solute transport model. *Journal of Hydrology 197* (1/4): 88–110.

Nitrate pollution of groundwater – national and regional studies

CHAPTER 14

Nitrate reduction and pyrite oxidation in the Netherlands

H.P. Broers
Netherlands Institute of Applied Geoscience TNO, P.O. Box 80015, 3508 TA, Utrecht,
The Netherlands

ABSTRACT: Nitrate reduction by pyrite oxidation is abundant in the Netherlands subsoil and limits the breakthrough of nitrate into deeper aquifers in parts of the country. Regional and local examples from the Netherlands show that the downward movement of an agricultural pollution front with nitrate might mobilize sulphate, trace metals and arsenic when pyrite is encountered. Therefore, European groundwater protection policy should not only focus on nitrate, but also on these reaction products.

INTRODUCTION

Nitrate contamination of shallow groundwater is widespread in the Netherlands. Intensive monitoring programmes exist for soil quality, shallow phreatic groundwater and deeper

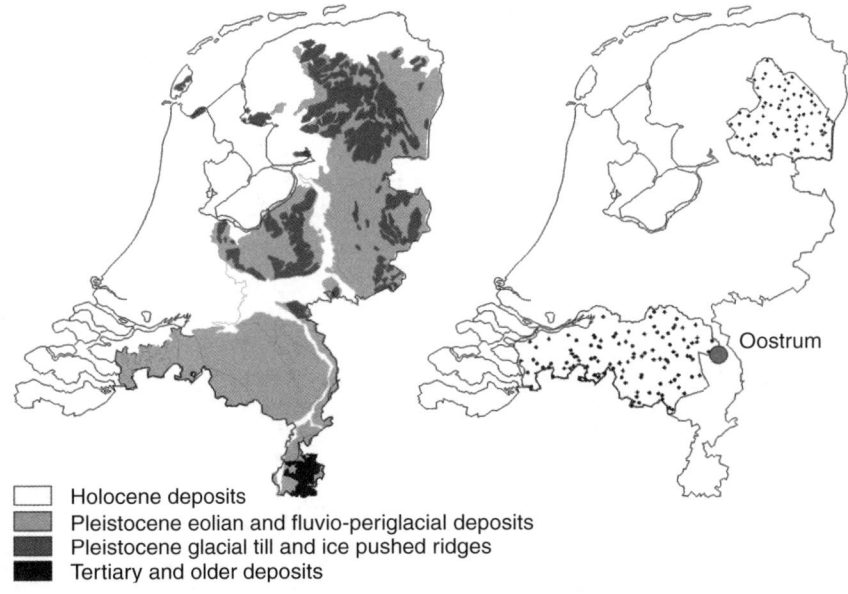

Holocene deposits
Pleistocene eolian and fluvio-periglacial deposits
Pleistocene glacial till and ice pushed ridges
Tertiary and older deposits

Figure 1. Location of the Pleistocene sandy areas in the Netherlands, the location of monitoring wells in the provincial networks of Drenthe (northern part) and Noord-Brabant (southern part) and the location of the phreatic well field Oostrum.

groundwater up to 30 m depth (Van der Aa, Van der Grift, Busink *et al.* 2002). These networks operate on a national and provincial scale and on the local scale of well fields. The monitoring results indicate that nitrate reduction by organic matter is important in areas with Holocene clay and peat soils and in areas with very shallow groundwater levels (<1.2 m).

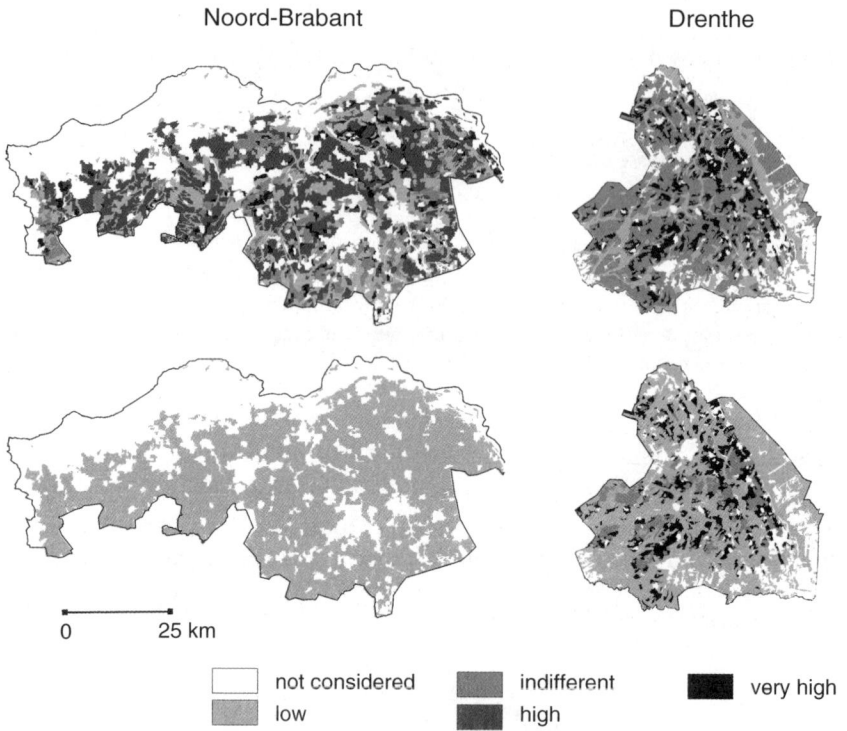

Figure 2. Classes of the proportion of nitrate-contaminated groundwater at 5–15 m depth (upper graphs) and 15–30 m depth (lower graphs) in the provinces of Noord-Brabant and Drente. The class low represents a proportion of less than 30% nitrate-contaminated groundwater at 95% confidence level (Broers 2002). Nitrate is virtually absent in deeper groundwater in Noord-Brabant, although the area is known for large inputs of manure.

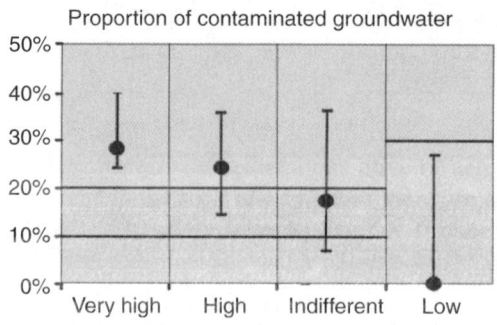

Figure 3. Definition of classes of contaminated groundwater used for the maps.

In the Pleistocene sandy regions of the Netherlands, nitrate contamination is abundant up to 15 m depth, especially in agriculturally used regional recharge areas. However, the vulnerability for nitrate leaching to deep parts of the aquifers is strongly dependent on the reactivity of the subsoil. This paper investigates the effects of pyrite oxidation on nitrate contamination of deeper aquifers, using examples from regional and local scale monitoring networks in the Netherlands.

EXAMPLES FROM TWO REGIONAL MONITORING NETWORKS

Examples from the regional monitoring networks in the provinces of Drenthe and Noord-Brabant show large differences in the depth of the nitrate front (Broers 2002). In the Drenthe recharge areas, large proportions of nitrate-contaminated groundwater are present at $15-30$ m depth (Figure 2). Contrarily, nitrate is absent in Noord-Brabant recharge areas at this depth and appears to be reduced by the oxidation of pyrite.

Classes were formed by comparing the 95% confidence interval on the proportion of contaminated groundwater with thresholds of 10%, 20% and 30% (Figure 3).

Sediment samples show that 0.1 to 0.2 weight% of pyrite is very common in the Noord-Brabant subsoil (Figure 4). In the aquifers, pyrite oxidation probably proceeds as:

$$5FeS_2 + 14NO_3^- + 4H^+ \rightarrow 5Fe^{2+} + 10SO_4^{2-} + 2H_2O + 7N_2$$

Although nitrate disappears with depth, the groundwater is still identifiable as being agriculturally polluted using indicators such as Oxidation Capacity (OXC, after Postma, Boesen, Kristiansen *et al.* 1991) and sum of cations (SUMCAT). Here, OXC was defined as:

$$OXC = 7[SO_4^{2-}] + 5[NO_3^-] \, \text{mmol} \, l^{-1}$$

OXC is transported conservatively if pyrite oxidation is the only process which controls the nitrate and sulphate concentrations. Using OXC as an indicator for agricultural pollution, the province of Noord-Brabant is much more affected than the province of Drenthe. Figure 5 shows spatial patterns of OXC contamination for two depths intervals.

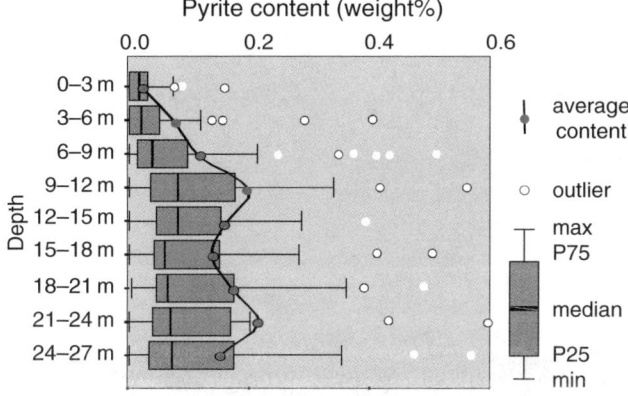

Figure 4. Average pyrite contents and frequency distribution (box plots) of 433 soil samples from 24 observation wells in the province of Noord-Brabant.

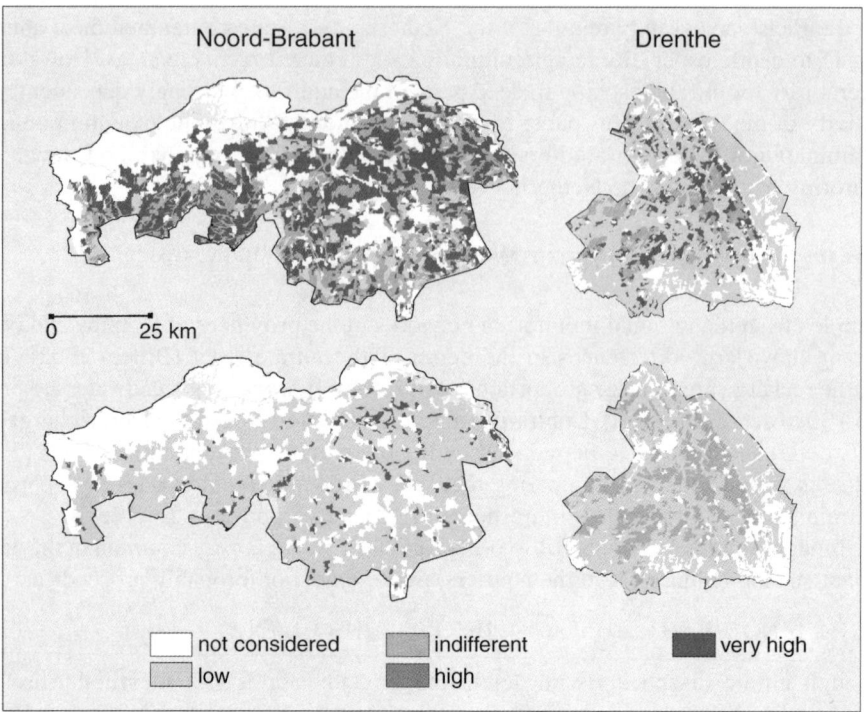

Figure 5. Classes of the proportion of OXC-contaminated groundwater at 5−15 m depth (upper graphs) and 15−30 m depth (lower graphs) in the provinces of Noord-Brabant and Drenthe. The class high represents a proportion of more than 20% nitrate-contaminated groundwater at 95% confidence level (Broers 2002). Breakthrough of OXC concentrations above 11 mmol l^{-1} is indicated for deeper groundwater in specific areas in the province Noord-Brabant.

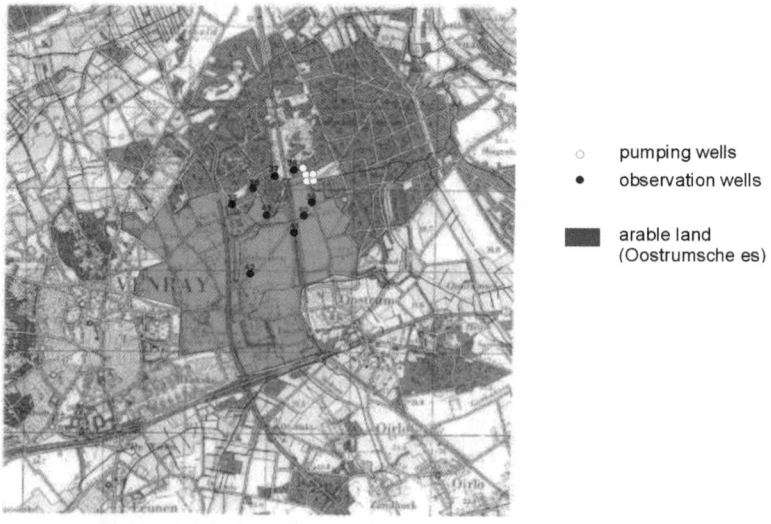

Figure 6. Location of pumping and observation wells at the Oostrum well field.

Groundwater was identified as OXC-contaminated if the concentrations were above the threshold of $11\,\text{mmol}\,l^{-1}$, which coincides with a potential exceedence of the Dutch sulphate drinking water standard of $150\,\text{mg}\,l^{-1}$. According to Figure 5, the concentrations of the reaction product sulphate have already increased above drinking water standards in deeper groundwater of specific areas in Noord-Brabant.

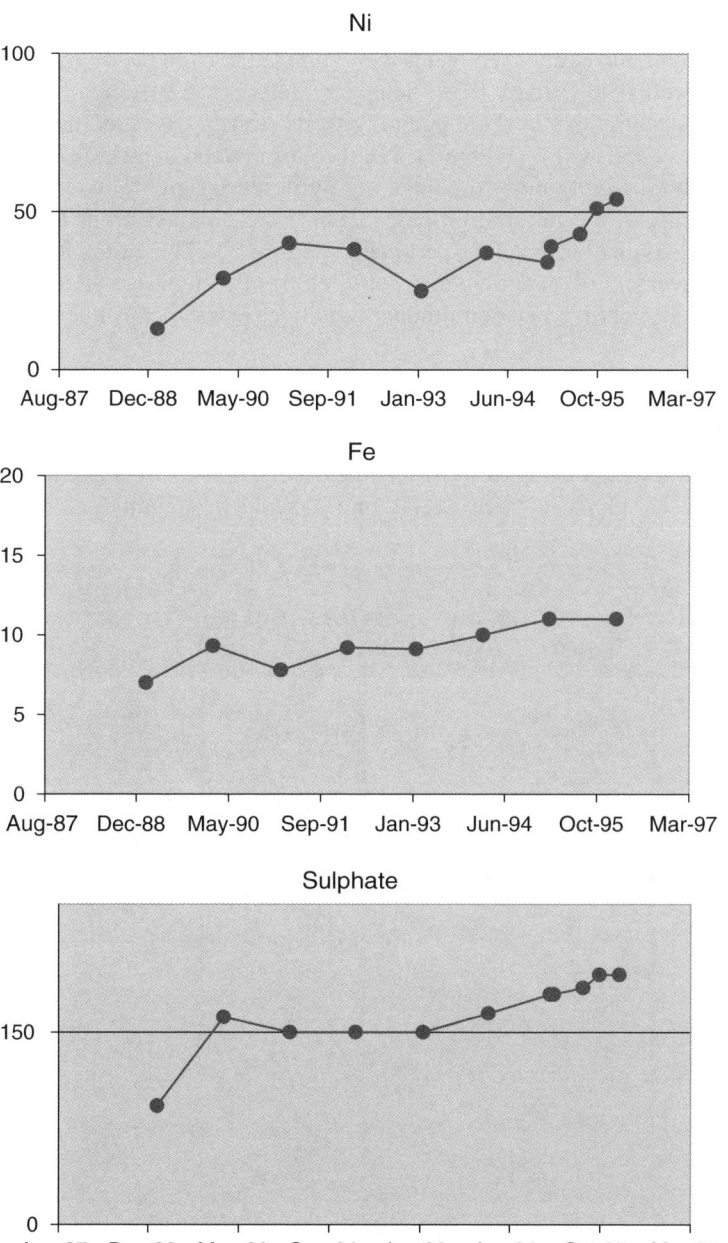

Figure 7. Increasing concentrations of nickel ($\mu g\,l^{-1}$), iron and sulphate ($mg\,l^{-1}$) in pumping well III.

EXAMPLES FROM WATER WELL FIELDS

Local studies of groundwater quality and sediment reactivity at phreatic well fields indicated that pyrite oxidation and exchange reactions induced by the infiltrating agricultural pollution fronts yield the mobilization of trace metals and arsenic from the sediments (Van Beek, Hettinga and Straatman 1989; Broers and Buijs 1997). Figure 6 shows the configuration of observation wells at the Oostrum well field in Noord-Limburg. Both groundwater and aquifer sediments were sampled to trace the origin and fate of trace metals at the well field (Broers 1998; Schipper, Helvoort, Appelo *et al.* 2000).

Sulphate concentrations in some pumping wells already exceeded the Dutch drinking water standard of $150 \, \text{mg} \, \text{l}^{-1}$ (Figure 7). The increasing trend in nickel concentrations was attributed to the dissolution of Ni-containing pyrite during pyrite oxidation. Indications for nickel release at the nitrate-iron redox cline were obtained from mini-screen wells (Figure 8) and oxidation batch experiments (Figure 9). The example shows that the downward movement of an agricultural pollution front with nitrate might induce a series of reactions that mobilize reaction products and release trace constituents.

CONCLUSIONS

The regional and local examples show that the excessive use of manure and fertilizer can indirectly cause sulphate, trace metal and arsenic contamination of groundwater

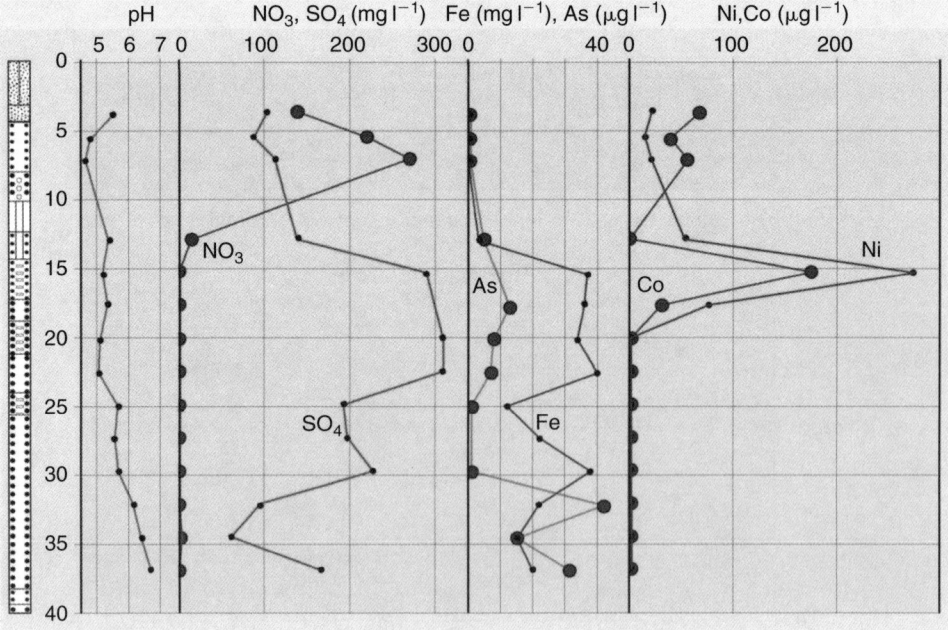

Figure 8. Groundwater quality profile of well 40 (see Figure 6 for location). Peak concentrations of Ni, Co and As were observed below the nitrate-iron redox cline.

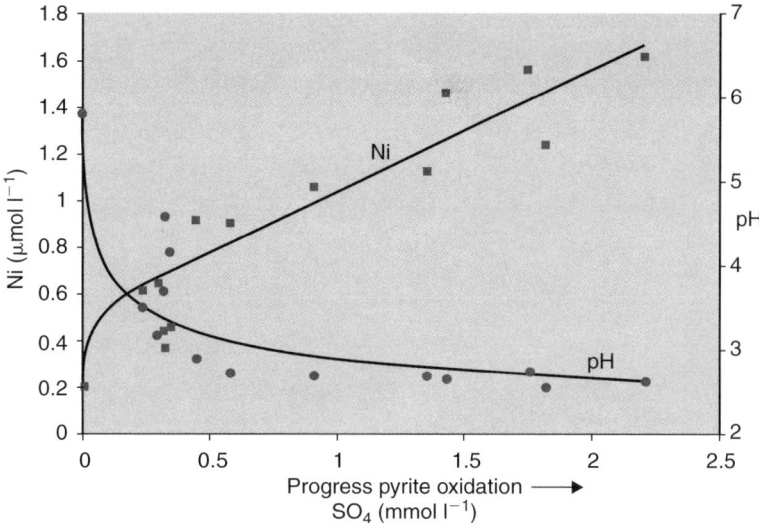

Figure 9. Ni-release at oxidation batch experiments. After an initial release of Ni from exchangeable sites, a constant Ni-sulphate dissolution ratio points to nickel release from pyrite (Van Helvoort, Broers, Schipper *et al.* 2000).

resources. Therefore, it is recommended that the European groundwater protection policy does not only focus on nitrate, but also on the reaction products that are mobilized during the propagation of agricultural pollution fronts.

REFERENCES

Broers, H.P. and Buijs, E.A. 1997: *Origin of trace metals and arsenic at the Oostrum well field.* Netherlands Institute of Applied Geoscience TNO, report NITG 97-189A (in Dutch).

Broers, H.P. 1998: *Release of Ni, Co and As from aquifer sediments at the Oostrum well field.* In: M. Herbert and K. Kovar (eds): Proc. Groundwater Quality 1998, Tübinger Geowissenschaftliche Arbeiten, Reihe C, no. 36. pp. 106–108.

Broers, H.P. 2002: *Strategies for regional groundwater quality monitoring.* Ph.D. Thesis. Netherlands Geographical Studies 306, KNAG/Faculteit Ruimtelijke Wetenschappen, Utrecht, p. 231.

Postma, D., Boesen, C., Kristiansen, H. and Larsen, F. 1991: *Nitrate reduction in an unconfined sandy aquifer: water chemistry, reduction processes, and geochemical modeling. Water Resources Research* 27 (8): 2027–2045.

Schipper, P., Helvoort, P.J., Appelo, C. and Broers, H.P. 2000: *Heavy metals in groundwater: pyrite oxidation and desorption: 2. Geochemical modelling* (in Dutch). *H₂O 24*: 19–22.

Van Beek, C.G.E.M., Hettinga, F.A.M. and Straatman, R. 1989: *Release of heavy metals in groundwater due to manure spreading.* Proceedings 6th International Symposium on Water–Rock Interaction, Malvern, U.K. 3–8 August 1989. Balkema, Rotterdam, Netherlands.

Van der Aa, M., Van der Grift, B., Busink, R. and Broers, H.P. 2002: *Regional monitoring of nitrate in phreatic and deeper groundwater.* Proceedings IAH Euromeeting 2002: Nitrate in groundwaters in Europe. This volume.

Van Helvoort, P.J., Broers, H.P., Schipper, P. and Appelo, C. 2000: *Heavy metals in groundwater: pyrite oxidation and desorption: 1. field and laboratory study Oostrum* (in Dutch). *H₂O 24*: 15–18.

CHAPTER 15

Nitrate and nitrite in shallow groundwater

M. Guzik, P. Liszka, M. Zembal and A. Pacholewski
Polish Geological Institute, Upper Silesian Branch, 41-200 Sosnowiec, ul,
Królowej Jadwigi1, Poland

ABSTRACT: The purpose of this investigation was to recognize the degree of pollution of
shallow groundwater by nitrate in the predominantly rural Lubliniec region.
   The results of water quality investigations conducted during the preparation of the Hydrogeological map of Poland at a scale 1:50,000, indicated high content of nitrogen compounds in
shallow groundwater. This problem is significant in the rural areas where high nitrate concentrations are connected with uncontrolled agricultural activities and improper water management.
   The most important hazard comes from the nitrate ion, which is active and subjected to sorption
only to a small extent and therefore can easily migrate to groundwater. The problem of groundwater contamination by nitrogen compounds, as a result of agricultural practices, has already been
monitored in the countries with intensive use of chemicals in agriculture for many years. The
Council of European Community issued the directive no. 91/676/EEC, in order to reduce the
contamination by nitrate from agricultural practices, which obliges the member states to monitor
waters, both underground and surface water. Poland, as a candidate to the European Union, should
also take into consideration the obligation to perform such investigations. This paper presents the
preliminary results of the investigations performed in 130 used domestic wells. This study was
financed by the Polish Committee of Scientific Research.

STUDY AREA

The Lubliniec region is situated between the Silesian – Cracow Upland and the Silesian
Lowland. Lubliniec is the major urban centre. This area is famous for its low urbanization
and typical agricultural character. Intensive farming activity is conducted in the western
and central parts of this area. Medium size animal-breeding farms dominate. In the
southern part of this region there is large forest area. Sources of groundwater pollution in
the Lubliniec region are presented in the Figure 1. Two aquifers have been recognized
here, Quaternary and Triassic.
   The shallow Quaternary aquifer is composed of sands and gravels. Its thickness does
not exceed 10 m. In the central part of the Lubliniec region, that is, in Woźniki–Lubliniec–
Pawonków zone, the usable aquifer also occurs in the carbonate deposits – the Woźniki
limestones of the Upper Triassic mudstone–siltstone complex. This is an unconfined
aquifer with the depth of water table ranging from 0.67 to 7.8 m. The average depth of
water table is 2.77 m.

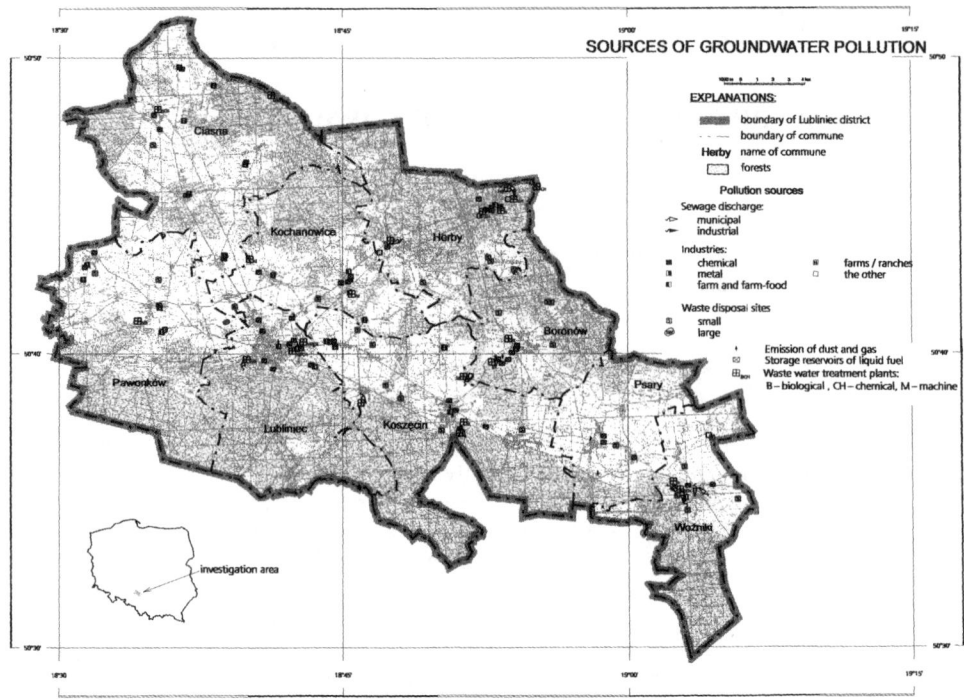

Figure 1. Sources of groundwater pollution in the Lubliniec region.

METHODS

During the field investigations in autumn of the year 2000, 130 domestic wells were analysed in order to determine the content of nitrogen compounds. The shallow wells are in use and situated in the areas of different types of cultivation. The field investigations included the colorimetric tests of nitrate and nitrite by the Riedel-de Haen test, measurements of pH, temperature and electrolytic conductivity of water.

RESULTS

The pollution of shallow groundwaters by nitrate in the Lubliniec region is considerable: maximum nitrate content in waters reaches $300 \, mg \, l^{-1}$, the average is $60.32 \, mg \, l^{-1}$. Maximum nitrite content in waters reaches $0.4 \, mg \, l^{-1}$, the average is $0.01 \, mg \, l^{-1}$. Selected statistical parameters are presented in the Table 1.

The content of NO_3^- in groundwaters is connected with the existing pollution sources, particularly with the animal-breeding farms, waste water treatment plants and waste storage sites located in the vicinity of water intakes. A considerable contribution to water pollution comes from leaky septic tanks and lack of sewage systems. In the vicinity of these sources, the nitrate content in the water increases to above the Polish Drinking Water Standard. Variability with depth in water chemistry is caused by the nature of the pollutant and possibilities of infiltration and sorption in the zone of aeration, as well as the

Table 1. Selected statistics parameters.

Parameters characteristic	Stratigraphy of the aquifer	Number of samples	Minimum value	Maximum value	Arithmetic mean	Median	Standard deviation
1	2	3	4	5	6	7	8
NO_2 (mg l^{-1})	Q + T	128	0.00	0.40	0.03	0.01	0.06
NO_3 (mg l^{-1})	Q + T	133	0.00	300.00	58.31	60.0	56.99
pH	Q + T	128	5.70	7.98	7.00	7.0	0.50
Electrolytic conductivity (μS cm^{-1})	Q + T	128	136.00	1600.00	586.70	528.5	281.80

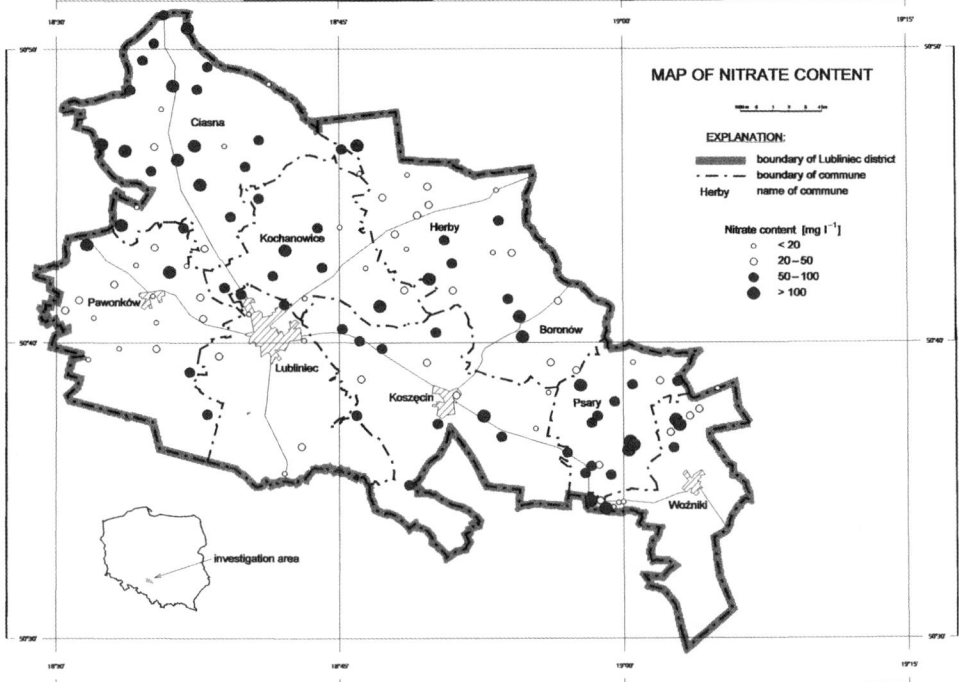

Figure 2. Map of nitrate content in the Lubliniec region.

dispersion conditions in the saturation zone. The most polluted waters occur in Psary, Kochanowice and Ciasna communes (Guzik, Liszka and Zembal 2001). The results of the investigation are shown in the Figure 2. The investigations have shown that 42% of domestic wells have water polluted by nitrate in excess of drinking water requirements. The Polish Standard of Drinking Water Quality is 50 mg l^{-1} NO_3^- and 0.1 mg l^{-1} NO_2^-. In the study area 8−9% of domestic wells have waters polluted with nitrite. Cumulative distribution curves and histograms of frequency distributions of nitrate and nitrite in shallow groundwater are presented in Figure 3.

In the study area, the nitrate content in waters increases with time. Although the application of fertilizers and pesticides causes water pollution by nitrate, the quality of

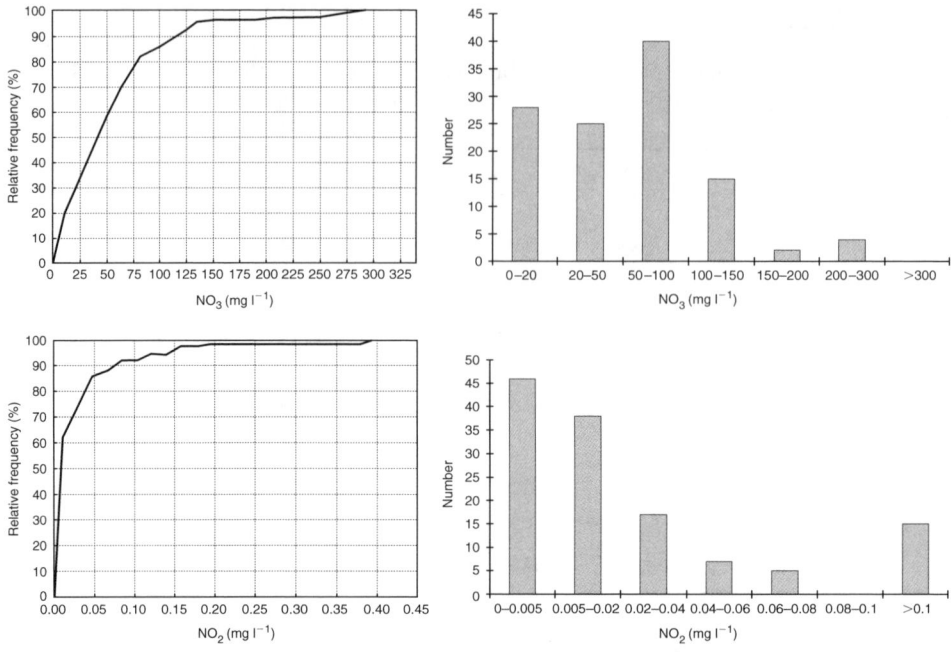

Figure 3. Cumulative distribution curves and histograms of frequency distribution of NO_3 and NO_2 in shallow groundwaters.

waters in the farming regions is generally better than in the dense housing areas (Kleczkowski 1984). This is a result of the disordered water- and sewage- management, and improperly protected septic tanks and dung heaps.

The nitrate ion is thought to be the biggest hazard, being also the most mobile and only subject to the sorption to a small degree (Witczak and Adamczyk 1995). It can easily migrate in groundwater.

The problem of groundwater contamination by nitrogen compounds, appearing as the result of agricultural activity, has been perceived in countries with intensive use of chemicals in agriculture for a dozen or so years.

Vulnerability of shallow groundwaters in this region is very high. The main factors which determine water pollution are:

• type and intensity of production and conditions of ground cultivation,
• hydrogeological and hydrodynamic conditions which occur at the study area,
• regional and local geomorphology which determine the infiltration.

The vertical variability of chemistry of shallow groundwater is especially visible near the pollution sources. The most significant for the migration of pollution is the thickness of the soil layer and unsaturated zone. The protection provided by the soil depends mainly on: humus and clay minerals content as well as the pH of the soil.

This process is determined by the load of pollution and capacities of infiltration and sorption in the unsaturated zone and also dispersion in the saturated zone (Górski 1989; Mikołajków 1995).

CONCLUSIONS

Shallow groundwater in the Lubliniec region is contaminated by nitrate and nitrite, high concentrations of which are harmful for the human health. Water contamination here is mainly caused by improper water management, leaky septic tanks and lack of sewage systems in the municipal area. In the rural area fertilizers are the main sources of pollution.

REFERENCES

Górski, J. 1989: *Recognition of hydrochemical background of Mosina town.* Reporting from realization of task 02.06.06 (in Polish). In frames CPBP 04.10.09, Archive IGPIK, Poznan.

Guzik, M., Liszka, P. and Zembal, M. 2001: *Nitrate contamination of shallow groundwaters in the Lubliniec district* (in Polish). Archive PGI, Sosnowiec.

Kleczkowski, A.S. (ed), 1984: Groundwater protection (in Polish). Wydawnictwa Geologiczne. Warsaw.

Mikołajków, J. 1995: Migration of nitrogen compounds in zone of aeration of the water pollution in fur agricultural areas (in Polish). *Present Problems of Hydrogeology.* Vol. II, part I, p. 323–329. Wydawnictwo Profile, Cracow.

Decree of minister of health from 4 September 2000 concerning the condition which should meet water for drinking and for industrial needs (Dz. At. No.82. item 937).

Witczak, S. and Adamczyk, A. 1995: *Catalogue of physical and chemical elements of groundwater – methods of analysing* (in Polish). Vol. II. PIOS, Monitoring Library of Environment, Warsaw.

CHAPTER 16

Nitrate variations in aquifers: a case study from the Neogene aquifers of Trifilia, SW Peloponnesus, Greece

G. Panagopoulos and N. Lambrakis
University of Patras, Geology Department, Section of Applied Geology and Geophysics, 26500, Rio, Greece

ABSTRACT: West Trifilia is an agricultural area with intense nitrogenous fertilization. A system of two aquifers developed in calcitic sandstones and separated by a thick marly and clayey bed is subject to nitrogen pollution. The unconfined aquifer exhibited a high concentration of nitrate ions with a mean of $68.6 \, mg \, l^{-1}$, while the confined aquifer exhibited low concentration with a mean of $12.0 \, mg \, l^{-1}$. The lack of nitrate ions in the confined aquifer is attributed to the role of the marly-clayey bed that prohibits their lateral advance. The similar spatial distribution of nitrate, ammonium and calcium ions especially in the high concentration areas, shows a common chemical process that relays these ions and which may be the cation exchange between Ca^{2+} from the sediments and NH_4^+ from groundwater.

INTRODUCTION

Sources of nitrate origin in groundwater are natural sources (nitrate from the nitrogen cycle), waste materials (municipal or industrial sludge, animal manure) and agrochemicals. A significant amount of nitrate in groundwater also originates from septic tank systems or household waste in localized areas. However, the most important source is the application of agrochemicals.

Chemical and microbiological processes such as nitrification and denitrification can influence the movements of various forms of nitrogen in groundwater. Nitrification is defined as the biological oxidation of ammonium-nitrogen to nitrate-nitrogen and takes place in two steps (NH_4^+ to NO_2^-) by nitrosomonas and (NO_2^- to NO_3^-) by nitrobacter (Reddy and Patric 1981). The overall process can be written as follows:

$$NH_4^+ + 2O_2 \; \rightarrow \; NO_3 + 2H^+ + H_2O$$

In this form of nitrate, nitrogen is transported under oxidized conditions in aquifers for long distances (Freeze and Cherry 1979; Antonakos and Lambrakis 2000). Denitrification is defined as the biological reduction of nitrate to gaseous nitrogen by the following steps: NO_3^- to NO_2^- to NO to N_2O to N_2 (Payne 1981). This very important process can characterize aquifers as 'bioreactors' (Korom 1992). The complete nitrate reduction equation can be written in two ways taking into consideration the reaction reflecting oxidation of organic matter (CH_2O) or the oxidation of a substrate mineral (FeS_2) as follows (Mariotti 1986):

$$CH_2O + 4/5NO_3^- + 4/5H^+ \leftrightarrow CO_{2(g)} + 2/5N_2 + 7/5H_2O$$

$$FeS_2 + 14NO_3^- + 4H^+ \leftrightarrow 7N_{2(g)} + 10SO_4^{2-} + 5Fe^{2+} + 2H_2O$$

However, the denitrification reaction can take place under four requirements (Firestone 1982): a) N oxides as terminal e^- acceptors, b) the presence of bacteria possessing the metabolic capacity, c) suitable e^- donors, and d) anaerobic conditions or restricted O_2 availability. Hence, an understanding of the subsurface process is fundamental to identifying natural or man-made sources of nitrate contamination and also to evaluate the factors responsible for this process.

Factors affecting the amounts of nitrate in groundwaters relate them with the time of fertilizer application, vegetation cover and soil porosity (Borneff and Adabe 1973). A number of hydrogeological factors can also influence the amount of nitrates in ground-water such as precipitation/runoff, soil and aquifer type, depth of groundwater level, the nature of the rocks hosting the aquifers, etcetera. Based on hydrogeological factors that are considered the most important factors controlling groundwater pollution, Aller, Bennett, Lehr *et al.* (1987) presented DRASTIC, a standardized numerical rating system for evaluating groundwater pollution potential. The acronym is derived from the factors 'Depth to groundwater', 'Recharge rate', 'Aquifer media', 'Soil media', 'Topography', 'Impact of the vadose zone', and 'Conductivity hydraulic of the aquifer'. Research conducted in the Peloponnese (Lambrakis, Tiniakos, Lazarou *et al.* 1997) showed that the most important factor affecting nitrate concentration in groundwater was the depth of the groundwater level and the aquifer media. Carbonate aquifers exhibited increased sensitivity to nitrogen pollution (provided mainly by animal manure), mainly due to luck of sorption and the high velocity groundwater recharge through the rock macropores (fissures and karstic conductors). Groundwater from recent deposits such as Quaternary sediments has a heavy load of nitrate ions (sometimes it is greater than $200\,mg\,l^{-1}$) because of intense fertilization, while in Neogene sediments with a thick clayey unsaturated zone, groundwater exhibited low concentrations of nitrate.

The aim of this article is to study the impact of the intense fertilization on groundwater of the Neogene sediments in Trifylia where the presence of a thick bed from marls and clay between the unconfined and confined aquifers of the area diversifies the quality of the water.

GEOLOGICAL AND HYDROGEOLOGICAL SETTING

The bedrock of the research area belongs to Tripoli series and is composed of a thick sequence of Eocene and dolomitic limestones overlain by an Oligocene flysch (ESSO-HELLENIC 1980). All these sediments are unconformably overlain by recent Pleistocene and Holocene deposits (Figure 1). The Pleistocene sediments comprise mainly alternating beds of calcitic sandstone, marly limestone, marls and clay schists, with an overall thickness ranging from 30 to 100 m. These sediments host two discrete aquifers that are separated by the marly and clayey beds (Figure 1).

The uppermost aquifer is unconfined with a mean thickness of 12 m and the lower one is confined and has a mean thickness of about 15 m. The hydraulic characteristics of these

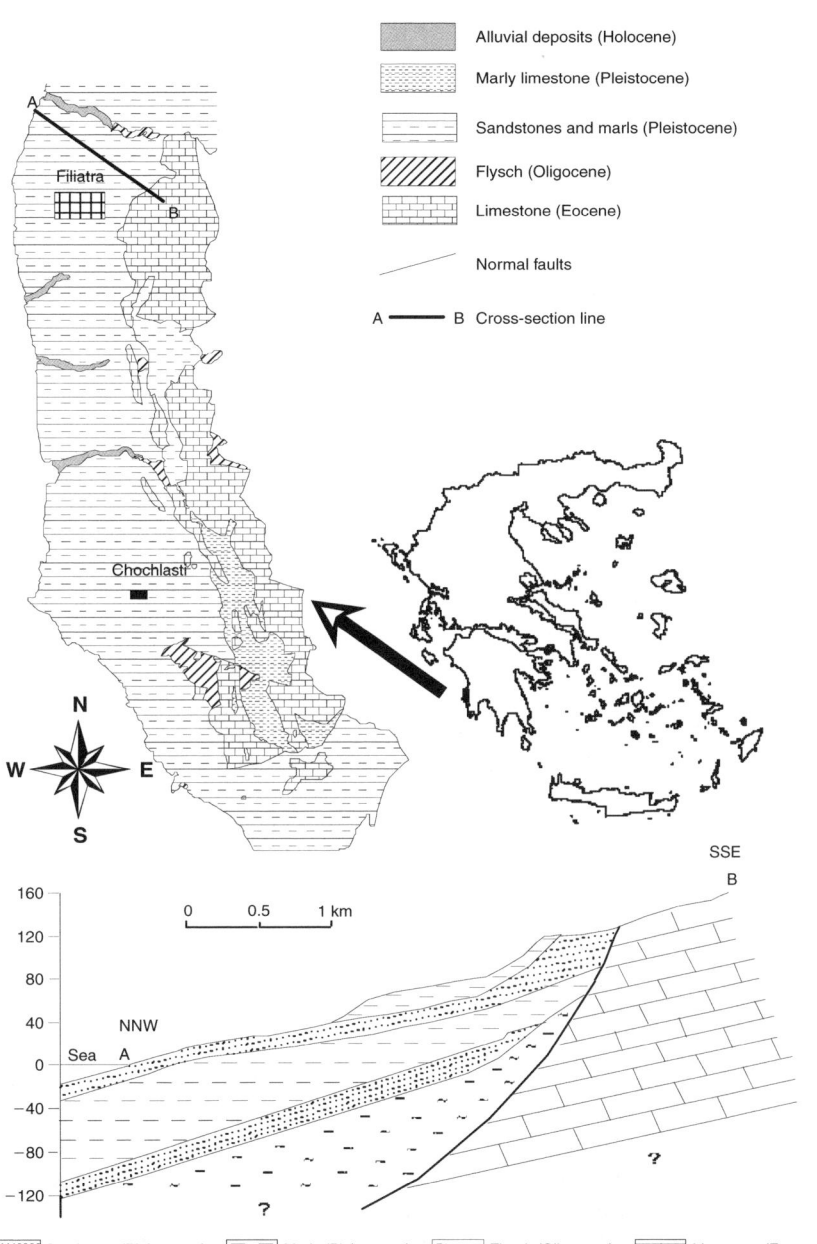

Figure 1. Geological map (ESSO-HELLENIC 1980) and cross-section of the research area.

aquifers were determined from pumping tests and showed different values for each aquifer. The unconfined aquifer has a transmisivity coefficient (T) ranging between 0.5×10^{-1} and $5.7 \times 10^{-3} \, \mathrm{m}^2 \, \mathrm{s}^{-1}$, a storage coefficient (S) varying between 1.14×10^{-3} and 5.35×10^{-3}, while the resultant hydraulic conductivity (K) is about $1.8 \times 10^{-3} \, \mathrm{m \, s}^{-1}$.

On the other hand, the confined aquifer has a mean transmisivity coefficient of $8.2 \times 10^{-4}\,\mathrm{m^2\,s^{-1}}$, a mean storage coefficient of 8.8×10^{-4}, while the hydraulic conductivity is about $5.5 \times 10^{-5}\,\mathrm{m\,s^{-1}}$.

The unconfined aquifer is recharged mainly from rainwater infiltration, while significant quantities of recharged water derive from the surface water percolation. The recharge conditions of the confined aquifer are similar but the recharge area is located outside and to the north of the research area. Aquifer exploitation is realized by a great number of shallow wells and deeper boreholes (more than 300).

ANALYTICAL METHODS

Among the 300 production wells and boreholes in the studied area, a network of 44 sampling points was defined, from which 32 points refer to the unconfined aquifer and 12 points to the confined (Figure 2a). Sampling runs were undertaken biannually, from May 1999 to October 2001, covering three hydrogeological cycles (six runs). The samples were collected at the well-pump outflow in two polyethylene bottles. All samples were filtered on site through 0.45-µm pore size Millipore filters. The unstable parameters such as pH and electric conductivity (E.C) were determined on site by using an ion/E.C metre (Consort® C533) with combined electrodes. The other analyses were performed immediately after collection in the Laboratory of Hydrogeology, University of Patras. The first sample of 0.5 l volume was acidified with 2 ml of 65% HNO_3 for cation analysis. The second non-acidified aliquot (1 l) was kept to determine bicarbonates (HCO_3^-) and chlorides (Cl^-) by Hach® titration kits. Sulphates (SO_4^{2-}), nitrates (NO_3^-) and ammonium (NH_4^+) were determined by a spectrophotometer (Hach®, DR/4000). Finally, the major cations Ca^{2+}, Mg^{2+}, Na^+ and K^+ were determined by the atomic absorption spectroscopy (GBC Avanta). The overall precision of the analyses is within $\pm5\%$. Processing of the data was carried out using a software package developed in the Laboratory of Hydrogeology, University of Patras (Lambrakis 1991). A statistical overview of the measured parameters is presented in Table 1.

HYDROCHEMISTRY

West Trifilia is an agricultural area mainly of olive trees (83%), irrigated crops (12%, mainly tomatoes and watermelons) and greenhouses (5%). Inorganic fertilizers, mainly nitrate ammonium, are widely used with a mean annual application rate in the order of $1,850\,\mathrm{kg\,ha^{-1}}$ (Sampatakakis, Makris, Gidonis *et al.* 1994). Because of the heavy fertilization in the area, groundwater contains an increased concentration of nitrates and ammonium (Figures 3a and b), with maximum values in the Chochlasti region (290.4 and $3.7\,\mathrm{mg\,l^{-1}}$ respectively). On the contrary, nitrates and ammonium concentrations are lower in the confined aquifer, due to the presence of the thick bed of impermeable marls and clays (Table 1).

The relatively high concentrations of nitrate and ammonium in the unconfined aquifer could be attributed to the application of nitrogen fertilizers, mostly NH_4NO_3, to the soil, and the thin unsaturated zone promotes this contamination. Thus, the applied fertilizers move through the unsaturated zone and infiltrate the aquifer.

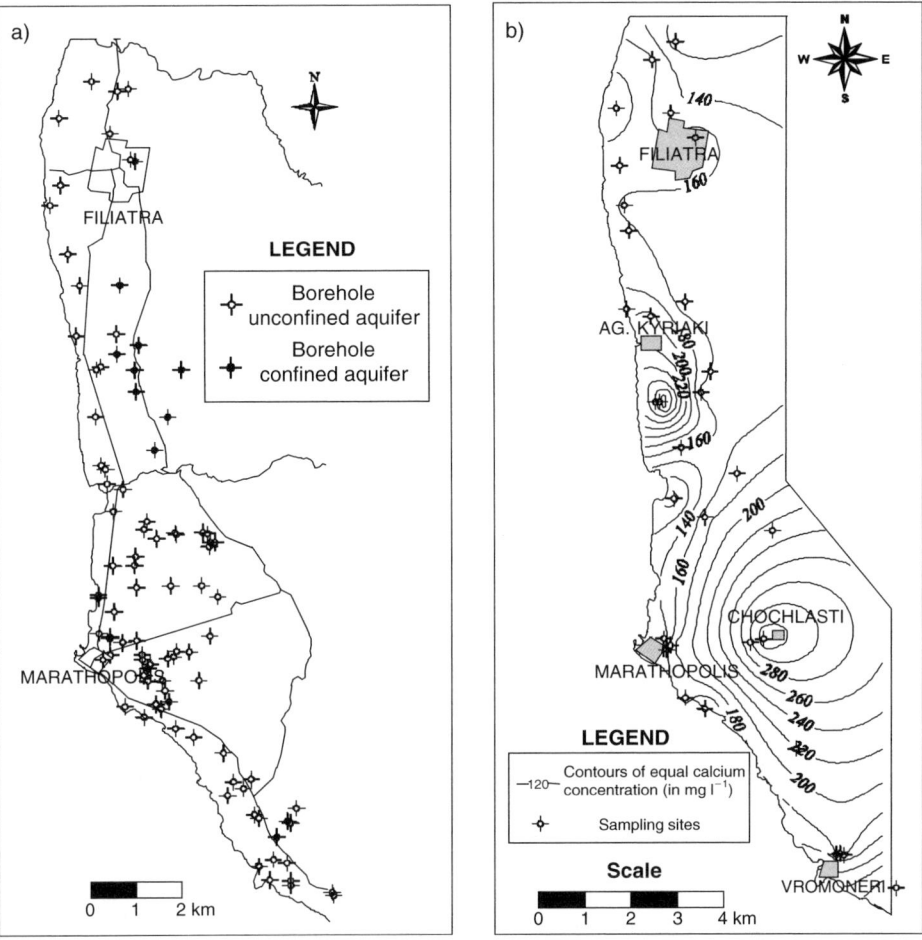

Figure 2. Water sampling network in the west Trifylia neogene aquifers (a) and calcium ions distribution map of the unconfined aquifer (b).

Table 1. Hydrochemical data of the unconfined and confined aquifer (wet season, May 2001). Averaged values are in $mg\,l^{-1}$ unless otherwise indicated.

Parameter	Unconfined aquifer ($n = 32$ samples)	Confined aquifer ($n = 12$ samples)
E.C ($\mu S\,cm^{-1}$)	1,664	1,146
pH	7.3	7.2
Ca^{2+}	179.7	158.7
Mg^{2+}	24.2	23.7
Na^{+}	125.8	48.5
K^{+}	5.4	4.7
HCO_3^{-}	308.0	372.6
Cl^{-}	218	74
SO_4^{2-}	177	129
NO_3^{-}	68.6	12.0
NH_4^{+}	0.9	0.5

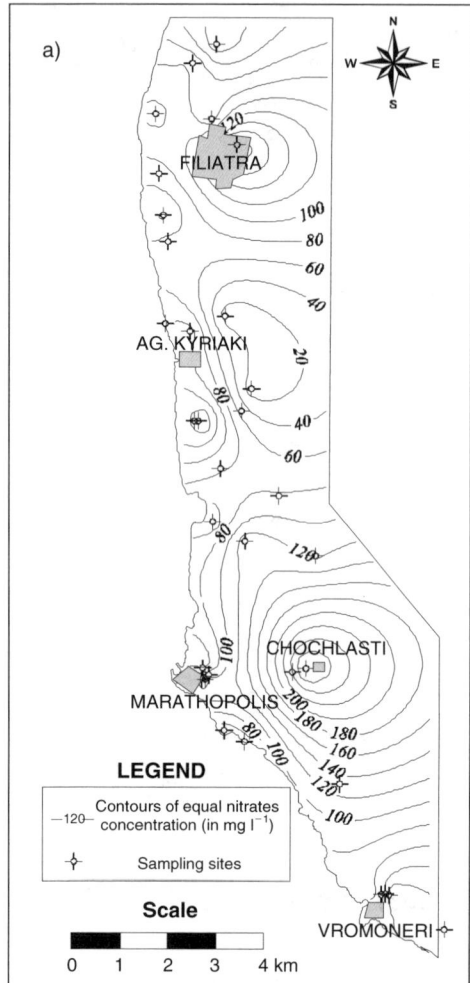 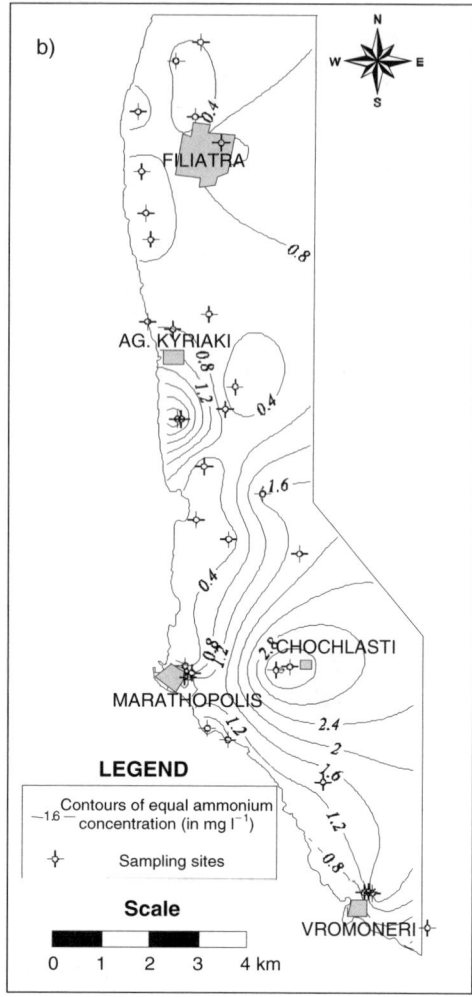

Figure 3. Distribution maps of nitrates (a) and ammonium (b) in the unconfined aquifer.

The nitrate salts are very soluble in water (more than $10\,\mathrm{g\,l^{-1}}$ at 20°C) and thus, dissociate according to the following reactions (Hatziioannou 1985):

$$NH_4NO_3 + H_2O \rightarrow NH_4OH + H^+ + NO_3^-$$

causing an increase of the nitrate concentration in the aquifer. NH_4OH is a weak base and dissociates according to the following reaction (Hatziioannou 1985):

$$NH_4OH \leftrightarrow NH_4^+ + OH^-$$

which causes an increase of the ammonium concentration in the aquifer.

As clearly shown in Figure 2b, the distribution pattern of calcium ions in the unconfined aquifer is similar to respective patterns of nitrate and ammonium, with maximum concentrations occurring in the same region of Chochlasti ($340.8\,\mathrm{mg\,l^{-1}}$). This could be attributed to cation exchange processes between water, rocks and clay minerals.

The Pleistocene sediments are rich in clay minerals, mainly smectite and illite. Ca^{2+} is the dominant cation in the interlayers of these minerals due to the presence of calcite in these sediments. The high ammonium concentration in the groundwater of the Chochlasti region could promote saturation of the clay minerals with NH_4^+ because of the high replacing power of NH_4^+ and its high concentration in the aquifer (Grim 1968). Ca^{2+}, which was originally in the interlayer of the clay minerals, was 'exchanged out' and liberated in the groundwater. In regions with intense fertilizer application, this procedure could explain the enrichment of the groundwater in Ca^{2+}.

CONCLUSIONS

From the hydrogeological–hydrochemical study in the West Trifilia coastal area, we conclude the following:

- The Pleistocene sediments host two discrete aquifers, with different hydraulic characteristics. The upper one is unconfined while the lower is confined and limited by marly impermeable strata.
- The marls layer of the confining unit prevents the leaching of nitrates into the groundwater.
- The high nitrate and ammonium concentrations (290.4 and 3.7 mg l^{-1} respectively) in the unconfined aquifer derive from the intensive application of inorganic fertilizers.
- The high calcium concentrations (340.8 mg l^{-1}) in the groundwater of Chochlasti area could be attributed to the cation exchange processes between water, rocks and clay minerals.

REFERENCES

Aller, L., Bennett, T., Lehr, J.H., Petty, R.J. and Hackett, G. 1987: DRASTIC: *A Standardized System for Evaluating Ground Water Pollution Potential Using Hydrogeologic Settings.* EPA 600/2−87/035, U.S. Environmental Protection Agency, Ada, Oklahoma.

Antonakos, A. and Lambrakis, N. 2000: Hydrodynamic characteristics and nitrate propagation in Sparta aquifer. *Water Research 34* (16): 3977−3986.

Borneff, J. and Adabe, B. 1973: Nitrate in Ground Water and its Relation to Fertilization. *Zentralbl. Bakteriol. Parasitenkd. Infektionskr. Hyg. Erste Abt. Orig. Reihe B. Hyg. Praev. Med. 157* (4): 337−345.

ESSO-HELLENIC 1980: Geological map of Greece. Filiatra sheet, I.G.M.E., Athens.

Firestone, M.K. 1982: Biological denitrification. In: F.J. Stevenson (ed.): *Nitrogen in Agriculture Soils.* American Society of Agronomy, Madison, Wis., pp. 289−326.

Freeze, R.A. and Cherry, J.A. 1979: Groundwater. Prentice-Hall, Englewood Cliffs, New Jersey.

Grim, R.E. 1968: International series in the Earth and Planetary Sciences. McGraw-Hill Book Company.

Hatziioannou, T.P. 1985: *Assay and chemical equilibrium.* O.P.D.B., Athens.

Korom, S. 1992: Natural denitrification in the saturated zone: a review. *Water Resources Research 28:* 1657−1668.

Lambrakis, N. 1991: Elaboration of chemical analysis data (in Greek). *Mineral Wealth 74:* 53−60.

Lambrakis, N., Tiniakos, L., Lazarou, A. and Kallergis, G. 1997: *Nitrate pollution of agricultural origin of groundwater of aquifers in Peloponnesus.* Proceedings of the fourth National Hydrogeological Congress, 14−16 November 1997, Thessaloniki, pp. 163−178.

Mariotti, A. 1986: La denitrification dans les eaux souteraines, principes et methods de son identification: une revue. *Journal of Hydrology 88:* 1−23.

Payne, W.J. 1981: Denitrification. John Wiley, New York.

Reddy, K.R. and Patric, W.H. 1981: Nitrogen transformation and loss in flooded soils and sediments. *CRC Critical Reviews in Environmental Control 13* (4): 273–303.

Sampatakakis, P., Makris, A., Gidonis, E. and Tzoulis, H. 1994: Phenomena of undergroundwater pollution in Trifilia coastal area from the use of agricultural fertilizers. In: *Bulletin of the Geological Society of Greece vol. XXX/4*: 115–125. Proceedings of the 7th Congress, Thessaloniki, May 1994.

CHAPTER 17

Nitrate pollution and quality degradation of Ionian coastal groundwater (Southern Italy)

M. Polemio, P.P. Limoni, D. Mitolo, F. Santaloia and R. Virga
C.N.R. I.R.P.I., Bari, Italy

ABSTRACT: The chemical and physical characterization of groundwater described by the paper resulted from a careful analysis of: the aquifers lying in the Ionian coastal plain; of the relations between surface waters and groundwater and of the effects of anthropogenic factors on water quality. Two main types of aquifers have been recognized in the study area: one involving the marine terraces and alluvial deposits in the inland sectors, and the second one corresponding with the alluvial, marine and coastal deposits of the Ionian coastal plain. The quality of the entire groundwater system has been degraded by anthropogenic activities. This degradation is significant in terms of nitrate concentrations, which have increased considerably, especially in the years between 1990 and 2002. Two areas show a complex and serious degradation. The first area is located between the Agri and Cavone rivers, while the second one is located between the Basento and Bradano rivers. The total and fecal coliforms have been reported in high concentrations in the latter area, along with nitrogen cycle species and carcinogenic elements.

INTRODUCTION

The study area lies in the southernmost part of the Basilicata (or *Lucania*) region in Southern Italy, along the Ionian coastal plain, known as the *Piana di Metaponto* (Metaponto Plain), stretching across the central and lower valleys of the Sinni, Agri, Cavone, Basento and Bradano rivers (Figure 1).

Throughout the twentieth century, land reclamation works, the construction of more than ten dams and the introduction of modern irrigation systems have deeply modified the water cycle along the coastal plain. The area is farmed intensively and the quality of the groundwater is vital for the expansion of the tourist industry and for farming activities throughout the Ionian plain.

Attention has been focused on the shallow aquifer of the coastal plain, which is the most important in terms of practical utilization, as highlighted by Radina (1956). This aquifer is also the most subject to seawater intrusion and the risk of groundwater degradation (Polemio, Limoni, Mitolo *et al.* 2002b).

The geological and hydrogeological profile of the study area, as well as chemical and physical features and the quality of groundwater, have been inferred from an analysis of historical data concerning 1,130 boreholes (Polemio *et al.* 2002c). A monitoring programme was realized in a selected area during 2002, in order to evidence the trend of nitrate pollution.

The characterization of nitrate concentration variations and the general effect of anthropogenic factors on groundwater in terms of pollution, emphasize how important the issue of nitrate pollution is for these groundwater resources.

Extensive research has indicated that agricultural practices may cause nitrate contamination to be so high as to exceed the maximum acceptable level for drinking-water (Böhlke 2002). As a consequence, nitrate contamination of groundwater has become a global problem (Spalding and Exner 1993), commonly related to a variety of causes, not just intensive farming but also high density housing, low-efficiency sewerage and purification systems, and irrigation using sewage effluents, particularly in a region of mixed agricultural land uses such as the study area Voterdy. The increase in groundwater nitrate concentrations is a source of real concern as it may be associated with loss of fertility of the overlying soil, eutrophication when the groundwater discharges into the surface water, and potential health risks for animals and human beings.

Soil and groundwater contamination due to agricultural and livestock management practices should already be regarded as a serious environmental threat to the area under examination.

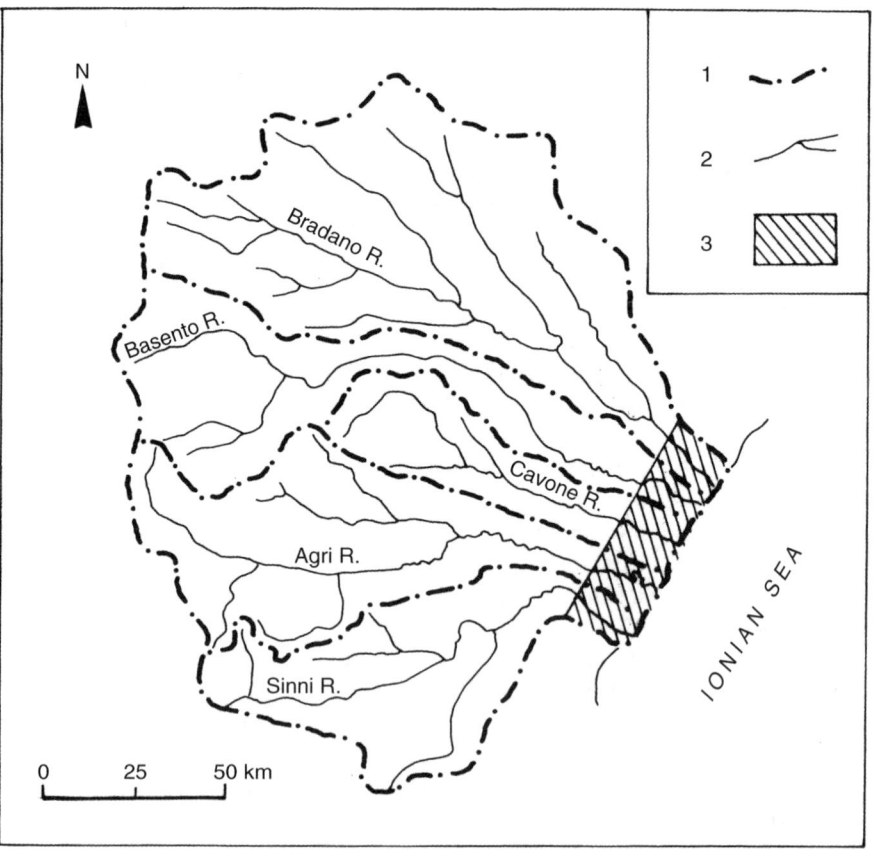

Figure 1. Study area and main rivers. 1) catchment's boundary; 2) main drainage network; 3) study area.

GEOLOGICAL AND HYDROGEOLOGICAL SET-UP

The Marine Terraced Deposits (regressive deposits consisting of sands, conglomerates and silts of the Middle–Upper Pleistocene age) overlying the Subapennine Clayey Formation (silty-clayey successions of the Late Pliocene?–Middle Pleistocene age) outcrop in the upland segments of the study area, while the alluvial, transitional, marine and coastal deposits (Holocene age) outcrop mainly in the coastal plain and along the rivers (Figure 2).

Based on lithological logs, the lithological profile of the shallow portions of the area has been accurately depicted (Figure 3). As regards the inland areas where marine terraces outcrop, three units have been identified below the topsoil (Figure 2). The upper unit consists of pebbles, locally cemented and dispersed in a sandy matrix, with sands and silty-clayey levels. The middle sandy unit, around 40 m thick, is composed of fine- to coarse-grained sands. Different clayey and silty-clayey strata, sandstone levels and gravelly lens are also widespread in this sandy unit. The third unit is represented by a clayey and silty-clayey succession.

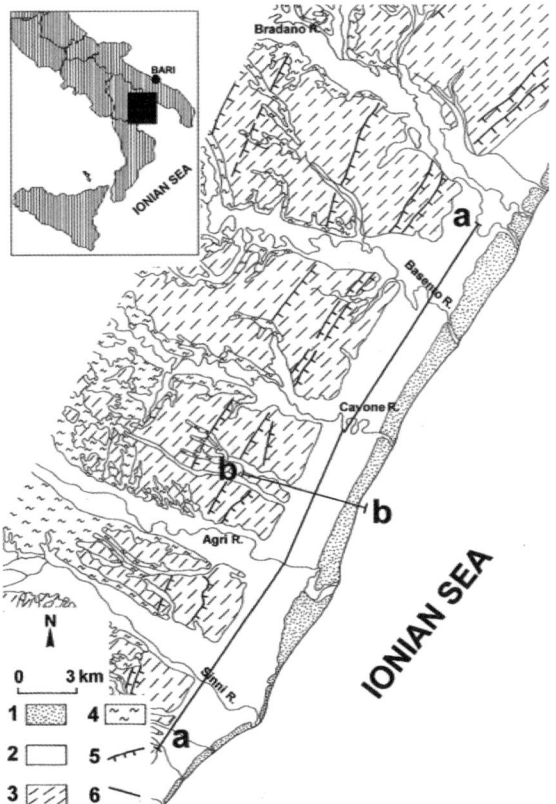

Figure 2. Schematic geological map of the study area. 1) coastal deposits; 2) alluvial, transitional and marine deposits; 3) Marine Terrace Deposits; 4) Subapennine Clays Formation; 5) marine terraced scarps; 6) lithological sections lines.

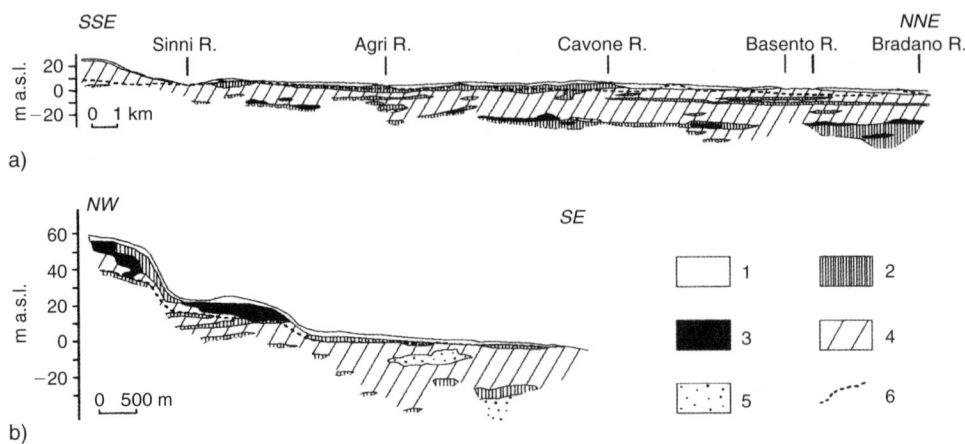

Figure 3. Schematic lithological sections. 1) soil; 2) clays or silty clays (yellow, brown, grey); 3) pebbles in a sandy and/or clayey matrix; 4) grey sands with clayey strata; 5) sands and/or silty sands; 6) piezometric surface (m a.s.l.).

With regard to the Ionian coastal plain, four principal units have been distinguished below the topsoil (Figure 3). The upper unit consists of grey and/or yellow clays, with a maximum thickness of around 10 metres. Below this, there is a sandy unit stretching down to 45–50 metres from ground level. Silty-clayey and clayey levels, gravelly sands and locally cemented pebbly lens are widespread in this sandy unit. The third unit consists essentially of grey silty-clays and clays, up to 30 m thick, locally with some pebbly lens. Finally, the fourth unit, reached only by few boreholes, consists of grey sands, from fine-grained to coarse-grained.

According to the texture, fertility and hydrogeological properties, three principal types of topsoil have been distinguished (Figure 4; AA.VV. 1997): sandy and sandy-silty topsoil, sandy-silty-clayey topsoil and silty, clayey and clayey-silty topsoil. The sandy and sandy-silty topsoil is characterized by medium to high hydraulic conductivity and low fertility, due to a scarcity of organic matter and scanty concentrations of nitrogen and phosphorus. The sandy-silty-clayey topsoil is fairly loose, with from low to medium hydraulic conductivity, moderate to high content of organic matter, low nitrogen and phosphorus concentrations and with high potassium concentration. The silty, brown clayey or clayey-silty topsoil has very low to low hydraulic conductivity, moderate organic matter content, and scant nitrogen and phosphoric anhydride concentrations.

Two main types of aquifers have been distinguished (Polemio *et al.* 2002a, 2002b, 2002c). The first encloses the aquifers of the marine terraces and the alluvial river valleys deposits. The marine terrace aquifers display medium to high hydraulic conductivity but river valleys regularly break their continuity across the area. The aquifers of the river valleys display low to medium hydraulic conductivity and do not generally permit an accumulation of significant groundwater resources. The second type of aquifer includes that of the coastal plain deposits and is more or less 40 km wide. It is the most important in terms of practical utilization, not for its hydraulic conductivity, which is not particularly high (mean value $2.28 \times 10^{-4}\,\mathrm{m\,s^{-1}}$), but because of its extension, thickness, continuity across the plain and, also, because of its location, where the water demand is the highest one.

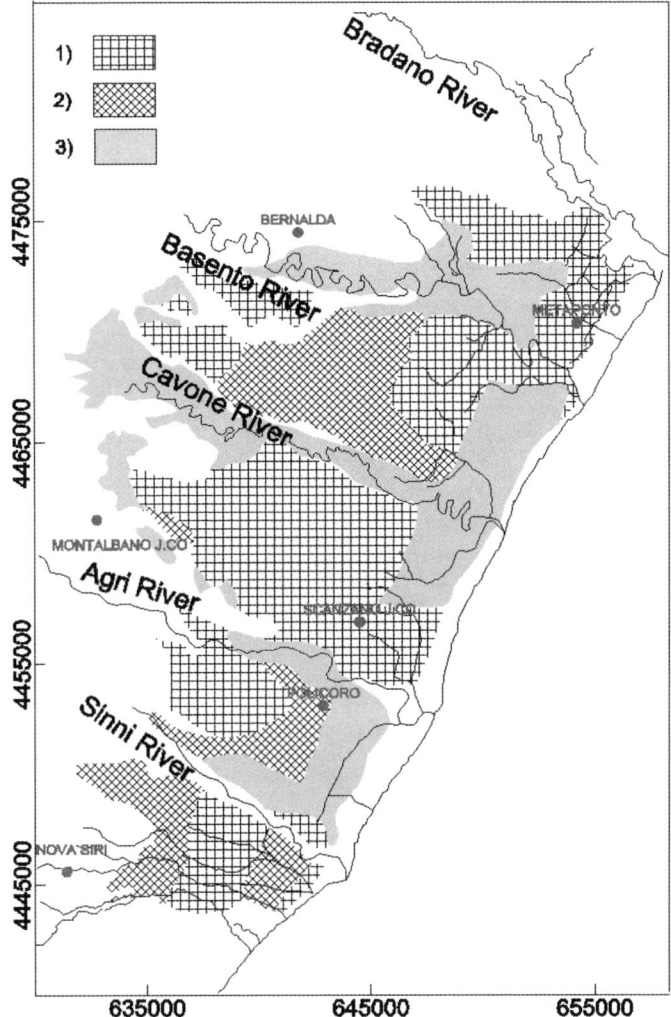

Figure 4. Map of the topsoil type. 1) sandy and sandy-silty topsoil; 2) sandy-silty-clayey topsoil; 3) silty, clayey and clayey-silty topsoil.

The groundwater of the coastal plain flows in a multilayered aquifer constituted by different sandy permeable strata. The shallow sandy aquifer, corresponding to the upper sandy unit, lying below the upper clayey unit, is generally the only one employed for practical uses. Moreover, the coastal aquifer does not outcrop everywhere due to the widespread presence of an upper, almost impervious stratum, 3 to 10 m thick. In particular, where this clayey stratum exists the aquifer is confined; otherwise it is generally unconfined.

The coastal aquifer is skirted downward by the Ionian Sea, and upward by the aquifers belonging to marine terraces or to the alluvial deposits of the river valleys. The shallow coastal aquifer is mainly recharged by the discharge from the upward aquifers (Figure 5) and by river leakage. As shown in Figure 3, the riverbeds in the coastal plain are almost

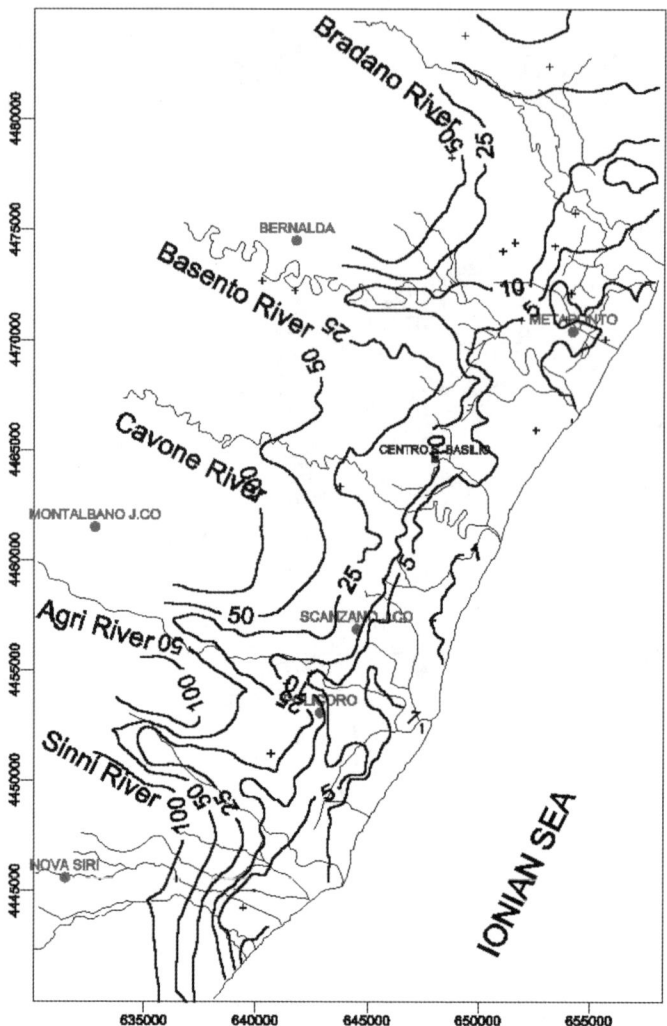

Figure 5. Piezometric map (m a.s.l.).

always deep enough to cut outcropping soils of low hydraulic conductivity, where they exist. Piezometric levels, near the riverbeds, are besides generally low enough to permit river leakage.

CHEMICAL AND PHYSICAL FEATURES OF THE GROUNDWATER

The chemical and physical characterization of the groundwater has been inferred from the analysis of different groundwater samples, taken from 158 wells which are uniformly distributed within the study area: 47 samples have been tapped from marine terraced deposits and 162 samples from alluvial and coastal deposits (coastal plain deposits). All the wells were sampled in 1990, and some were sampled also in 1999.

According to the distribution of the major ions, two dominant types of groundwater have been identified: HCO_3-Ca and SO_4-Cl-Na type. The former is mainly typical for the groundwater flowing in the marine terraces and alluvial deposits, while the latter is characteristic of the coastal plain deposit samples. SO_4-Cl-Ca type is less frequent except in the area surrounding the Basento river, where it is associated with the pollution detected there. Moreover, the water drained into the five catchment areas (Figure 1) supplies the discharge of the five principal rivers flowing in the study area. The discharge yield is particularly high for rivers in the basin of which impervious lithotypes outcrop extensively, as in the case of the Basento River (85% of the catchment area has impervious outcropping lithotypes). Considered the peak location of population density and of polluting industrial activities located in the region, the Basento river area represents a major hazard. On the other hand, leakage of the Basento river water is possible in the coastal plain (Figure 3).

The high concentration ranges of the major ions could not simply be due to differences in the sources of public data, as the ranges remain largely constant even on the basis of a single data source (the validation and references of available public data are described by Polemio *et al.* 2002c); nor could the pollution be due only to the river leakage, though it appears to be a local factor. The reasons for this variability should be connected to other factors, which are briefly summarized below.

First of all, the conditions, which characterize both the depositional environments of the various geological units and the lithogenetic processes, have influenced the geochemical composition of the different lithotypes constituting the aquifers, and thus the chemical features of their groundwater. At the same time, terraced deposit aquifers constitute recharge areas in which rainfall infiltration is relatively rapid and direct, and in which groundwater flows in a space where clayey levels are encountered frequently.

Neither of these two circumstances exists in the coastal plain aquifer. River leakage and geochemical and hydrogeological factors cannot justify the concentration ranges of the coastal plain groundwater, which are generally wider than those of the marine terrace groundwater (Figures 6 and 7).

Coastal plain groundwater is deeply influenced by other factors, primarily seawater intrusion. As shown by Polemio *et al.* (2002b), seawater intrusion affects a coastal strip that is $1-1.5$ km wide on average. Anthropogenic factors could also be inferred to explain the high variability of the ion concentration. A relevant contribution is due to the widespread irrational procedures of irrigation and of fertilizer utilization and to the consequent water irrigation and chemical surpluses.

Irrigation is important in this semi-arid area because of the high rate of potential evaporation–transpiration, corresponding to a mean annual value of 860 mm (161% of the mean annual rainfall). The irrigation water is supplied by dams, built in the inner sectors of the main river basins and dating to the 1960s. Before using aqueducts, supplied by dams, the irrigation water used to be tapped by wells and today this practice still persists during drought periods, when the artificial lakes are largely empty (as is the currently the case).

In both cases an increasing load of salt mass, which is introduced by the irrigation water, and pollutants, which are added to the topsoil, mainly in the form of fertilizers but also pesticides, may cause a degradation of the quality of the fresh, pure groundwater system. During intensive groundwater exploitation (that is, in drought periods) the situation becomes more serious as the irrigation – distributing salt mass – is at least partly the cause of seawater intrusion (Polemio *et al.* 2002a, 2002b).

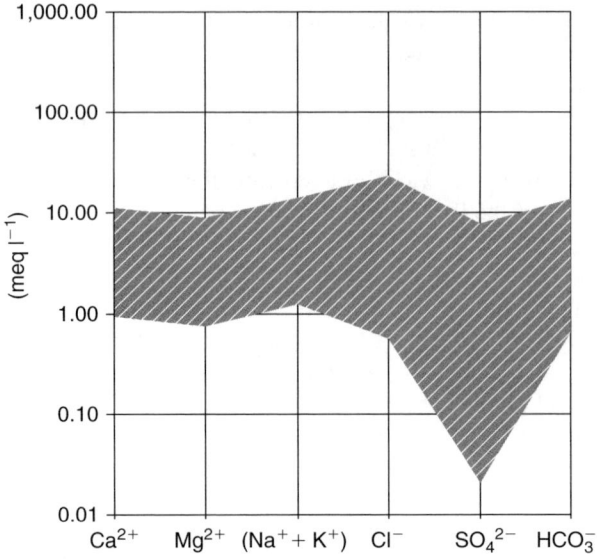

Figure 6. Schoeller diagram of the whole groundwater samples in the terraced marine aquifers.

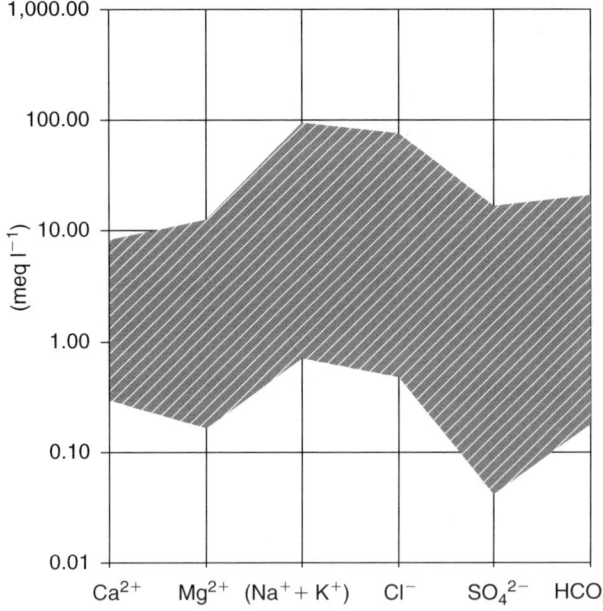

Figure 7. Schoeller diagram of the whole groundwater samples in the alluvial and coastal aquifers.

Another factor could be also related to the extremely low drainage capacity of the coastal plain, which is subjected to frequent and extensive seasonal pounding due to river flooding. The evaporation of the pounded waters may also be a further source of salts and pollutants, added to the system, as pointed out by Lopez *et al.* (1986).

Nitrate distribution and trend

The environmental impact of agricultural activities is becoming a serious problem, especially in industrialized countries. Eutrophication, for instance, as caused by high concentrations of nutrients in waters, represents one of the possible environmental impacts of agriculture and animal husbandry practices. In Italy, in the Northern Adriatic Sea, algal blooms have acquired the dimension of an ecological catastrophe since 1988 (AA.VV. 1997).

Nitrates represent one of the most important causes of water quality degradation, and nitrate losses from agricultural areas are generally larger than the losses from natural ecosystems.

In the study area, the direct contribution of anthropogenic factors to groundwater quality degradation is significant in terms of nitrate concentrations and other contaminants, as summarized by Table 1.

The major contaminant considered is nitrate, as it is generally the most significant, mobile and persistent agricultural contaminant (Böhkle 2002). The spatial trend of nitrate concentrations is shown in Figure 8; the trend is fairly high for the study area, which is characterized primarily by arable land and the intensive use of fertilizers. The mean value of the nitrate concentration is about $16\,mg\,l^{-1}$, but locally this chemical parameter may exceed $150\,mg\,l^{-1}$.

A close relationship has been detected between topsoil properties and spatial trend nitrate concentrations, and between nitrate concentrations and soil hydraulic conductivity. In fact, there are two areas most significantly subjected to nitrate contamination (Figure 8). The first is located between the Agri and Cavone rivers (along the coast between the mouth of the Agri river and the village of Scanzano Jonico, and inland between the village and the Cavone river). The second area is between the Basento and Bradano rivers in the coastal plain. In both these areas the topsoil, constituted by sandy and sandy-silty soils, shows the highest hydraulic conductivity (Figure 5). On the other hand, the above-mentioned natural scanty nitrogen concentrations in the topsoil do not justify the high nitrogenous compounds measured in the groundwater under examination.

The spatial trend for ammonia concentrations, reported in Figure 9, highlights the existence of two areas with concentrations in excess of $0.5\,mg\,l^{-1}$. The first area, characterized also by high nitrate concentrations, lies between the Sinni and the Agri rivers where two important towns are situated (Scanzano Jonico and Policoro). The second one is located near Metaponto.

The presence of nitrate seems to be related to an intense use of agricultural fertilizers and, in limited areas, to urban activities. In some groundwater samples, the high NH_4 concentrations measured seem to be distinctive of sewage leakages and polluting urban activities.

To evaluate the evolution over time of nitrate pollution, a test area was selected, between the Basento and Cavone rivers, and groundwater samples were taken from 34 wells of the coastal plain aquifer. The survey was realised during 2002 (Figure 10).

The 34 wells of the 2002 survey are uniformly distributed in the costal aquifer of the test area, while some of the 40 wells in the 1990 survey are also located in the terraced deposits.

As shown by nitrate maps (Figures 8 and 10), the nitrate ion is detected consistently and abundantly in the test area. The $25\,mg\,l^{-1}$ contour line of 1990 several times cuts the

Table 1. Statistical outline, listing the minimum, the maximum and the mean values of parameter selected to highlight the anthropogenic groundwater pollution of the whole Metaponto plain (1990 survey) and the test area (2002 survey). <DL: Less than detection limit.

Quality parameters	Terraced deposits – 1990			Coastal plain deposits – 1990			Coastal plain deposits – 2002		
	min	max	mean	min	max	mean	min	max	mean
Ammonia (mg l^{-1})	<DL	24	1	<DL	50	3	1.30	21.27	8.40
Nitrate (mg l^{-1})	<DL	68	16	<DL	183	16	2.95	92.72	29.70
Nitrate (mg l^{-1})	<DL	21	1	<DL	<DL	<DL	/	/	/
Colonies at 22°C	800	3,200	1,201	800	3,500	2,375	/	/	/
Colonies at 36°C	120	2,500	686	450	2,500	1,463	/	/	/
Total coliforms (MPN/100 ml)	16	16	13	<DL	16	7	/	/	/
Fecal coliforms (MPN/100 ml)	2	6	4	<DL	16	4	/	/	/
Fecal streptococci (MPN/100 ml)	6	16	9	<DL	16	4	/	/	/
Strontium (mg l^{-1})	<DL	21	1	<DL	<DL	<DL	/	/	/
Lithium (mg l^{-1})	<DL	21	1	<DL	<DL	<DL	/	/	/
Fluorine (mg l^{-1})	<DL	21	1	<DL	<DL	<DL	5.13	270.00	38.33
Iron (mg l^{-1})	<DL	24	1	<DL	10	4	/	/	/
Manganese (mg l^{-1})	<DL	21	1	<DL	<DL	<DL	/	/	/
Copper (mg l^{-1})	21	35	20	15	52	48	/	/	/
Zinc (mg l^{-1})	4.15	6.50	3.70	0.04	1.25	0.44	/	/	/
Arsenic (mg l^{-1})	0.009	0.009	0.006	0.010	0.016	0.012	/	/	/

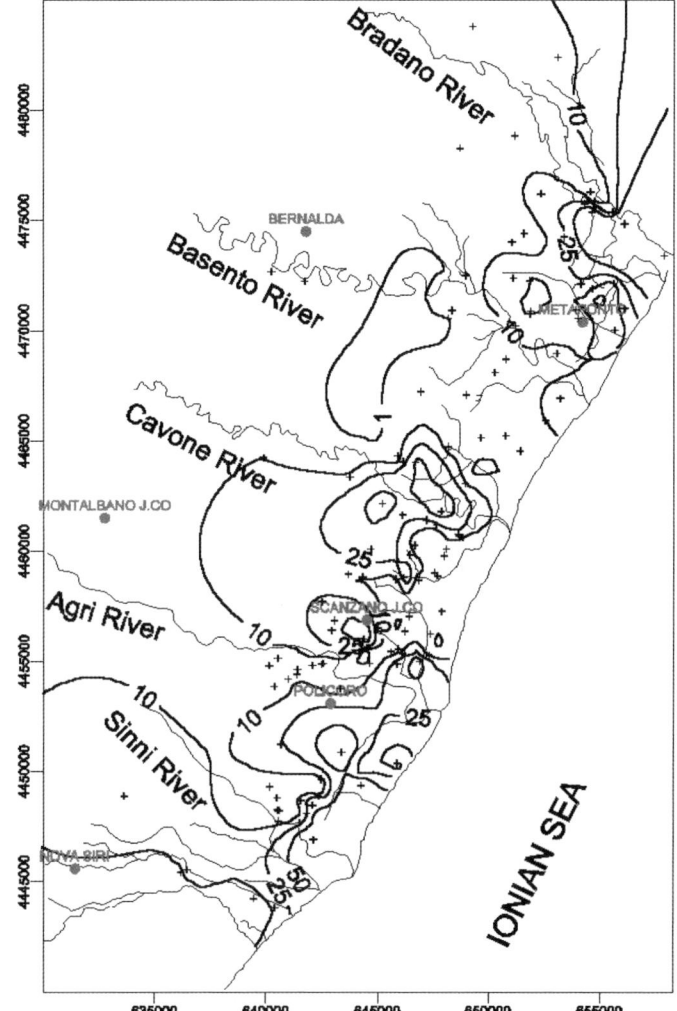

Figure 8. Map of nitrate groundwater concentration, year 1990 (mg l^{-1}).

limit between the terraced deposits and the coastal plain deposits, roughly corresponding to the railway line (Figure 10). Upward, the concentration is higher in the southern portion and lower in the remaining portion; downward, in the coastal plain aquifer, the concentration is lower.

The 25 mg l^{-1} contour line of 2002 overlaps with the 1990 line only in the northern portion of the test site; in the central portion it has moved downward, while in the southern portion it is replaced by the 50 mg l^{-1} contour line. In each point of the test area, the coastal aquifer display nitrate concentrations that are generally increased, somewhere significantly.

This result is a cause for considerable concern due to the dramatic increase in groundwater nitrate concentrations. The concentrations seem to have more than doubled

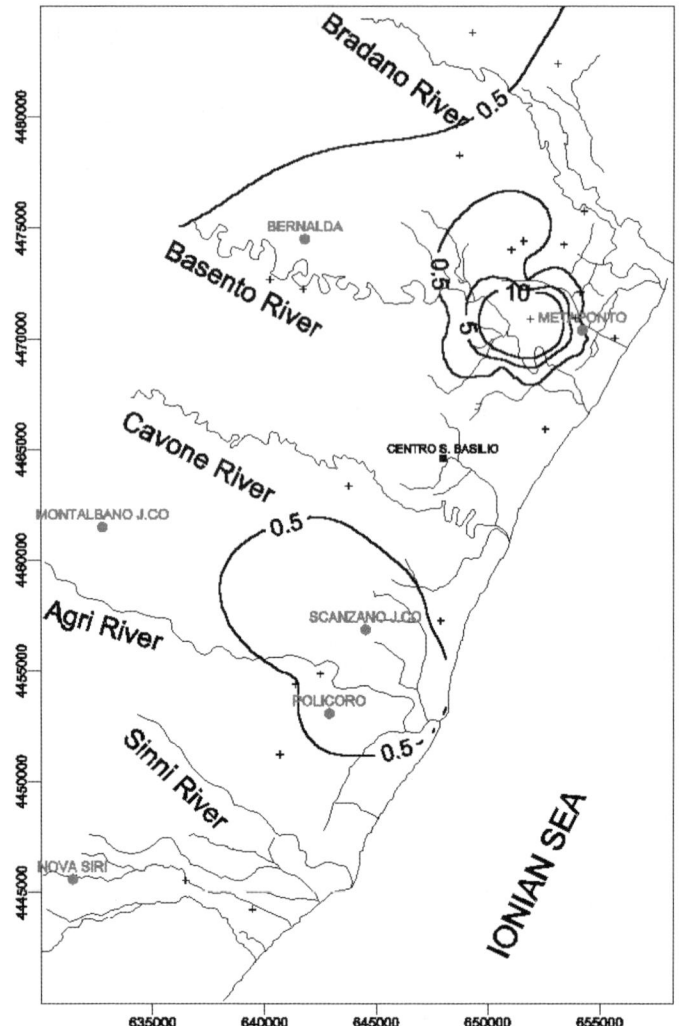

Figure 9. Map of ammonium groundwater concentration, year 1990 (mg l^{-1}).

in the coastal plain, not only where the topsoil is more permeable (Figure 4). If this trend were extrapolated to the whole study area the result would be devastating.

In the test area the statistical parameters (minimum, mean and maximum) rose significantly not only in terms of nitrate but also ammonia, between 1990 and 2002 (Figures 8–10 and Table 1).

Moreover, nitrate and ammonia are not the only source of risk. Widespread bacteriological pollution of the studied groundwater was also observed in the 1990 survey (Table 1). This phenomenon may be explained by the presence of several towns and villages, in addition to extensive livestock management practices, in the coastal plain.

The presence of metals such as Fe, Mn, Cu, As and Zn, detected in the groundwater, may be associated with urban waste water, polluted river leakage, direct recharge, mineral dissolution and anthropogenic activities.

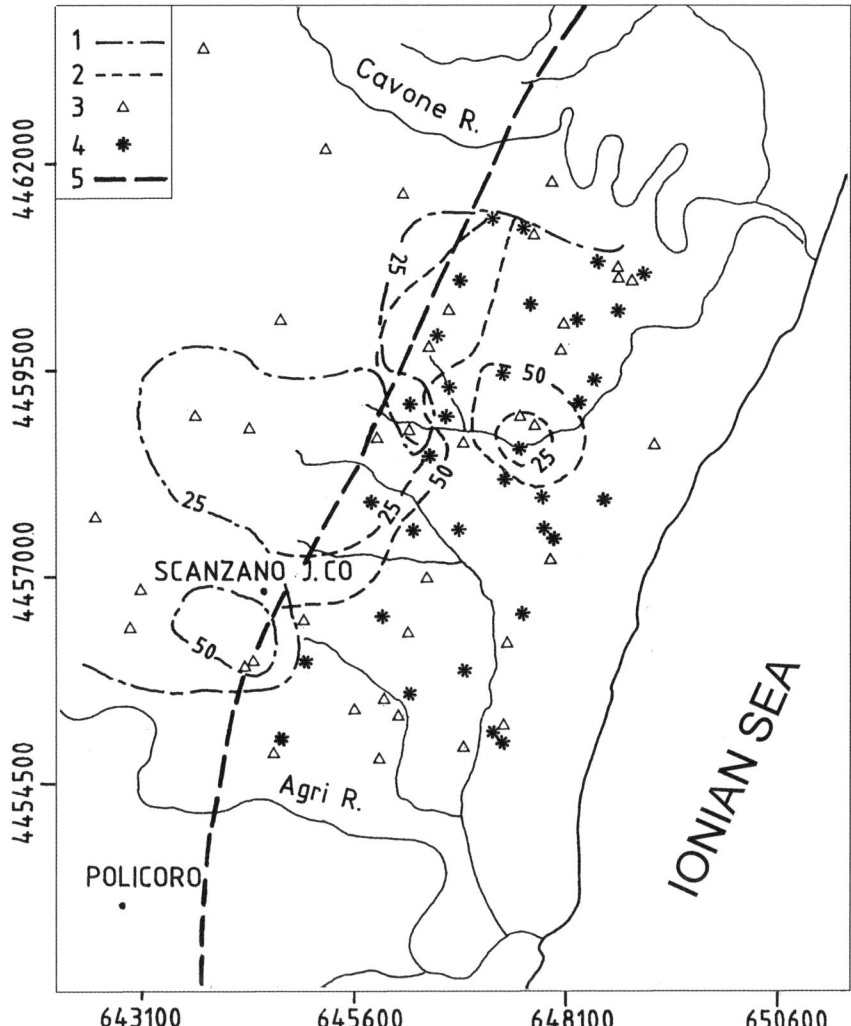

Figure 10. Map of nitrate concentration from 1990 to 2002 in the test area. 1) 1990 nitrate concentration (mg l^{-1}), 2) 2002 nitrate concentration (mg l^{-1}), 3) 1990 monitoring well, 4) 2002 monitoring well, 5) railway.

Moreover, the portion of coastal aquifer between the Bradano and the Basento rivers is characterized by a widely diversified concentration of contaminants. Total and fecal coliforms have been reported in high concentrations (exceeding 15 MPN/100 ml) along with nitrogen cycle species and carcinogenic elements, such as arsenic, zinc and copper. This phenomenon is suggestive of the poor groundwater quality caused by human activities. Indeed, the groundwater of this stretch of the Ionian coastal plain has been increasingly polluted by both the urban and industrial waste water discharged into the Basento river, which is interconnected to the downward groundwater system, and by a dense drainage network of reclamation works.

CONCLUSIONS

The analyses of the hydrogeological and chemical data suggest that the risk of groundwater quality degradation is not negligible in the Metaponto plain. The analysis show also the extreme vulnerability of the Ionian-Lucanian groundwater system. In particular, if the nitrate concentration is the principal parameter considered, the pollution level is high and the quality of the groundwater is dramatically poor.

The direct effect of anthropogenic factors to the deterioration of the groundwater quality is relevant also in terms of nitrate concentration, especially in two areas. The first is located between the Agri and Cavone rivers, the second is between the Basento and Bradano rivers; here total and fecal coliforms are present in high concentrations in the form of nitrogen cycle species and carcinogenic elements, such as arsenic, zinc and copper. In both these areas the topsoil, comprised of sandy and sandy-silty soils, shows the highest hydraulic conductivity. In these areas the ammonia concentration is higher than $0.5 \, \mathrm{mg} \, \mathrm{l}^{-1}$.

The detection in groundwater of toxic substances which are generally absent or present in minimal quantities, and the deteriorating quality of groundwater elsewhere, call for a more judicious management of groundwater resources, together with the entire water cycle, and a more careful utilization of chemicals in agricultural activities.

REFERENCES

AA.VV. 1997: *Studio agronomico del comprensorio irriguo del Consorzio di bonifica di Bradano e Metaponto per la gestione delle risorse idriche a fini irrigui*. Consorzio di bonifica di Bradano e Metaponto, Matera, Italy.

Böhlke, J.K. 2002: Groundwater recharge and agricultural contamination. *Hydrogeology Journal 10* (1): 153–179.

Lopez, G., Ascione, S. and Marrone, G. 1986: Caratteri fisico – chimici dei suoli e delle acque di falda di un ambiente del litorale ionico lucano. *Annali Istituto Sperimentale Agronomico 17*: 49–65.

Polemio, M., Limoni, P.P., Mitolo, D. and Santaloia, F. 2002a: *Characterisation of Ionian-lucanian coastal plain aquifer*. In: E. Bocanegra, D. Martinez and H. Massone (eds): Groundwater and human development, Mar del Plata, Argentina, pp. 874–883.

Polemio, M., Limoni, P.P., Mitolo, D. and Santaloia, F. 2002b: *Characterisation of Ionian-lucanian coastal aquifer and seawater intrusion hazard*. Proceedings of 17th Salt Water Intrusion Meeting, Deft, Netherlands, pp. 422–434.

Polemio, M., Dragone, V., Limoni, P.P., Mitolo, D. and Santaloia, F. 2002c: *Extended report of CNR-CERIST unit on deliverables: first year of activity*. European research project 'CRYSTECHSALIN', VPQ of European Commission (RST-EESD), Brussels.

Radina, B. 1956: Alcune considerazioni geoidrologiche sulla fascia costiera jonica compresa fra i fiumi Bradano e Sinni. *Geotecnica 1*: 3–11.

Spalding, R.F. and Exner, M.E. 1993: Occurence of nitrate in groundwater. *Journal of Environmental Quality 22*: 392–402.

CHAPTER 18

Nitrates in water of the vadose and phreatic zones, Cracow Jurassic — Poland

J. Różkowski

University of Silesia, Faculty of Earth Sciences, Będzińska 60, 41-200 Sosnowiec, Poland

J. Motyka and K. Różkowski

University of Science and Technology AGH, Faculty of Mining and Geoengineering,
A. Mickiewicza 30, 30-059 Cracow, Poland

ABSTRACT: Unconfined karstic aquifers in agricultural areas are vulnerable to human impacts. In karstic environment processes influencing water chemistry in both the vadose and phreatic zone can be observed. Mobile, slowly degraded, toxic nitrates present in such environments are the best indicators of groundwater anthropogenic contamination. Gaseous pollutants, polluted precipitation, agricultural chemicals and the rural economy are the main artificial sources of nitrates. The investigations took place in the karstic area of the Cracow Upland (in southern Poland) where a useable aquifer in the Upper Jurassic limestones occurs. This agricultural area is under strong human impact from the adjacent industrial-urban agglomerations of Cracow and Upper Silesia. At present the nitrate hydrochemical background of the vadose and phreatic zone is respectively 0.2–8.5 and 1.2–7.5 mg N-NO$_3$/l, while its range is seven to eight times as great as the natural one (0–1 mg N-NO$_3$/l). Degradation of water quality is observed with an absence of the highest quality water (<1 mg N-NO$_3$/l) within hydrochemical background, shift of hydrochemical background range (from 0.2–6.0 mg N-NO$_3$/l in the 60th), considerable share of supra-sanitary standards concentrations (>10 mg N-NO$_3$/l) up to 110 N-NO$_3$/l within vadose zone and to 22 mg N-NO$_3$/l in phreatic zone (especially during thawing period and in autumn). Mean nitrate nitrogen concentrations in groundwater are: urban areas — 7.4 mg N-NO$_3$/l, rural areas — 6.6 mg N-NO$_3$/l, forests — 3.6 mg N-NO$_3$/l. On the basis of groundwater flow modelling, as well as hydrochemical research, the estimated nitrogen load carried away from the usable Upper Jurassic aquifer (area of 652 km^2) amounts from 0.97 to 2.58 kg N-NO$_3$/24 h·km^2.

INTRODUCTION

Nitrate anion NO$_3^-$ is one of the basic forms of nitrogen found in soil, unsaturated (vadose) zone and in shallow open water-bearing strata where oxidation environment occurs. Nitrate comes from precipitation, decay and mineralization of organic substances and human activity. High solubility and anion-forms sorbed on a small scale (retardation of migration R is less than 2) are reasons for easy nitrate migration in groundwater. In natural conditions nitrates degrade slowly. Half-life, a measure of nitrate decomposition in denitrification process, is variously estimated by researchers in the range from 1 to 2 years (Duynisveld, Strebel, Boettcher *et al.* 1989) to even about 20 (Kozlovsky 1988).

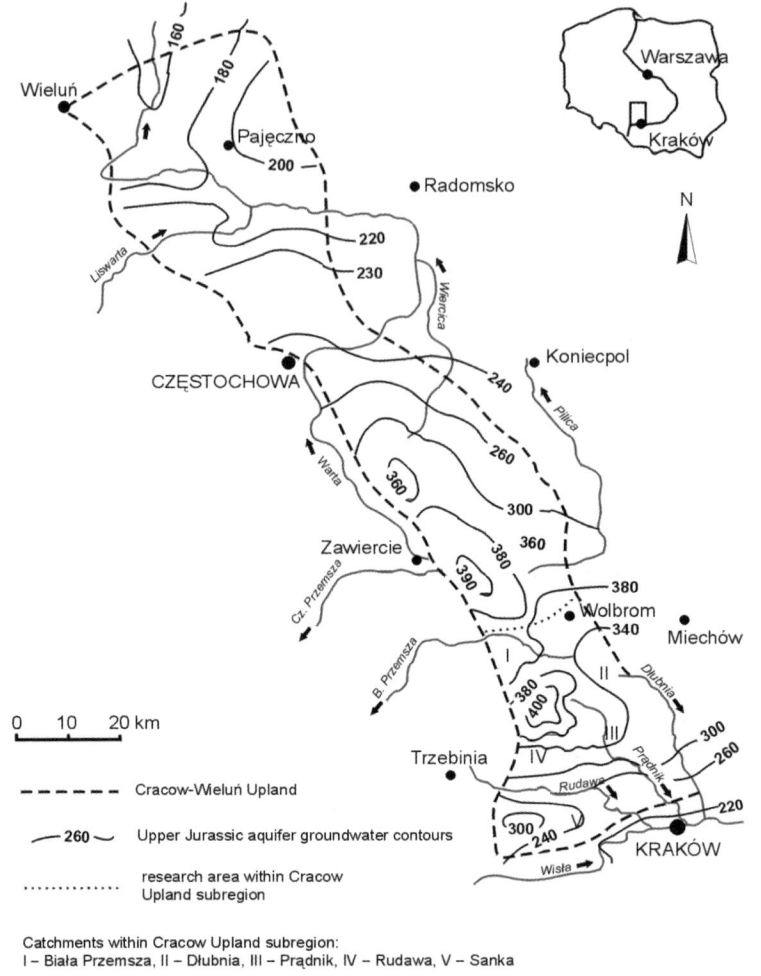

Figure 1. Hydrogeological map of Cracow-Wieluń Upland.

In southern Poland, north from Cracow, the main useable groundwater aquifer occurs in the carbonate Upper Jurassic sediments (Figure 1). It is under increased human impact. The unconfined nature of the aquifer favors intensive penetration of pollution to groundwater, while the hydraulic fissure-karstic structure favours rapid spreading. The opportunity for water purification within the compact reservoir rock is limited. The karstic waters in this agricultural area are characterized by regional scale pollution by nitrates derived from human activities. Nitrate presence in the karstic environment of the Cracow Upland is presented in Figure 2.

The problem is important and critical at a regional scale. Water pollution analysis of usable aquifers in Poland shows that they are mainly polluted by agriculture, farming and rural settlement (33%). The most common substances polluting groundwater are nitrogen compounds of agricultural origin. They occur in 45% of the examined point sources of pollution, in 38% of local sources and in 22% of regional sources (Macioszczyk and Mitręga 1997). Karstic aquifers of open character and vulnerable to pollution are the best

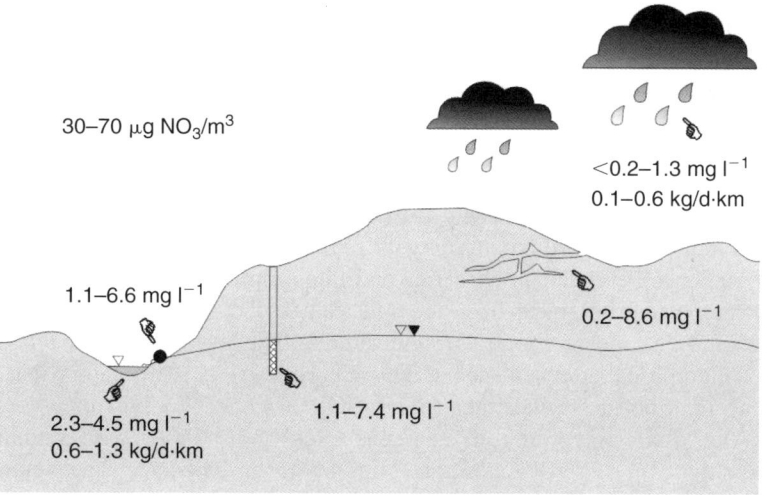

30–70 μg NO_3/m^3

<0.2–1.3 mg l^{-1}
0.1–0.6 kg/d·km

1.1–6.6 mg l^{-1}

0.2–8.6 mg l^{-1}

2.3–4.5 mg l^{-1}
0.6–1.3 kg/d·km

1.1–7.4 mg l^{-1}

Figure 2. Nitrate nitrogen in the hydrosphere of Cracow Upland (mean concentrations and loads).

testing ground for studying agricultural pollution migration in time and space, in vadose zone (caves) and phreatic zone (wells, springs).

SOURCES OF WATER QUALITY DETERIORATION

The natural environment of the Cracow Upland is influenced by pollution from the adjacent urban-industrial agglomerations of Cracow and Upper Silesia. This pollution source is one of the biggest in Poland. The average annual value of air pollution in the area under study reaches $50-80 \mu g \, m^{-3}$ of nitrogen oxides. Rainfall chemistry in this area depends on a phenomenon of 'rainout' in which the concentration of pollutants in precipitation depends on influx of rain-bearing clouds from distant areas. The most frequent winds from W and SW (65% of population) bring strongly polluted air from the Upper Silesian Industrial Area. Average rainfall in the central part of the Cracow Upland during a period of 30 years was 718 mm. In 1989–1995, mean nitrate nitrogen concentration in rainfall was 0.7 mg N-NO_3/l (maximal value 5.0 mg N-NO_3/l) while in NE part of Poland that is considered as a not polluted area, it was about 0.2 mg N-NO_3/l. The mean annual N-NO_3 load for elementary rainfall was in that time 22 kg m^{-3} (maximal load 157 kg m^{-3}; Leśniok 1996). In 1996–2000, mean annual nitrogen concentration in rainfall was 0.8 mg N-NO_3/l (Leśniok and Partyka 2001). The main source of nitrates in rainfall is the emission of nitrogen oxides from fuel combustion. They are the main components of atmospheric gaseous pollutants. The nitrate load reaching the vadose zone from rainfall is an input of lesser significance. The principal nitrate pollution source in karstic environments comes from agricultural and communal activity. The agricultural economy of the Cracow Upland area causes an average hazard (relative to the whole country) to the natural environment including groundwater. Agricultural spatial pollution is connected mainly with chemicalization. Point sources of pollution are of communal origin or they are connected with the infrastructure of rural management (poor water-supply and sewage-disposal management, ineffective mechanical–biological sewage treatment plants, lack of properly managed waste dumps, fattening houses and direct storage of fertilizers on the ground; Kleczkowski 1984). Excessive use of artificial and natural fertilizers exceeding the plant's needs and the absorption capacity of the soil can cause pollution of groundwater by washing out nitrate from cultivated fields. Usage of nitrogen from mineral fertilizers in the area of the Cracow Upland varies from 30 to $80 \, kg \, ha^{-1}$. Among the NPK components of mineral fertilizers, the greatest loss is from nitrate because the plants can hardly use them (about 50%). In the case of intensive farming, an additional hazard is caused by the organic mass left after the crop is harvested. From November to February, in climatic conditions similar to that in Poland, about 60% of annual nitrate load migrates from the soil (Sapek 1990). There is a distinct relation between the nitrate concentration in groundwater and the level of nitrogen fertilizer use in the field. An empirical study (Benes, Pekny, Skorepa *et al.* 1989) shows that estimated nitrate concentration in groundwater (mg N-NO$_3$/l) is equal to 1/10 of use of nitrogen fertilizers (kg N/ha). According to Juergens-Gschwind (1989), with rational use of fertilizers, only 10% of fertilizer nitrate migrates to groundwater. The amount of soil nitrogen is, on average, a few dozen times higher than the amount of nitrogen introduced to the ground with fertilizers. Cracow Upland regosols with a shallow parent rock favour quick nitrate migration through the soil profile, which has important significance for the nitrate load size washed-out from soil (Juergens-Gschwind 1989; Witczak and Żurek 1994). Natural nitrate hydrochemical background according to Witczak and Adamczyk (1995) ranges from 0 to 1 mg N-NO$_3$/l. Abnormally high concentrations reaching several hundreds mg N-NO$_3$/l are observed in shallow groundwater in agricultural areas, in dug wells within rural settlements with improper sewage disposal managagement, and within zones of pollution down-gradient from industrial and municipal waste sites.

NITRATES IN THE WATER OF A VADOSE ZONE

Caves in the Cracow Upland are an excellent testing ground for studying processes influencing the chemistry of precipitation infiltrating through a vadose zone. In the caves the physical properties and chemical composition of water from drips, calcareous sinter gours and from pools can be studied. Some of them occur on the boundary of the vadose and phreatic zones. The studied caves are shallow, with a carbonate rock overburden thickness in the range from several to a maximum of a dozen or so metres. The hydraulic system of a fissure-karstic vadose zone is very complex (Bakalowicz 1995), composed of subcutaneous karst zone (Mangin 1975) and a zone of free vertical percolation (Ford and Williams 1989). Within the described fissure-karstic aquifer, the differentiation of hydraulic resistance is high. From the beginning of the 1990s, the chemistry of cave waters in the southern part of the Cracow Upland was studied by a group of karst specialists (Różkowski and Różkowski 1999; Goc *et al.* 2000).

The range of nitrate concentration in cave waters of the studied karstic area is wide − from trace values (<1 mg N-NO$_3$/l) to over 100 mg N-NO$_3$/l. The highest concentrations of this ion (20−110 mg N-NO$_3$/l) were observed in the caves of Cracow city area (Smocza Jama cave, horst of Zakrzówek). Smocza Jama cave is situated directly beneath the Wawel hill and is probably influenced by the historical settlement. On the other hand, at the area of Zakrzówek horst, an old explosives store might be situated (Goc *et al.* 2000). Generally nitrate concentrations in cave water are in the range 0.2−12 mg N-NO$_3$/l and

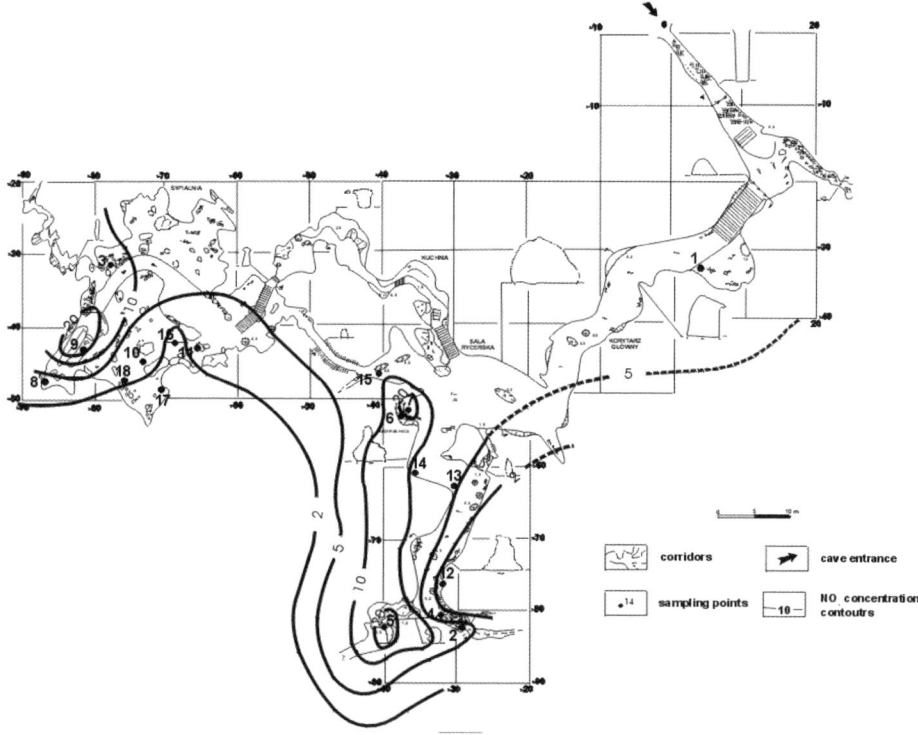

Figure 3. Spatial nitrate concentration diversity in drips within Łokietek Cave.

hydrochemical background varies from 0.2 to 8.5 mg N-NO$_3$/l. Spatial differentiation of nitrate concentrations in percolating water is observed even within single caves (Figure 3). The study carried out in the Ojców National Park showed the highest seasonal variability of nitrate concentration among all major components (in the range from 1.5 to 5.2 mg N-NO$_3$/l) which confirms its anthropogenic origin. Maximum nitrate concentrations are observed during thawing period, in summer and late autumn.

NITRATES IN FISSURE-KARSTIC WATER OF THE PHREATIC ZONE

Groundwater pollution with nitrates within the studied area is of regional and permanent character (Różkowski 1996; Tyc 1997). The wide range of hydrochemical background (6–8 mg N-NO$_3$/l) corresponds to individual decades. Degradation of water quality occurred in the succeeding decades. As a result, there is lack of the highest quality waters (<1 mg N-NO$_3$/l) within hydrochemical background, a shift of hydrochemical background range from 0.2–6.0 to 1.2–7.5 N-NO$_3$/l (Figure 4), the occurrence of supra-standard contamination in all population and an increase of positional parameters values of statistical distribution (mode and median). In groundwater of the Quaternary aquifer situated above the Upper Jurassic one, the range of nitrate concentration variability is higher (0.02–33 mg N-NO$_3$/l) in comparison to the last one. Occurrence of high nitrate concentrations in groundwater of the Quaternary aquifer (>20 mg N-NO$_3$/l) is observed,

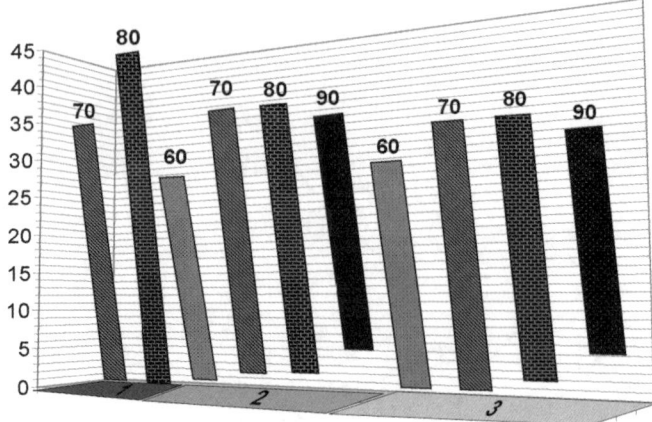

Figure 4. Range of nitrate hydrochemical background in groundwater, years 1960−2000 (mg NO$_3$/l). 1 − shallow dug wells; 2 − deep drilled wells; 3 − karstic springs.

which is not in the water of the Upper Jurassic aquifer. However the higher share of supra-standard determinations (15%) corresponds to this observation. Such regularity results from direct and permanent agricultural and communal pollution of shallow groundwater. At the same time a higher share of low nitrate concentration is observed as well. It manifests with point character of pollution from the surface of the Quaternary aquifer.

In natural watercourses receiving water from springs, linear outflows or fissure-karstic flow from adjacent catchments, a consistent nitrogen concentration (2.3−4.5 mg N-NO$_3$/l, Figure 2) is observed. Nitrate nitrogen concentrations in the Upper Jurassic aquifer observed in the 90th were in range <0.1−22.4 mg N-NO$_3$/l. Low concentrations of nitrate in water (<1 mg N-NO$_3$/l) correspond to the natural hydrochemical background. Generally elevated nitrate concentrations affect only the high (not the highest) quality waters and, seasonally (in the thawing period and in autumn), those of low quality.The level of nitrate concentration in fissure-karstic waters is correlated to a form of the land use. Mean nitrate nitrogen concentrations in groundwater within rural areas, arable land and forests did not significantly vary in the 90th, amounting respectively to 7.4, 6.6 and 3.6 mg N-NO$_3$/l. Scale effect causes that in regional scale Jurassic fissure − karstic − porous aquifer is homogenous. Mean concentrations of nitrate nitrogen are similar in individual groundwater catchments: I) Biała Przemsza 3.4−11.3, II) Dłubnia 1.3−6.6, III) Prądnik 1.1−4.3, IV) Rudawa 1.8−5.9, V) Sanka 1.3−7.4 (mg N-NO$_3$/l, Figure 1). The lowest concentrations of nitrates were measured within an area protected by law − the catchment of the Prądnik River (Ojców National Park).

In 1999 within the confines of Cracow Upland a numerical model of Jurassic aquifer was constructed using the MODFLOW programme. The aim of the modelling study was determination of the groundwater circulation system, calculation of the groundwater balance, assessment of groundwater renewal and resources in the Upper Jurassic aquifer. Within this modelling study a numeric model of the area of 652 km^2 was prepared.

On the basis of groundwater flow balance determined from the modelling, it can be stated that the total Upper Jurassic aquifer runoff within Cracow Upland amounts to 310,000 m^3/24 h, which is correlated to mean drainage modulus 475 m^3/24 h·km^2. The runoff recharged mainly by precipitation infiltration equals 269,910 m^3/24 h

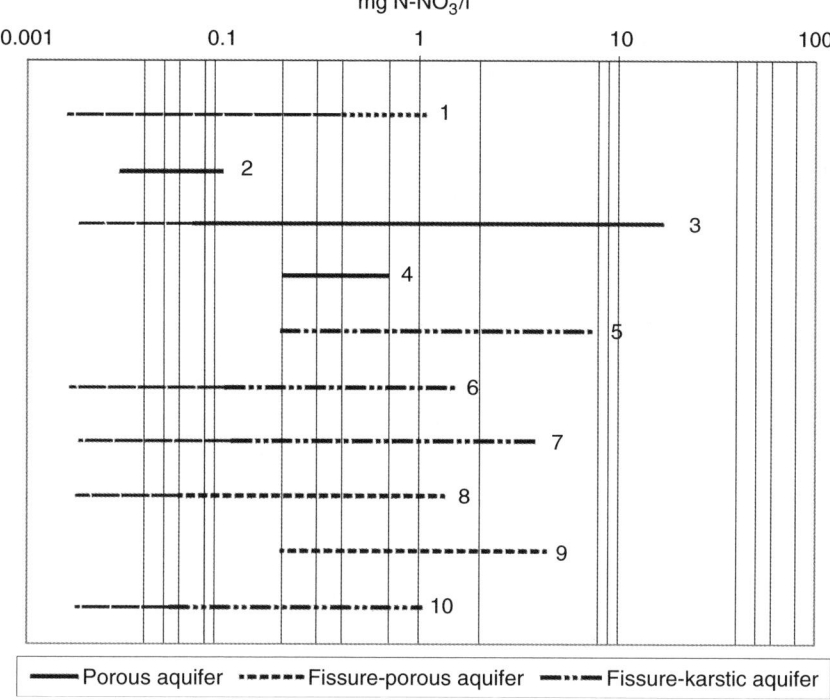

Figure 5. Hydrochemical nitrate background (N-NO₃) at the selected areas of main groundwater aquifers in Poland (according to Macioszczyk 1991, with supplement). 1) precipitation at the Cracow Upland, confined basin; 2 – Cretaceous of Gdañsk, semi-confined basins; 6 – Triassic of Lubliniec-Myszków; 7 – Triassic of Gliwice, unconfined basins; 3 – Quaternary, Kurpie, sandr sands; 4 – Quaternary, Wroclaw, paleovalley of the Odra river; 5 – Jurassic of Cracow; 8 – Carboniferous, Silesia; 9 – the Sudeten Mts., Palaeozoic, crystalline massif; 10 – the Tatra Mts., Tertiary, Triassic.

$(414 \, m^3/24 \, h \cdot km^2)$. The groundwater river drainage reaches $181{,}907 \, m^3/24 \, h$ $(279 \, m^3/ 24 \, h \cdot km^2$, Różkowski, Kowalczyk, Rubin *et al.* 2003). The nitrate load introduced by precipitation of average nitrate concentration in range from 0.2 to 1.3 mg N-NO₃/l equals $0.06 - 0.37 \, Mg \, N\text{-}NO_3/24 \, h$ $(0.09 - 0.56 \, kg \, N\text{-}NO_3/24 \, h \cdot km^2)$. The nitrate load carried by the rivers out of the research area with average nitrate concentration in surface water in the range from 2.3 to 4.5 Mg N-NO₃/l, equals $0.41 - 0.82 \, Mg \, N\text{-}NO_3/24 \, h$ $(0.63 - 1.26 \, kg$ $N\text{-}NO_3/24 \, h \cdot km^2)$. The nitrate quantity taken out from the Cracow Upland catchment within the confines of total runoff is estimated from 0.63 to $1.68 \, Mg \, N\text{-}NO_3/24 \, h$ $(0.97 - 2.58 \, kg \, N\text{-}NO_3/24 \, h \cdot km^2)$, with the nitrate concentration modal value in groundwater ranging from 2.0 to 5.4 mg N-NO₃/l.The nitrate hydrochemical background for selected areas of main groundwater aquifers in Poland is shown in Figure 5.

FINAL REMARKS

Hydrochemical investigations of karstic aquifers in agricultural areas such as the Cracow Upland concern the environments that are very sensitive to human impact. Within fracture-karstic environment, mobile, slowly degraded and toxic nitrates provide the greatest local and regional evidence of anthropogenic groundwater pollution. At the same time in carbonate massif, processes influencing the chemistry of infiltrational waters in vadose zone (caves) and phreatic zone (wells, springs) can be studied. In the studied area, the components of the input function derived from human pressures include atmospheric pollution in the form of dry and wet precipitation, agricultural chemicalization and rural economy. Mean annual nitrate concentration in atmospheric air reaches $30-70\,\mu g\,m^{-3}$ in the central part of the Cracow Upland and $0.6-1.0\,mg\,N\text{-}NO_3/l$ in atmospheric precipitation. Use of nitrogen in mineral fertilizers is now at the level of $30-80\,kg\,N/ha$. The human impacts cause a regional increase in the contamination of fissure-karstic waters of the Cracow Upland with nitrates, which influences water quality. The current hydrochemical background of nitrates in cave waters $(0.2-8.5\,mg\,N\text{-}NO_3/l)$ and phreatic zone waters $(1.2-7.5\,mg\,N\text{-}NO_3/l)$ shows seven to eight times as wide a value as the natural background $(0-1\,mg\,N\text{-}NO_3/l)$. Spatial differentiation of nitrate concentrations in percolating water is observed even within single caves. Mean nitrate concentrations in groundwater from varied land use do not differ significantly and amount in rural areas to $7.4\,mg\,N\text{-}NO_3/l$, arable land $-6.6\,mg\,N\text{-}NO_3/l$ and forests $-3.6\,mg\,N\text{-}NO_3/l$. On the basis of groundwater flow modelling as well as hydrochemical research, the nitrate load introduced with precipitation was assessed to $0.09-0.56\,kg\,N\text{-}NO_3/24\,h\cdot km^2$, while nitrate load carried away from the useful Upper Jurassic aquifer estimated from 0.97 to $2.58\,kg$ $N\text{-}NO_3/24\,h\cdot km^2$, including rivers $0.63-1.26\,kg\,N\text{-}NO_3/24\,h\cdot km^2$.

REFERENCES

Bakalowicz, M. 1995: La zone d'infiltration des aquiferes karstiques. Methodes d'étude. Structure et fonctionnement. *Hydrogèologie 4*: 3–21.
Benes, V., Pekny, V., Skorepa, J. and Vrba, J. 1989: Impact of diffuse nitrate pollution sources on groundwater quality – some examples from Czechoslovakia. *Environmental Health Perspectives 83*: 5–24.
Duynisveld, W.H.M., Strebel, O., Boettcher, J. and Kinzelgach, W. 1989: *Long term prognosis of the groundwater quality as influenced by land use and land use changes in the Fuhrberger Feld using numerical solute transport models.* In: H.E. Kobus and W. Kinzelbach (eds): Contaminant transport in groundwater. Balkema, Rotterdam, pp. 89–96.
Ford, D. and Williams, P. 1989: *Karst geomorphology and hydrology.* Unwin Hyman, London.
Goc, P., Górny, A., Klojzy-Karczmarczyk, B. and Motyka, J. 2000: Nitrates in cave waters of southern part of Cracow Upland (in Polish). *Kras i Speleologia 10* (XIX): 67–83. Wydawnictwo Uniwersytet Śląski, Katowice.
Juergens-Gschwind, S. 1989: *Groundwater nitrate in other developed countries* (Europe) – *relationships to land use patterns.* In: R.F. Folett (ed.): Nitrogen management and groundwater protection. Ser. Developments in Agricultural and Managed – Forest Ecology 21. Elsevier, Amsterdam, pp. 75–125.
Kleczkowski, A. (ed.) 1984: *Groundwater protection* (in Polish). Wydawnictwa Geologiczne, Warszawa.
Kozlovsky, E.A. (ed.) 1988 *Geology and the Environment*, vol. I, Water Management and the Geoenvironment. Unesco Paris, France. UNEP Nairobi Kenya, p. 179.
Leśniok , M. 1996: *Pollution of precipitation in the Silesia – Cracow Upland* (in Polish). Wydawnictwo Uniwersytet Śląski, Katowice.
Leśniok, M., Partyka, J. and Pawlicka, A. 2001: *Atmospheric pollution at the Ojców National Park area.* (in Polish) In: Mat. konf. Zmeny geografickeho prostredi v pohranicnich oblastech Hornoslezskeho a Ostravskeho regionu. Ostravski univerzita v Ostrave, pp. 164–168.

Macioszczyk, A. 1991: *Hydrogeochemical background and quality of groundwater exploited in Poland.* In: Ochrona wód podziemnych w Polsce, stan i kierunki badań (in Polish). Publ. CPBP 04.10. vol. 56. Wyd. SGGW-AR, Warszawa, pp. 253–273.

Macioszczyk, A. and Mitręga, J. 1997: *State of groundwater pollution in productive aquifers of Poland.* In: Mat. konf. Współczesne Problemy Hydrogeologii (in Polish), vol. VIII: 361–364. Poznań.

Mangin, A. 1975: Contribution a l'étude hydrodynamique dse aquiferes karstiques. *Ann. De Speleol. 29* (3): 283–332; (4): 495–601; *30* (1): 21–124.

Różkowski, J. 1996: Transformations in chemical composition of karst water in the southern part of the Cracow Upland (Rudawa and Prądnik drainage areas) (in Polish). *Kras i Speleologia* nr specj. 1. Wyd. Uniw. Śląski, Katowice.

Różkowski, J. and Różkowski, K. 1999: Human impact on the groundwater chemistry of the Tenczynek Horst karstic area. In: Mat. konf. *Współczesne Problemy Hydrogeologii* (in Polish), vol. IX. Warszawa–Kielce: PIG, pp. 315–322.

Różkowski, J., Kowalczyk, A., Rubin, K. and Wróbel, J. 2003: Groundwater circulation balance, renewal and resources in the Cracow Jurassic karstic aquifer in the light of modeling study. *Kras i speleologia*, vol. 11 (XX). Wydawnictwo Uniwersytet Śląski, Katowice.

Sapek, A. 1990: Evaluation of CREAMS model usefulness for nitrate washout forecasting according to fertilising, cultivation and soil parameters in Poland. In: Mat. uzupeln. do sprawozd. z projektu Pg-Po-376 (PL-ARS-129). Inst. Melioracji i Użytków Zielonych, Falenty, p. 91.

Tyc, A. 1997: Anthropogenic impact on karst processes in the Silesian-Cracow Upland (Olkusz–Zawiercie area as example) (in Polish). *Kras i Speleologia* nr specj. 2(1997). Wydawnictwo Uniwersytet Śląski, Katowice.

Witczak, S. and Adamczyk, A. 1995: *Catalogue of specific physical and chemical groundwater pollution indicators and measurement methods*, vol. II (in Polish). Bibl. Monitoringu Środowiska, PIOŚ, Warszawa.

Witczak, S. and Żurek, A. 1994: *Use of soil – agricultural maps in the evaluation of protective role of soil for groundwater.* In: A.S. Kleczkowski (ed.): Metodyczne podstawy ochrony wód podziemnych (in Polish), pp. 155–180.

CHAPTER 19

Nitrate in Danish groundwater

J. Stockmarr and P. Nyegaard
Geological Survey of Denmark and Greenland (GEUS), Øster Voldgade,
DK-1350 Copenhagen, Denmark

ABSTRACT: Drinking water supply in Denmark is totally based on groundwater and since the early 1980s Danish researchers have been aware of the heavy nitrate pollution of Danish groundwater.

Groundwater monitoring builds on data from groundwater monitoring areas, agricultural watershed catchment areas, and water abstraction wells that together give a comprehensive picture of the groundwater chemistry and state of pollution. Nitrate pollution primarily from agriculture is the most serious problem to the Danish groundwater resource. During the last decades many minor waterworks have been closed. Sixteen per cent of the monitoring wells have nitrate concentration above the maximum admissible concentration of $50 \, \text{mg} \, l^{-1} \, NO_3^-$ for drinking water. However the main part of the water supply wells, 68% contain nitrate below $1 \, \text{mg} \, l^{-1}$.

The development of the nitrate content during the last 50–60 years generally follows the use of fertilizers. In 1987 an Action Plan for the Aquatic Environment was approved by the Danish Parliament to reduce the nitrate leaching by 50% within 5 years. However the effect on the nitrate content in groundwater is very little as the age of groundwater in the majority of well screens predates the Action Plan.

GROUNDWATER MONITORING PROGRAMME

As the drinking water supply in Denmark is totally based on groundwater, quality monitoring of the groundwater is extremely important to the Danish community. Groundwater quality monitoring in Denmark is described in proceedings from the IWA Conference in Buenos Aires (Henriksen and Stockmarr 1999), GEUS (2002), Stockmarr, Grant and Jørgensen (in press), and in GEUS homepage www.groundwater.dk (English) and www.grundvandsovervaagning.dk (Danish).

The Danish annual groundwater quality monitoring reports are based on data from approximately 6,000 water supply wells, approximately 1,100 well screens from the groundwater monitoring network and 100 shallow screens from the five agricultural watersheds (GEUS 2002). More or less evenly distributed, the groundwater monitoring constitutes 67 groundwater monitoring catchment areas (Figure 1); each with approximately 17 monitoring screens in the main aquifer and in upper secondary aquifers. The flow in the catchment is conducted by a minor water supply well; see the principle in Figure 2.

The Danish groundwater monitoring programme, which is a part of the National Monitoring Programme for the Aquatic Environment, NOVA 2003, was originally started

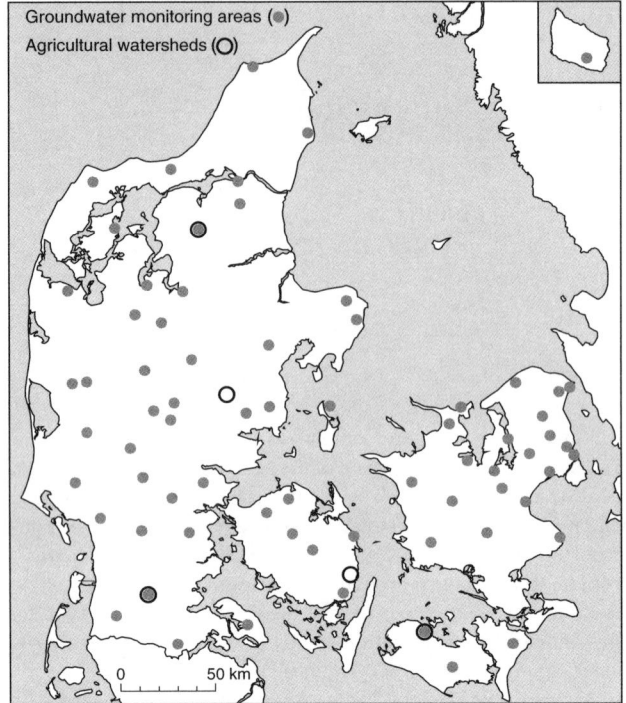

Figure 1. Groundwater monitoring areas and agricultural watersheds in Denmark (GEUS 2001).

in 1987 based on a political plan for the Aquatic Environment. The main objective for the Plan was the registration of the high pollution with nitrogen and phosphorus compounds from agriculture and the effects of the changes in the nutrient impact. Further, the objective for the groundwater monitoring was generally to examine the development in groundwater quality and quantity in order to secure sufficient drinking water of good quality. One final objective was to describe the quality of the water that constitutes the basic flow to the surface water.

WATER SUPPLY WELLS

Since 1989 it has been compulsory to all waterworks producing more that 3,000 m^3 per year to monitor the groundwater quality in every water supply well. Until now about 20,000 analyses from about 8,000 water supply wells have been reported to GEUS groundwater quality database (Figure 3).

The groundwater monitoring frequency is, however, dependent on the size of the water supply. All waterworks supplying more than 1.5 million m^3 per year must monitor the groundwater quality every three years. Waterworks supplying between 35,000 m^3 per year and 1.5 million m^3 per year must monitor the quality every four years and for the minor waterworks down to 3,000 m^3 per year it is every five years. The monitoring programmes comprise about 20 common main physical and chemical parameters. To this are added some inorganic trace elements such as nickel, arsenic and aluminium, and finally a list of

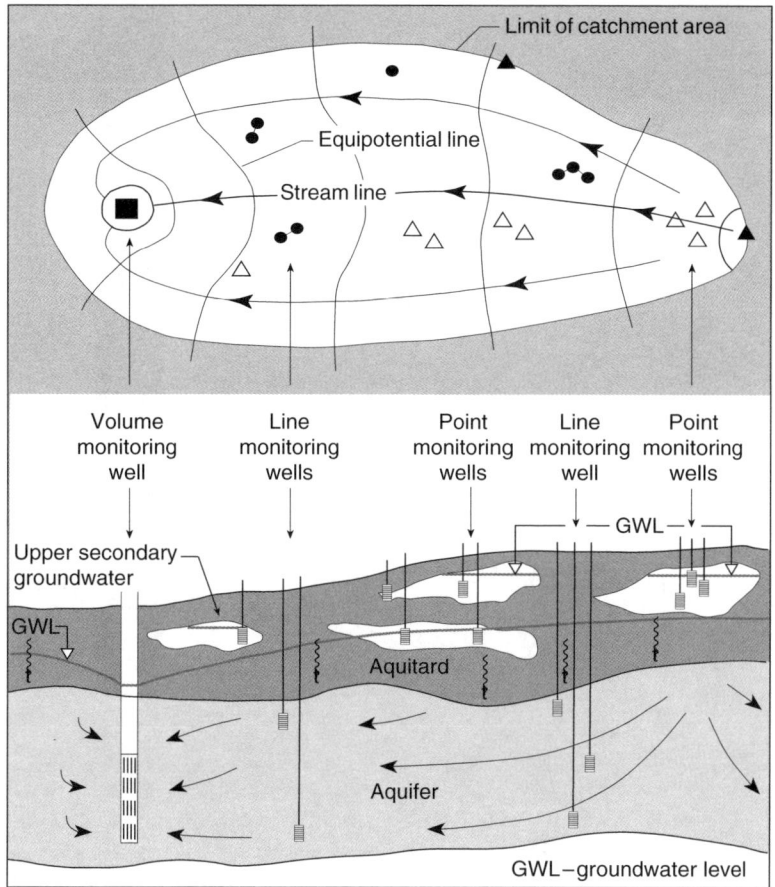

Figure 2. General principle for an ordinary Danish groundwater monitoring area (GEUS 2001).

pesticides and metabolites (at present 23) and organic micro pollutants, depending on pollution threats in the vicinity and up-stream.

Reporting of groundwater monitoring data

Every year since 1989 GEUS has published a report on the state of the groundwater quality and quantity based on monitoring data collected by GEUS and the counties. The most comprehensive was the 1995-report (GEUS 1995) and the most recent is the 2001-report (GEUS 2001).

NITRATE IN DANISH GROUNDWATER

The water supply wells generally have long screens and are assessed to give a representative distribution of the nitrate contents in the primary groundwater reservoirs.

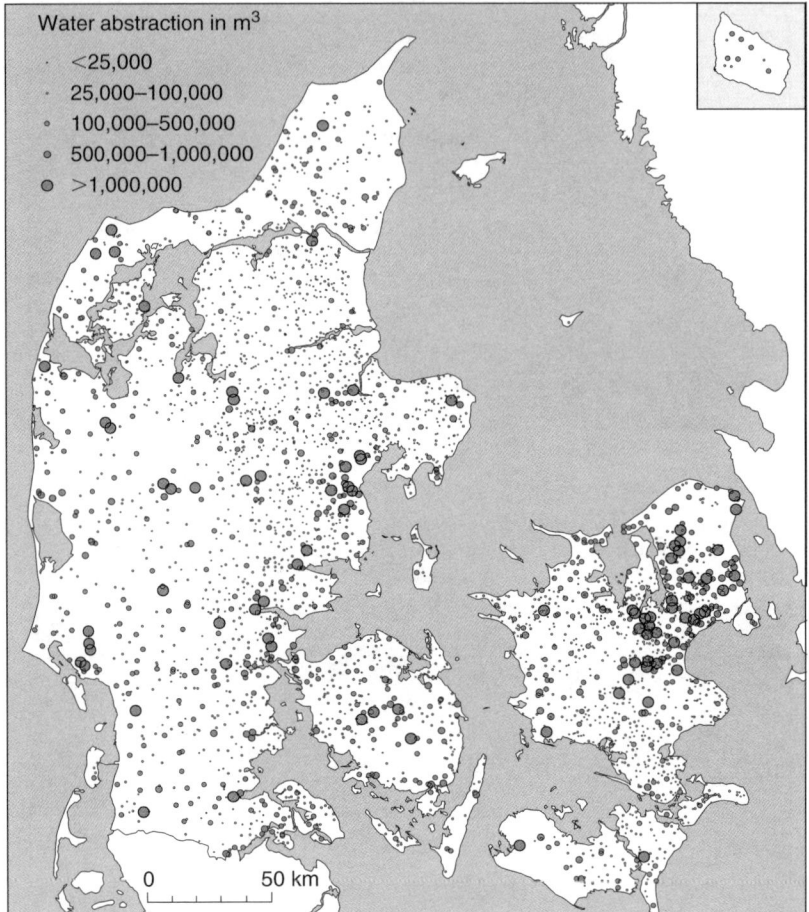

Figure 3. Distribution of waterworks in Denmark with abstraction size classification (GEUS 1997).

On the other hand, the data from the water abstraction wells are biased, as the water works must produce drinking water with nitrate concentration below the maximum admissible concentration at $50\,\mathrm{mg\,l^{-1}}\,NO_3^-$ (Figure 4). Groundwater monitoring wells give a more accurate picture of the general nitrate pollution in the Danish groundwater. Of them, 16.4% have nitrate contents above the MAC level and about 60% have no nitrate (below $1\,\mathrm{mg\,l^{-1}}$).

The spatial distributions of nitrate in the groundwater reservoirs (Figure 5) vary from west to east Denmark. West of the 'main stationary line' of the Weichselian glaciation, the sedimentary formations are sandy meltwater deposits with unconfined sandy reservoirs overlying deeper confined Quaternary and Miocene sand reservoirs. East of the main stationary line the sandy meltwater deposits and the Pre-Quaternary limestone reservoirs are often covered with clayey till deposits, and the reservoirs are mostly confined. Differences in the vertical distribution show that unconfined reservoirs have a much higher fraction of screens with more than $25\,\mathrm{mg\,l^{-1}}$ nitrate.

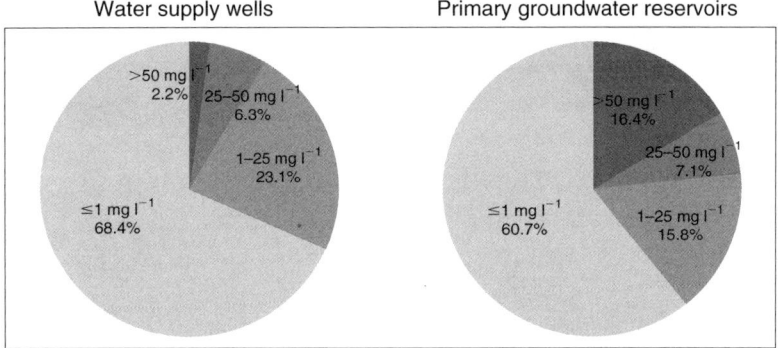

Figure 4. Distribution of nitrate in groundwater from water supply wells and monitoring wells in 4 classes: $<1\,\mathrm{mg\,l^{-1}}$ nitrate, $1-25\,\mathrm{mg\,l^{-1}}$ nitrate, $25-50\,\mathrm{mg\,l^{-1}}$ nitrate and $>50\,\mathrm{mg\,l^{-1}}$ nitrate. Median values for 1990–2000 (GEUS 2001).

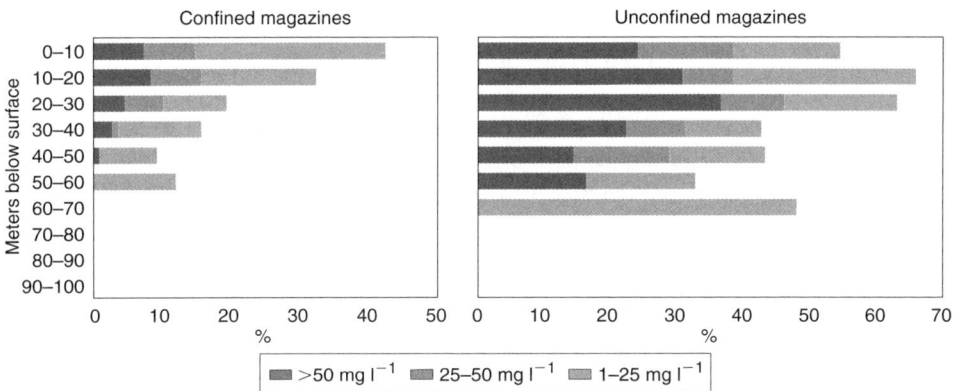

Figure 5. Depth distribution of nitrate in groundwater monitoring (Stockmarr *et al.* in press).

The high concentrations of nitrate have also reached a deeper level in areas with unconfined reservoirs. The reason for this is a much lower nitrate reduction capacity in the sandy meltwater deposits than in the clayey till deposits. The western sandy areas have also been exposed to oxidizing conditions for a long time. An additional factor is the higher rate of precipitation in the western part of Denmark than in the eastern, which leads to a higher degree of nitrate leaching from the topsoil. Further the western part is under greater pressure from the livestock production.

In the last half century farming has intensified and thereby the use of fertilizers. The groundwater from the monitoring screens has been dated by the CFC content (GEUS 2001: Hinsby, Laier, Thomsen *et al.* 1997; Laier, this volume), and in Figure 6 the nitrate content is given as a function of the CFC age. The use of fertilizers in kilograms per hectare (Danish Plant Directorate 2001) is shown for comparison. The nitrate contents in the age-dated groundwater follow the increase in the use of fertilizers.

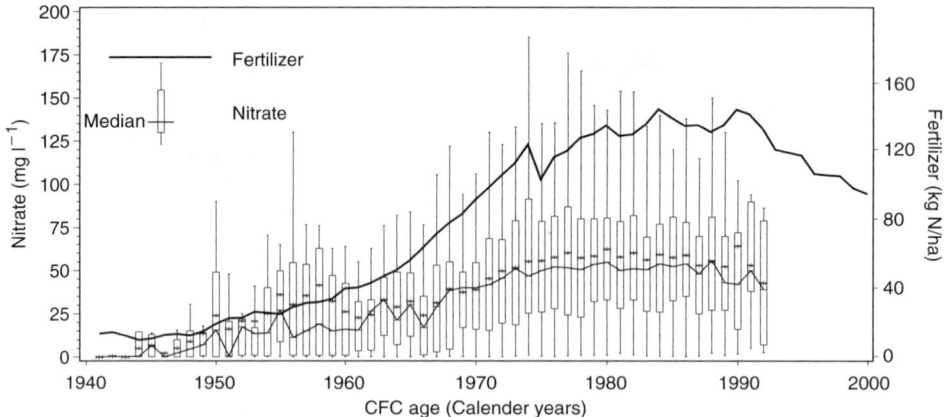

Figure 6. Nitrate content versus the CFC-age in groundwater monitoring together with the use of fertilizers in kg ha^{-1} (Danish Plant Directorate 2001). Only samples with nitrate and oxygen above 1 mg l^{-1} are included. As only very few samples are younger than 1992 they are not included.

NITRATE DEVELOPMENT IN THE GROUNDWATER MONITORING AREAS

The evaluation of the nitrate development in the groundwater monitoring areas is based on all nitrate monitoring data from the period 1990–2000 with a median value for the period 1990–2000 above 1 mg l^{-1} nitrate (ca. 40% of the well screens, see Figure 4). This means that data can be divided into groundwater under oxic conditions and groundwater under anoxic conditions. The maximum number of well screens is 254 screens in the oxic groundwater and 154 in the anoxic groundwater. The development in these two redox categories is shown in Figure 7.

In the groundwater monitoring areas the nitrate content of the two redox zones vary in every single screen, but the median value for the period 1990–2000 show little variation with a high degree of deviation. In general, no change in the nitrate concentration is seen even in the superficial aquifers where the nitrate concentration is mainly dependent of winter precipitation.

NITRATE IN YOUNG GROUNDWATER

The nitrate content in the youngest groundwater is monitored in the five agricultural watersheds both in sandy and clayey areas, and the variations in the nitrate content together with the winter precipitation from the last 11 years are illustrated in Figure 8. The screens in these watersheds are located from 1.5 to 5 metres below the surface. The median values of the nitrate content is 2–3 times higher in the sandy areas than in the clayey areas, but for both types the trend in nitrate content roughly follows the trend of the winter precipitation.

A high deviation can be seen on the nitrate data and the concentrations are somewhat higher in sandy soils than in clayey soils. By comparing the median values for the nitrate

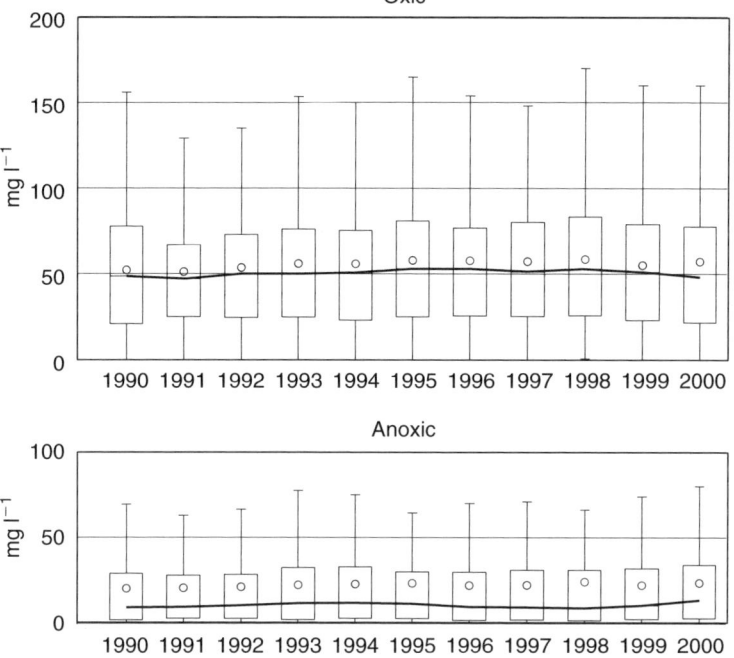

Figure 7. Nitrate development in mg l^{-1} from the period 1990–2000 based on data from all monitoring screens with more than 1 mg l^{-1} in the entire period. The redox zones are oxic (with oxygen) with up to 254 monitoring screens and anoxic (with nitrate) with up to 154 monitoring screens (GEUS 2001).

concentration with the winter precipitation a clear coherence is seen. The first high winter precipitation shows a high nitrate infiltration, which again the next year results in a decrease in nitrate concentration probably due to the thinning effect during the next winter with high precipitation. There is a quicker reaction in sandy sediments than in clay sediments. Consequently it is indicated that the leaching of nitrate from the topsoil is strongly influenced by the magnitude of the precipitation, and that it might be very difficult to demonstrate minor changes in the leaching from agricultural soils due to minor changes in the land-use.

MULTI-SCREENED REDOX WELLS

To get more detailed information on the eventual movements of the nitrate 'front', four multi-screened monitoring wells, called 'redox wells', were established in unconfined sandy aquifers mainly of glacial origin, and more are to follow.

The redox wells are established with at least 15 screens each. The screens are only a few centimetres and the spacing between the screens is generally below one metre in the anoxic zone (nitrate zone). The analyses frequency is six times a year and we already know that the concentrations results vary, but still the results are not sufficiently well understood for presentation.

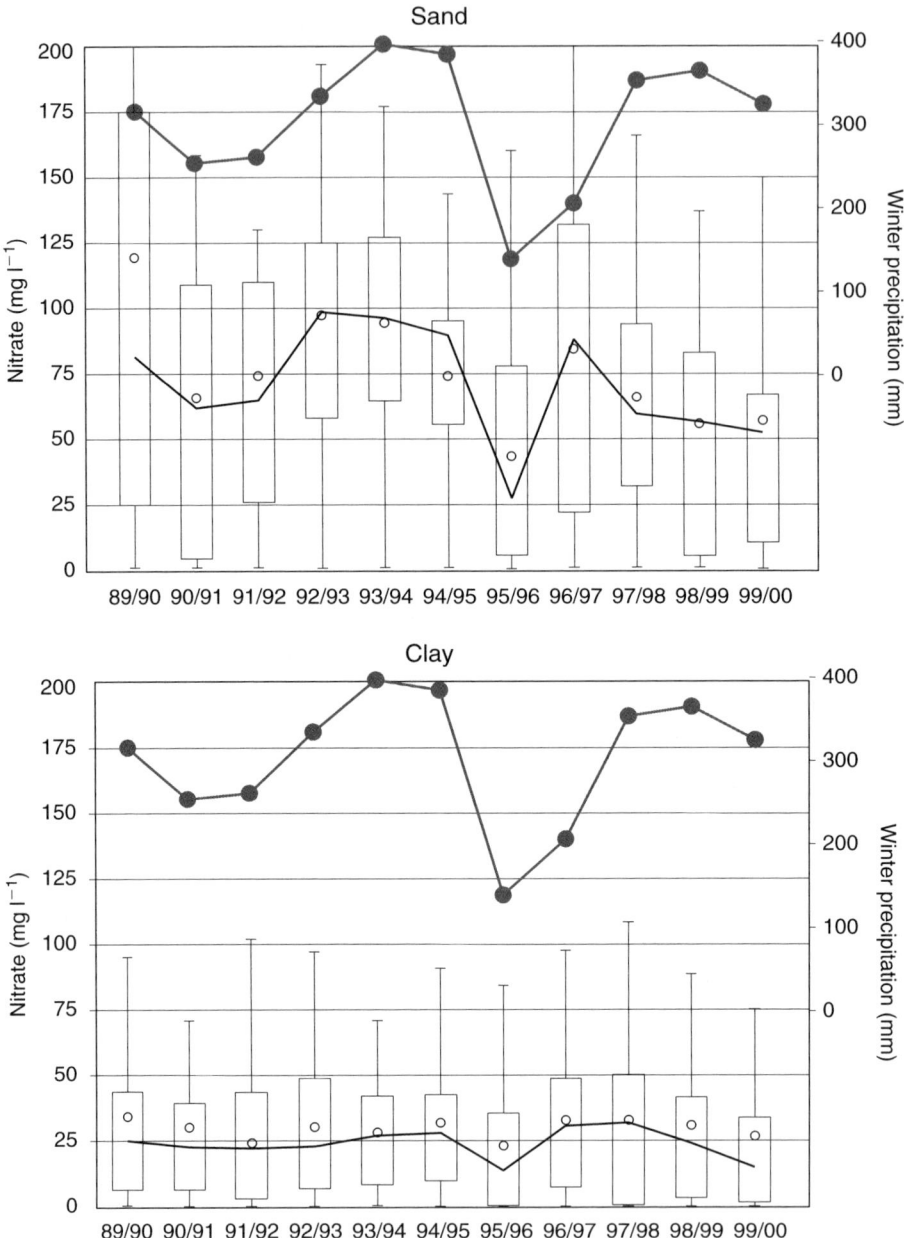

Figure 8. Nitrate in the agricultural watershed areas divided in sandy and clayey soils, compared to winter precipitation from the 4. and 1. quarter of the years (upper curve) (GEUS 2001).

The division between the different redox-zones is always dependent on the quality of the analyses. The oxygen analyses must be made in the field and the detection limit should be as low as $1 \, mg \, l^{-1}$. The lower limit for the anoxic zone (nitrate zone) is nitrate $< 1 \, mg \, l^{-1}$, and below this the reduced zone or iron-sulphate-zone is normally developed.

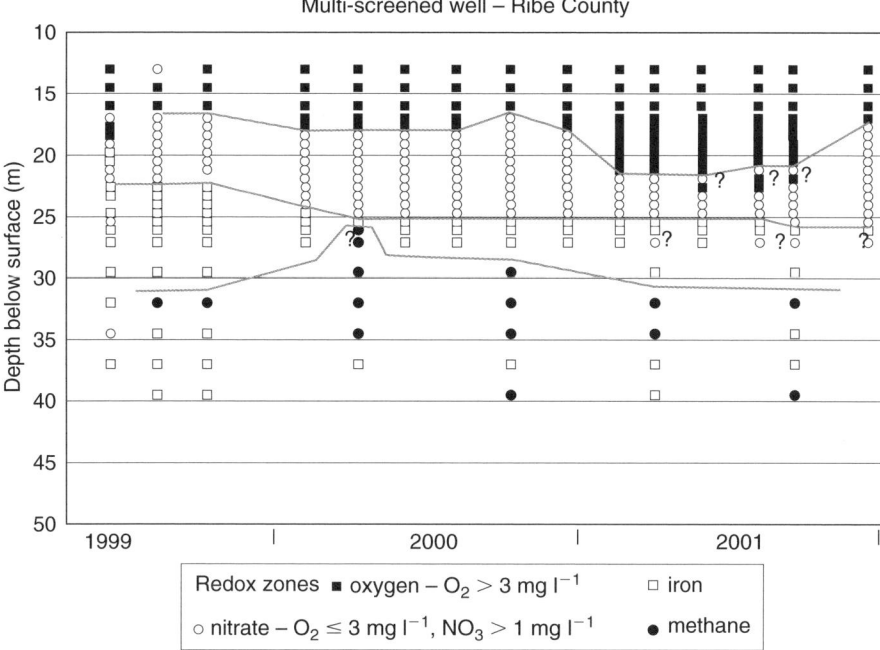

Figure 9. Multi-screened well in Grindsted, Central Jutland showing changes during the period 1999–2001. The oxic zone is well developed and varies between 16 and 21 m depth. The anoxic zone varies between 22 and 25 m depth, and below the iron-sulphate and methane zones are developed. It is of special interest that there seems to be seasonality in the zones (GEUS 2002).

The redox-well at Ribe County, Central Jutland (Figure 9) was established in 1999 with 23 monitoring screens from 13 to 39.5 m depth. The screens are placed in glacial meltwater sand and below that in tertiary quartz and mica sand with lignite.

The oxic zone extended in 1999 down to 16 m depth and the oxygen content in the non-reduced part of the oxic zone is about $9 \, mg \, l^{-1}$. During the period up to 2001 the zone extend down to 21.5 m depth before a slight increase began. The nitrate zone began in 1999 at 22 m depth but expanded rather soon down to 25 m depth (establishment effects?). By the end of 2001 there seems to be a further deepening of the nitrate front. Below the nitrate front is the iron-sulphate zone with some ammonium and, from about 40 m depth, the concentration of methane increases indicating the beginning of the methane redox zone.

COHERENCE BETWEEN VARIATIONS IN GROUNDWATER CHEMISTRY AND GROUNDWATER LEVEL

One of the premises for doing meaningful analyses of time series is that the phenomena you describe are dependent on time in a transparent way. This means that when samples are taken the age of the water should increase regularly to depth and that the catchment of a screen does not change.

These premises are not fulfilled if the groundwater potential changes and then induces changes in the flow pattern (Figure 10). So it is very important to know all the factors that

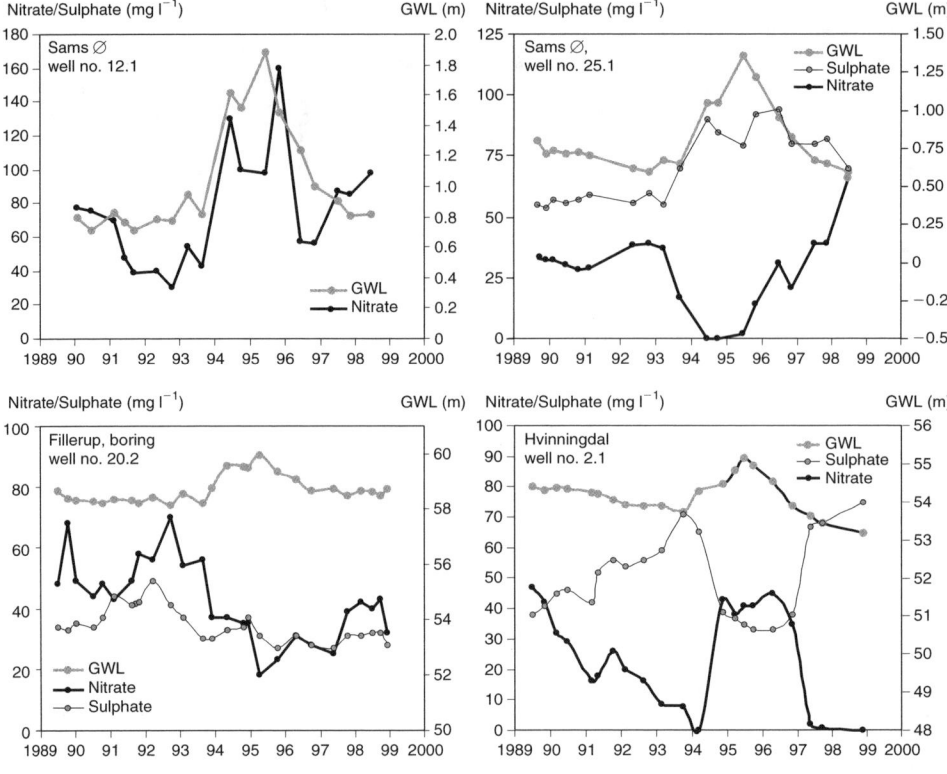

Figure 10. Nitrate and sulphate concentrations compared to groundwater level in unstable monitoring wells in Århus Amt (1999).

may interfere when the groundwater quality changes. It may be due to changes in land-use, but it may also be changes in the natural conditions.

Four wells in Figure 10 show a clear coherence between the variations in groundwater level and changes in groundwater chemistry. In the four cases the peak in the groundwater level from 1994 to 1997 induces increase or decrease of concentration of nitrate or sulphate, which may be due to changes in the groundwater or the flow pattern to the screens.

If the aquifer is homogenous and unconfined, increasing nitrate with a decrease in groundwater level should indicate higher nitrate concentration in the upper aquifer (Samsø well no. 25.1 and Fillerup well no. 20.2). However in the two other cases (Samsø well no. 12.1 and Hvinningdal well no. 2.1) the reaction is opposite, indicating that increasing groundwater level also increases the nitrate concentration for later reduction. Changes between clear oxic conditions and clear reducing conditions are also seen.

NITRATE IN GROUNDWATER MONITORING AND INCREASING AND DECREASING TENDENCIES

In order to illustrate if nitrate data from groundwater monitoring in water supplies show a development (increasing or decreasing), data are divided in 3-year groups, 1990–1992,

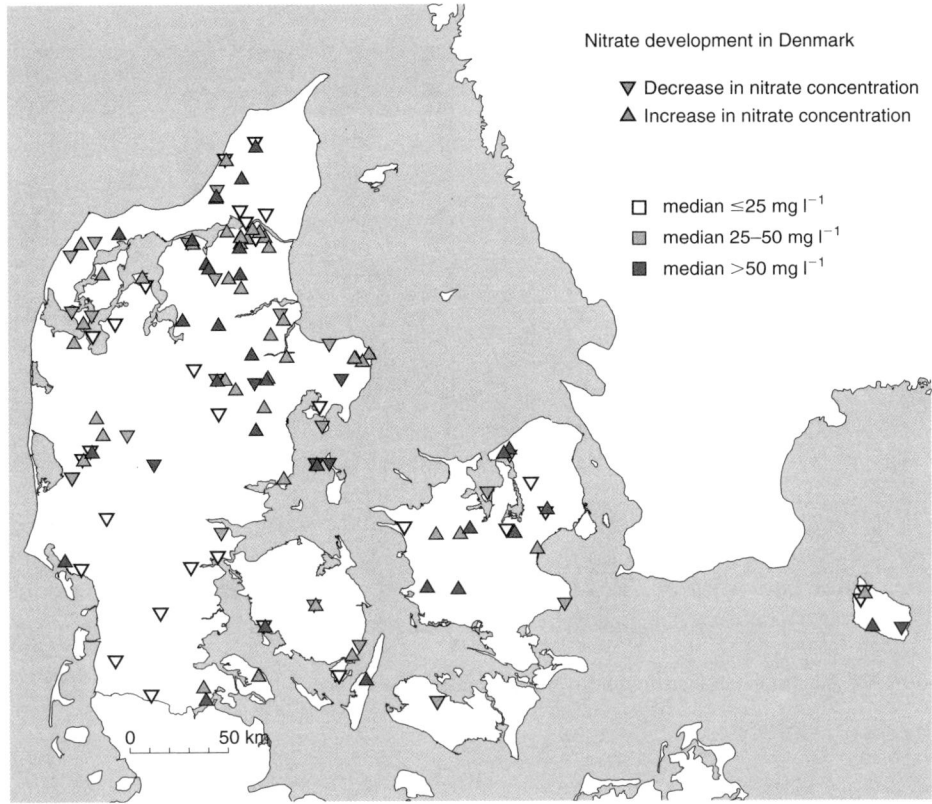

Figure 11. Nitrate variations in 3-year periods for water supplies. Nitrate concentrations are based on all nitrate data from 1990 to 2001. Only wells with more than 25 mg l^{-1} in a 3-year period and only increases and decreases are shown (GEUS 2002).

1993–1995, etc. Median values are calculated for all wells in the periods. Only data with more than 25 mg l^{-1} for at least one 3-year period are included (Figure 11) (GEUS 2002). Most water supplies have unchanged nitrate content. However there are slightly more wells with an increase in nitrate compared to wells with a decrease in nitrate (Figure 12). The period 1990–2000 covers 5,381 wells with less than 1 mg l^{-1} nitrate. 1,867 wells have between 1 and 25 mg l^{-1}, 517 wells have nitrate content between 25 and 50 mg l^{-1} and 173 wells have above 50 mg l^{-1}. Even though wells with above 50 mg l^{-1} are generally closed and replaced by wells without nitrate (GEUS 2001).

The majority of the water supply wells, 68%, represent groundwater with no nitrate, that is, less than 1 mg l^{-1} nitrate. The amount of wells with nitrate above 25 mg l^{-1} does not vary significantly.

Generally the nitrate content of the abstracted water has not changed during the last 11 years, eventhough many wells with more than 50 mg l^{-1} nitrates are closed and new wells without nitrate are established. The high nitrate values in the water supply wells, above 25 mg l^{-1}, mainly occur in the so-called 'nitrate belt' of North West Jutland. However, groundwater abstracted in several other regions of Denmark which lack deeper aquifers also shows bad water quality.

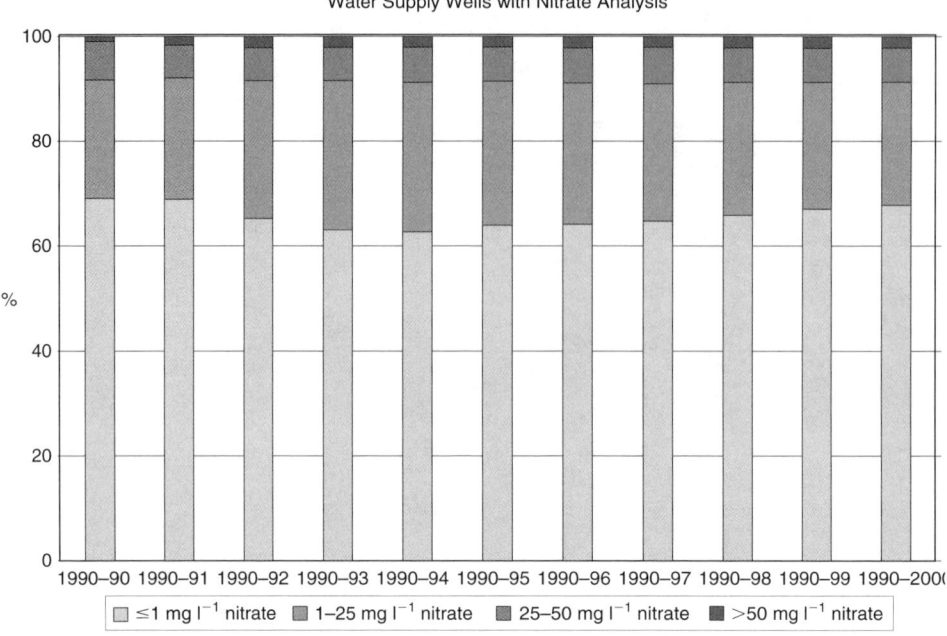

Figure 12. Accumulated distribution of nitrate analyses from water supply wells.

FINAL STATEMENT

In Danish groundwater we have two serious major pollution problems. One is nitrate and the other is pesticides and their metabolites. Giving up the use of plant protection agents or pesticides may over a timespan of tens of years eliminate pesticides as a major problem for groundwater. However, only by giving up agricultural production can nitrate pollution be eliminated, and, even if the Danish (and European) politicians some day might be ready to abandon the use of pesticides, they will never agree to give up agricultural production.

During the last decade many changes in farming behaviour have been implemented and the decrease in nitrate leaching to surface water is now significant. Groundwater formation takes time and still no decrease in nitrate is seen in the groundwater. However, eventhough there is a decrease in nitrate concentration in surface water, farmers have still not fulfilled the Action Plan for the Aquatic Environment from 1987 and more changes in farming behaviour will be introduced in the near future.

The most viable way is to reduce the use of manure to an acceptable level, depending on the hydrogeology, and the use of fertilizer. A proposal could be to make it unacceptable to add more manure than is equal to 1.3 animal unit per hectare or even less, and that the additional use of fertilizer never must exceed 70–80% of the optimal amount of nitrate needed (including manure), according to agricultural norms. To day the maximum is 90% of the optimum.

REFERENCES

Århus Amt 1999: Statusrapport 1998. Grundvandsovervågning i Århus Amt Århus, 69p. Danish Plant Directorate 2001: Forbruget af handelsgødning i 1999/00 – (www.pdir.dk).

GEUS 1995: Grundvandsovervågning 1995. GEUS, Special edition, København, 209p.

GEUS 1997: Grundvandsovervågning 1997. GEUS, Special edition, København, 101p.

GEUS 2001: Grundvandsovervågning 2001. GEUS, Special edition, København, 114p.

GEUS 2002: Grundvandsovervågning 2002. GEUS, Special edition, København, 122p. (www.grundvandsovervaagning.dk and www.groundwater.dk).

Henriksen, H.J. and Stockmarr, J. 1999: *Groundwater Resources in Denmark. Modelling and Monitoring.* IWA World Water Congress 1999, 9 Special Subject 1–11, Buenos Aires, Argentina.

Hinsby, K., Laier, T., Thomsen, A., Engesgaard, P., Jensen, K.H., Larsen, F., Jakobsen, R., Busenberg, E. and Plummer, L.N. 1997: *CFC-dating, and transport and degradation of CFC-gases in different redox environments: two case studies from Denmark.* Geological Society of America Annual Meeting, Salt Lake City, Utah.

Laier, this volume.

Miljøstyrelsen 1983: *Nitrat i drikkevand og Grundvand i Danmark. – Redegørelse fra Miljøstyrelsen.* København, 125p.

Stockmarr, J., Grant, R. and Jørgensen, U. (in press): *Monitoring effectiveness of the EU Nitrates Directive Action Programmes: Approach by Denmark.* Draft input to International Workshop Mon-NO_3, Den Haag 2003. 20 pp.

CHAPTER 20

Nitrates in fresh water in the Miechów Region, Poland

J. Wagner, B. Gajowiec and R. Formowicz
Polish Geological Institute, Upper Silesian Branch, Department of Deposit Geology,
Hydrogeology and Environment Protection, 1 Królowej Jadwigi Str.,
41-200 Sosnowiec, Poland

ABSTRACT: The region of Miechów is located in southern Poland. This paper describes the current quality of groundwater in the region, as well as the variation in nitrogen, potassium and phosphorus levels determined from 1992 to 1999. This is based on data acquired while preparing the Hydrogeological Map of Poland by the Polish Geological Institute.

INTRODUCTION

This paper deals with the quality of groundwater in the Miechów region in terms of the variation in nitrogen, potassium and phosphorus levels as determined from 1992 to 1999. The data acquisition was done while preparing the Hydrogeological Map of Poland by the Polish Geological Institute. The sheets of the Hydrogeological Map of Poland were in preparation for five years.

The main problem in the study area is caused by groundwater pollution originating from the usage of manure and fertilizers in the agricultural industry. The aim of this paper is to show the variation in the levels of nitrogen, potassium and phosphorus compounds in groundwater, which indicates groundwater pollution resulting from human activity.

STUDY AREA

The study area is located in the Miechów administrative district in Małopolska Province found in the Southern Poland (Figure 1). Geographically this area is located in the sub-region of the Nida Basin (Miechów Trough), within the sub-province of the Małopolska Highlands (Wyżyna Małopolska).

Geology

The elevation of the Miechów region is very diversified, varying from +220 m to +390 m above sea level. The southern limit of the Central Polish glaciation runs close to the centre of the basin. In the northern part of the basin, the Quaternary cover is thicker and the relief is significantly related to Quaternary geomorphology consisting of moraines, outwash

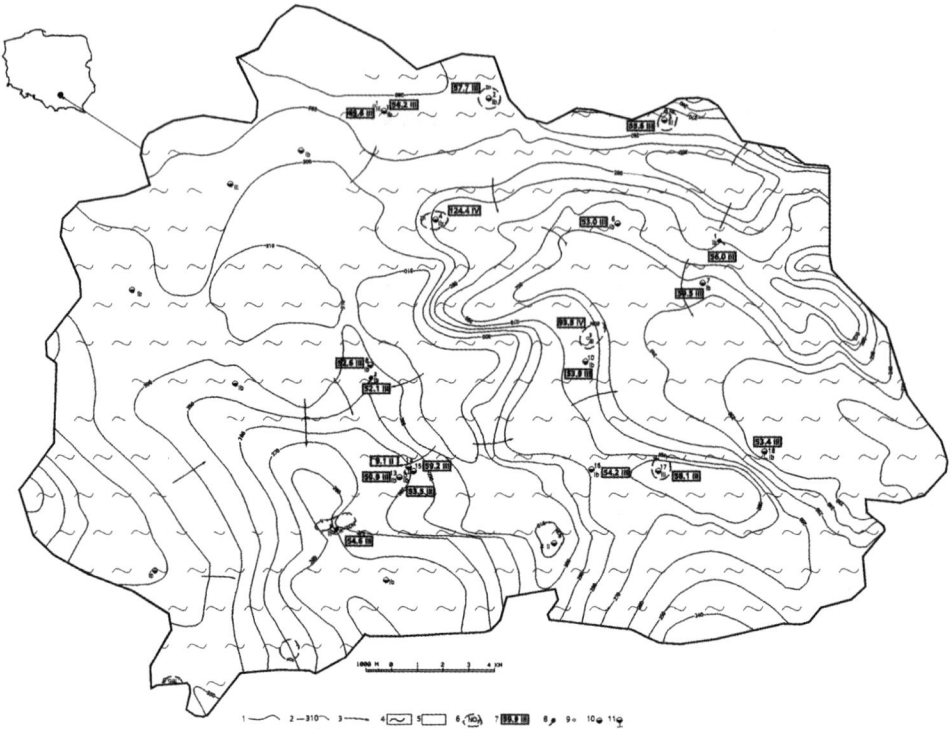

Figure 1. Quality of fresh water in Cretaceous aquifer in the Miechów administrative district. 1) boundary of the district; 2) water table contour of Cretaceous aquifer; 3) groundwater flow direction; 4) Ib class, water of good, but possibly unstable quality due to the surface exposure, requires no treatment; 5) II class, water of moderate quality, requires simple treatment; 6) extent of area where the groundwater quality exceeds drinking water standards. Symbols of excessive elements: Al, Cr, Fe, Mn, Zn, N-NO$_2$, N-NO$_3$. Deterioration of groundwater by fertilizers. 7) integrated index of water quality deterioration by fertilizers and degree of water quality deterioration by fertilizers, sampling sites of groundwater quality for Hydrogeological Map of Poland, 1: 50,000 scale, after Table 2; 8) springs (numbers 1s, 2s, 3s); 9) domestic wells (number 1dw, 2dw, 3dw, 4dw); 10) drilled wells (numbers from 1 to 8, 10, from 13 to 18); 11) sites of Regional Groundwater Monitoring (numbers 14, 18).

plains, dunes, loess deposits and valleys. In the southern part of the basin the deposits from the Southern Polish glaciation are considerably reduced and the morphology becomes predominantly Upper Cretaceous in origin.

In the southern part of Miechów Trough there are sediments of the Alpine orogenic cycle, with Cretaceous carbonates covered by Quaternary sand and gravel. Cretaceous strata are composed of marls and limestones.

Hydrogeology

The Miechów region is drained by the rivers Szreniawa and Nidzica, which are northern tributaries of the Vistula river. The surface water can be classified as class II and class III in respect of water purity, as well as the water whose degree of pollution exceeds the water purity classification limits.

The area of investigation is located in the Vistula river basin. The main usable aquifer is found in the Upper Cretaceous (Senonian), which is composed mainly of marls and gaizes. The aquifer recharge takes place through the outcrops of Cretaceous carbonate rocks or through the Quaternary sands overlying the Cretaceous strata. The area is drained by river valleys (water courses and springs), which are located along major tectonic lines of disturbance, or through water wells.

The Cretaceous aquifer is fissured. The fissuring disappears at depths between 80 and 120 m. Being for the most part unconfined, this aquifer is locally confined at low pressure (up to 100 kPa). The yield of water wells ranges from $10-70\,m^3\,h^{-1}$, locally reaching $150\,m^3\,h^{-1}$, with a specific capacity ranging from 0.1 to 83.3 m^3/h·m. Permeability values are found to be within the range of $4-331$ m/day. The water table of a confined piezometric surface ranges in depth from $0.5-32.8$ m. The discharges of numerous springs occurring in the area of investigation reach up to $135\,l\,s^{-1}$. The prevailing type of groundwater composition is HCO_3-Ca.

Agriculture

Owing to the type of soil, agriculture has become the main industry of the Miechów region. The following types of soil are found: podzols, black earths, sandy soils and muds in the river valleys. The underlying bedrock is mainly composed of marls and gaizes. The type of soil is a primary factor controlling cultivation and animal breeding. The main crops grown in the area are potatoes, corn (mainly wheat), mangolds, beets, tobacco and vegetables. The agricultural activity needs fertilization of soil, which is achieved through the introduction of nitrogen, potassium and phosphorus compounds to soil and, consequently, to underground water.

RESEARCH DESCRIPTION

The first phase of the investigations, which was conducted from 1995 until 1998 (Guzik and Pacholewski 1997; Różkowski 1997; Zembal and Liszka 1997), was followed by the second phase completed between 1998 and 2000 (Wagner and Gajowiec 2000; Włostowski 2000). The hydrogeological investigations were guided by 'Instructions for the Generation and Computerized Edition of the Hydrogeological Map of Poland' (1999). A group of hydrogeologists from the Polish Geological Institute (PGI) were assigned to perform the task. The following sheets of the map were completed: Wolbrom (Guzik and Pacholewski 1997), Skała (Różkowski 1997), Słomniki (Zembal and Liszka 1997), Miechów (Wagner and Gajowiec 2000) and Działoszyce (Włostowski 2000). The area covered by these sheets is shown on the map (Figure 1). These investigations resulted in the characterization of the Cretaceous aquifer in respect of underground water conditions, water quality and water resources.

The second phase of investigations (sheet Miechów conducted in the years 1998–2000) depicted the quantity and quality of groundwater in the area of extensive agricultural activities. Relatively high concentrations of nitrogen, phosphorus and potassium found in the water resulted from agricultural practices in the Miechów region. In the meantime, physical and chemical characteristics of the Upper Cretaceous underground water were investigated within a framework of the regional (Prażak, Janecka-Styrcz, Kowalczewska *et al*. 1996) and national (Hordejuk and Gawin 1998) underground water monitoring. The authors of sheet Miechów carried out their tests in 20 representative, hydrological data collection sites: three springs, four domestic water wells and 13 drilled water wells. In some cases the water quality exceeded the standards for drinking water.

The underground water samples collected between 1998 and 1999 represent mostly HCO_3-Ca chemical type of water, with a pH ranging from 6.7 to 7.6. The water was found to be hard, with total dissolved solids (TDS) within the range of $292-606$ mg l^{-1}. In terms of drinking water standards the water is of good quality and does not need treatment. However, higher concentrations of such elements as Fe_{og}, Mn^{2+}, Zn^{2+}, Al^{3+}, Sr^{2+}, NO_3^-, NH_4^+ and Pb^{2+} (referred to as anthropogenic pollutants) downgrade the quality of analysed water (to a lower class of water purity). This was the case mainly for springs and drilled wells. At some sites, which occurred erratically, groundwater collected from domestic wells exceeded drinking water quality standards (The Ordinance of the Minister of Health...). Higher concentrations of nitrite and nitrate-nitrogen were detected in the water of village settlements. In the neighbourhood of the municipal waste storage site, groundwater was found to be polluted by NO_3^-, NH_4^+ and Pb. The occurrence of Cl^- and NO_3^- ions detected at certain sites is related to a local contamination source rather than to a regional one. The contamination of Sr^{2+} and SO_4^{2-} is regarded to originate as a result of natural geological processes; these pollutants are referred to as 'geogenic pollutants' (Witczak and Adamczyk 1994).

RESULTS

In order to provide a detailed characterization of groundwater quality, the map of water quality deterioration in the Upper Cretaceous aquifer was prepared. As a result, the assessment of underground water quality deterioration was made. The scale of underground water quality deterioration from the use of fertilizers was then established (Table 1).

Table 1. Compensated scale of groundwater quality deterioration by fertilizers.

						Conversion factors (compensating) for indexes of agricultural contamination
Indexes of	N_{NO_3}	1	10	10	50	1.00
agricultural	N_{NO_2}	1	2	3	10	100.00
contamination	N_{NH_4}	1	5	7	15	10.00
	P	1	10	50	250	153.33
	K	1	2	2.4	4	0.20
R_{NPK} integrated index of water quality deterioration by fertilizers		5	29	72.4	329	
Degree of water quality deterioration by fertilizers		I	II	III	IV	

Table 2. Integrated indexes of water quality deterioration R_{NPK} and the degrees of water quality deterioration by fertilizers.

Site ID	Date	N_{NO_2}	R_{N/NO_2}	N_{NO_3}	R_{N/NO_3}	N_{NH_4}	R_{N/NH_4}	K	R_K	P	R_P	R_{NPK}	Degree of deterioration
1	99.05.17	0.003	0.30	5.78	5.78	0.039	0.39	1.000	0.20	0.323	49.52	56.2	III
2	99.08.13	0.003	0.30	6.80	6.80	0.070	0.70	2.000	0.40	0.323	49.52	57.7	III
3	99.04.22	0.003	0.30	4.81	4.81	0.078	0.78	1.000	0.20	0.323	49.52	55.6	III
4	99.08.16	0.697	69.70	2.58	2.58	0.241	2.41	1.000	0.20	0.323	49.52	124.4	IV
6	99.08.13	0.003	0.30	1.69	1.69	0.085	0.85	3.000	0.60	0.323	49.52	53.0	III
7	99.08.13	0.003	0.30	7.55	7.55	0.116	1.16	5.000	1.00	0.323	49.52	59.5	III
8	99.08.16	0.003	0.30	1.17	1.17	0.085	0.85	4.000	0.80	0.323	49.52	52.6	III
10	99.04.21	0.003	0.30	2.85	2.85	0.078	0.78	2.000	0.40	0.323	49.52	53.9	III
13	99.08.16	0.003	0.30	7.57	7.57	0.070	0.70	9.000	1.80	0.323	49.52	59.9	III
14	98.07.22	0.003	0.30	6.70	6.70	0.050	0.50	8.000	1.60	0.000	0.00	9.1	II
15	99.04.22	0.024	2.40	4.72	4.72	0.078	0.78	9.000	1.80	0.323	49.52	59.2	III
16	99.08.16	0.003	0.30	3.32	3.32	0.047	0.47	3.000	0.60	0.323	49.52	54.2	III
17	99.08.16	0.003	0.30	6.01	6.01	0.124	1.24	5.000	1.00	0.323	49.52	58.1	III
18	99.08.13	0.003	0.30	2.37	2.37	0.039	0.39	4.000	0.80	0.323	49.52	53.4	III
1s	99.08.13	0.003	0.30	4.79	4.79	0.101	1.01	2.000	0.40	0.323	49.52	56.0	III
1dw	99.05.17	0.003	0.30	18.34	18.34	0.039	0.39	5.000	1.00	0.323	49.52	69.6	III
2s	99.08.16	0.003	0.30	1.07	1.07	0.062	0.62	3.000	0.60	0.323	49.52	52.1	III
2dw	99.05.18	0.003	0.30	5.74	5.74	0.039	0.39	1.000	0.20	0.323	49.52	56.2	III
3s	99.08.16	0.003	0.30	1.98	1.98	0.039	0.39	7.000	1.40	0.323	49.52	54.6	III
3dw	99.04.21	0.003	0.30	42.69	42.70	0.078	0.78	1.000	0.20	0.323	49.52	93.5	IV
4dw	99.04.22	0.003	0.30	2.23	2.23	0.008	0.08	7.000	1.40	0.323	49.52	53.5	III

Explanation: collection sites: drilled wells – 1, 2, 3, ..., 18; springs – 1s, 2s, 3s; domestic wells – 1dw, 2dw, 3dw, 4dw.

The degree of groundwater quality deterioration by fertilizers can be determined using the compensated scale of groundwater quality deterioration by fertilizers, which was created by the authors (Table 1). The new scale is based on the usable water quality classification adopted for the groundwater monitoring (Staniewicz-Dubois 1995). Due to large differences in the absolute values of the concentrations of chemical substances that are used as water quality indexes (for instance N_{NO_3} occur in quantities on the order of $n \cdot 10^0 \, mg \, l^{-1}$, whereas P in quantities on the order of $n \cdot 10^{-2} \, mg \, l^{-1}$), the class limitation values of N_{NO_3}, N_{NO_2}, N_{NH_4}, P and K have to be counterbalanced. This scale is also modified in respect of the NH_4 limitation for Class II, which is assumed as $0.7 \, mg \, l^{-1}$ in accordance with 'Instructions for the Generation and Computerized Edition of the Hydrogeological Map of Poland' (1999). The limit concentration of pure phosphorus is established from orthophosphate (HPO_4^{2-}). All the limitations for the water quality indexes (Staniewicz-Dubois 1995) are brought down to a starting point on the scale, which is assumed as a dimensionless value of one. The conversion is then made by multiplying the limitation value of Class Ia by its reciprocal (given in Table 1) used as a conversion factor (compensation) for agricultural contamination index values. Hence, this process of relativising the scale of water quality index values enables us to propose a new scale of groundwater quality deterioration by fertilizers. This four-degree scale is determined by the limitation values of the integrated water quality deterioration index R_{NPK}.

The underground water classification was also prepared using the integrated index of water quality deterioration R_{NPK}. In this classification the degree of underground water quality deterioration by fertilizers can be classified into four categories: Class I — very pure water ($R_{NPK} < 5$), Class II — pure water ($5 < R_{NPK} < 29$), Class III — polluted water ($29 < R_{NPK} < 72.4$), Class IV — strongly polluted water ($72.4 < R_{NPK} < 329$) (or 1st, 2nd, 3rd, 4th degree of underground water quality deterioration by fertilizers). In most cases R_P (phosphoric) and R_{NO_3} (nitrate) are the key indexes that determine the degree of underground water quality deterioration by fertilizers (Wagner and Gajowiec 2000).

The completed map of underground water quality deterioration depicts the conditions of underground water deterioration by fertilizers, showing the occurrence of waters ranging from Class II (pure water) through Class IV (strongly polluted water). However, these two classes do not reveal continuous successions, but Class III (polluted water) is predominant.

The investigations were additionally focused on the Upper Cretaceous underground water with a view to locating poor quality water and determining its origin. These efforts resulted in the preliminary assessment of groundwater quality persistence in time and occurrence, as well as the interpretation of test results from the two sites of the Regional Groundwater Monitoring over a period of ten years 1990–1999.

CONCLUSIONS

The findings of investigations conducted on a regional as well as on a local scale indicate that the excessive use of manure and fertilizers in the Miechów region is an indirect cause of groundwater contamination, mainly by increasing the levels of nitrogen, phosphorus and potassium compounds.

REFERENCES

Guzik, M. and Pacholewski, A. 1997: *Hydrogeological Map of Poland*, 1:50,000 scale. Wolbrom Sheet (914) (in Polish). Polish Geological Institute, Warsaw.

Różkowski, J. 1997: *Hydrogeological Map of Poland*, 1:50,000 scale. Skała Sheet (946) (in Polish). Polish Geological Institute, Warsaw.

Zembal, M. and Liszka, P. 1997: *Hydrogeological Map of Poland*, 1:50,000 scale. Słomniki Sheet (947) (in Polish). Polish Geological Institute, Warsaw.

Wagner, J. and Gajowiec, B. 2000: *Hydrogeological Map of Poland*, 1:50,000 scale. Miechów Sheet (915) (in Polish). Polish Geological Institute, Warsaw.

Włostowski, J. 2000: *Hydrogeological Map of Poland*, 1:50,000 scale. Działoszyce Sheet (916) (in Polish). Polish Geological Institute, Warsaw.

Prażak, J., Janecka-Styrcz, K., Kowalczewska, G. and Paciura, W. 1996: *Report on the quality of underground usable water in the Kielce province on the basis of the monitoring investigations conducted between 1991 and 1995* (in Polish). The Library of Environmental Monitoring, Warsaw.

Hordejuk, T. and Gawin, A.M. 1998: *Quality of underground waters on the basis of the monitoring investigations conducted between 1991 and 1995* (in Polish). The Library of Environmental Monitoring, Warsaw.

Staniewicz-Dubois, H. 1995: *Guidelines on the methodology of creating the regional and local monitoring of underground waters. Modified Second Edition.* The Library of Environmental Monitoring. The State Inspection of Environmental Protection, Warsaw.

Witczak, S. and Adamczyk, A.F. 1994: *Catalogue of selected physical and chemical indexes of underground water pollution, as well as the procedures of their determination.* Volumes I and II (in Polish). The State Inspection of Environmental Protection. The Library of Environmental Monitoring, Warsaw.

The Ordinance of the Minister of Health of 4 September 2000, specifying the standards for drinking water, commercial water use, contact recreation water, as well as the rules for water quality inspections by the Sanitary Inspection representatives (in Polish). *Legal Gazette* No. 82, item 937.

Instructions for the Generation and Computerized Edition of the Hydrogeological Map of Poland in 1:50,000 scale (in Polish). Ministry of Environmental Protection Natural Resources and Forestry. National Fund of Environmental Protection and Water Management. Polish Geological Institute 1999, Warsaw.

CHAPTER 21

Nitrate pollution of groundwater in Flanders (Belgium)

K. Walraevens and R. Eppinger*

Laboratory for Applied Geology and Hydrogeology, Ghent University,
Krijgslaan 281 − S8, 9000 Ghent, Belgium

ABSTRACT: Nitrate pollution of shallow groundwater in Flanders is widespread. The assessment of its distribution is made difficult however because of the redox zoning in groundwater. Indeed, nitrates are stable only in the oxic groundwater zone, in the presence of oxygen. After oxygen has been (largely) consumed, mainly by oxidation of organic matter and sulphides, the redox potential has dropped to a level allowing for nitrate reduction. This nitrate-reducing zone is found at a certain depth, depending on the flow regime and the reduction capacity of the aquifer matrix. After nitrates have been reduced, at still larger depth and lower redox potential, Mn- and Fe-oxides become unstable and reduced Mn^{2+} and Fe^{2+} appear in the solution. At still lower redox potential, sulphates are reduced. It is evident that nitrate monitoring wells should have their screens in the oxic zone, in order to allow conclusions as to whether or not nitrate pollution of groundwater occurs. They should not have their screens below the nitrate-reducing zone, except to confirm the redox zone, and possibly to monitor the shift of redox zones towards larger depths as a function of time. The reduction capacity of aquifers will indeed be depleted as oxygen, nitrates and Fe^{3+} are reduced, resulting in a shift of redox zones towards depth. Especially in lowly reactive aquifers to which highly nitrate-loaded infiltration water is recharged, this may lead to a noticeable shift over the years.

An assessment of the results of the preliminary Flemish observation network for nitrate pollution of groundwater, demonstrated that for only 37% (97 wells) of the 260 wells, groundwater was pumped from the oxic zone, allowing the confirmation of the presence or absence of groundwater pollution. The remaining wells were too deep, or were pumping mixed waters from several redox zones. From the suitable 97 wells, 70% are polluted at a level $>50 \, \text{mg} \, \text{l}^{-1}$.

The issue of vulnerability of groundwater to nitrate pollution, in view of EC guidelines, was approached in relation to the maximum potential contamination depth, stating that the sensitivity increases with increasing depth. The Flemish territory was subdivided into hydrogeologically homogeneous zones with respect to nitrate pollution, in which the characteristics of the shallow aquifer can be considered to be comparable in terms of hydrogeological constitution, hydraulic conductivity, hydraulic gradient, thickness of unsaturated zone, and total reduction capacity (TRC), constituted mainly by organic matter and sulphide content.

For each hydrogeologically homogeneous zone, the maximum potential contamination depth with nitrates was assessed, irrespective of the nitrate load at the water table, and only taking reservoir characteristics into account. The zones with the largest contamination depths are considered to be the most nitrate sensitive.

---

* now at Flemish Community, Ministry of Environment and Agriculture, Department Water, Emile Jacqmainlaan 20, PB 5, 1000 Brussels, Belgium

INTRODUCTION

Nitrate pollution of shallow groundwater in Flanders is widespread. The assessment of its distribution is, however, made difficult because of the redox zoning in groundwater. The precise extent of the nitrate problem is not known by now, due to the absence of an adequate monitoring network.

The Ministry of the Flemish Community is facing the problem. According to the EC Nitrate Directive (91/676/EEC), areas within the Member States where groundwater is containing or could contain more than $50 \, \text{mg} \, \text{l}^{-1}$ nitrates without appropriate measures being taken, should be designated as vulnerable zones. In fact, in the whole of Flemish territory, groundwater could contain more than $50 \, \text{mg} \, \text{l}^{-1}$. But different degrees of vulnerability exist, related to the maximal potential contamination depth, which depends on the characteristics of the shallow groundwater reservoir. In the absence of a well-developed monitoring network, we have proposed a methodology for distinguishing between different classes of nitrate sensitivity. Moreover, the proposed subdivision into hydrogeologically homogeneous zones offers the starting point for a regionally differentiated treatment of the problem.

The assessment of the distribution of nitrate pollution of shallow groundwater is complicated by the redox zoning in groundwater. The analytical results of the preliminary Flemish observation network have been evaluated, taking the redox zoning into account.

HYDROGEOLOGICALLY HOMOGENEOUS UNITS AND ZONES

The occurrence of nitrates in groundwater depends on 3 main factors:
1) the input of nitrates to the water table, mainly by flushing of nutrients from the root zone of agricultural land;
2) the spread of nitrates by transport processes in groundwater;
3) chemical reactions, with denitrification, reducing the spread of nitrates in groundwater.

While the first factor depends largely on land-use practices, the second and third factors (in natural groundwater flow conditions) are independent of anthropogenic influence and are specific for the characteristics of the concerned shallow aquifer system, mainly its flow regime and its chemical reactivity. This provides the basis for subdividing the territory into regions, considering the characteristics of the shallow groundwater reservoir to be homogeneous within a region. On the larger scale, the territory of Flanders was subdivided into hydrogeologically homogeneous units (Figure 1). These are defined as areas in which the structure and main characteristics of the shallow groundwater reservoir with relevance to nitrate distribution are assumed to be similar.

In some parts of Flanders, the shallow aquifer system is mainly or predominantly composed of Quaternary sediments, while in other parts, the Quaternary cover is less dominant because of its limited thickness. This has delivered a first criterion for subdivision:
- in areas where the thickness of the Quaternary sediments is >10 m, only the Quaternary sediments are considered for the delimitation of hydrogeologically homogeneous units;
- in areas with a thin Quaternary cover (<10 m), the underlying Tertiary (or sometimes older) rocks are also taken into account.

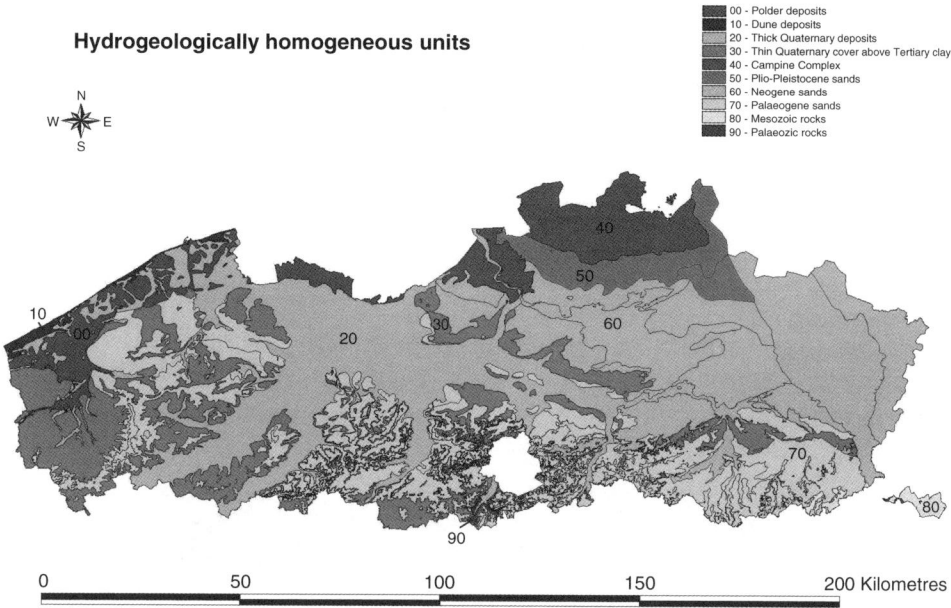

Figure 1. Map showing the hydrogeologically homogeneous units in Flanders.

The following hydrogeologically homogeneous units have been distinguished:
  unit 00: areas with saline groundwater in the shallow aquifer (Quaternary polder areas)
  unit 10: Quaternary dune areas
  unit 20: Pleistocene valleys with thick Quaternary cover
  unit 30: Palaeogene clay layers with thin Quaternary cover
  unit 40: Pleistocene Campine Complex
  unit 50: Plio-Pleistocene sands with thin Quaternary cover
  unit 60: Neogene sands with thin Quaternary cover
  unit 70: Palaeogene sands with thin Quaternary cover
  unit 80: Mesozoic marl and chalk with thin Quaternary cover
  unit 90: Palaeozoic Basement Complex with thin Quaternary cover.

Within a unit, a further subdivision was made, based on geological differences (for example, different ages of Palaeogene clay layers) or other distinguishing features. This has resulted in several hydrogeologically homogeneous zones within a unit (Figure 2). For example, unit 20 (Pleistocene valleys with a thick Quaternary cover) has been subdivided into 3 zones:
  zone 21: Flemish Valley and coastal plain
  zone 22: Meuse-Rhine deposits
  zone 23: High Terrace deposits.

The mapping of Flanders into hydrogeologically homogeneous units and zones was based on existing information. The sources of information used were:
• geological maps
• topographic maps

Hydrogeologically homogeneous zones

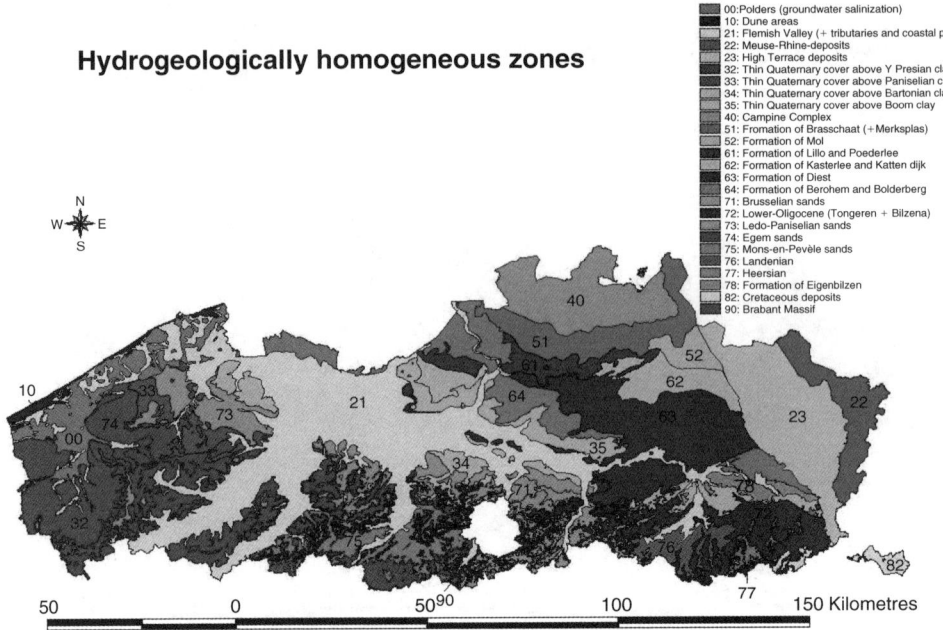

Figure 2. Map showing the hydrogeologically homogeneous units in Flanders.

- map of the depth of the fresh/salt-water interface (De Breuck *et al.* 1974)
- map with thickness of Quaternary deposits
- map of Quaternary deposits of the Flemish Valley (De Moor and Heyse 1978)
- report on the geology of the Campine area (Wouters and Vandenberghe 1994)
- drilling descriptions of different archives.

Five (out of ten) representative hydrogeologically homogeneous units were selected for detailed investigation of the processes controlling nitrate distribution in groundwater (starting from the water table), by means of the implementation of test sites (Eppinger and Walraevens, this paper). The criteria used for the selection of the units were as follows:

1) regional extent: the conclusions should be representative for the major part of Flanders, so regionally important units should be chosen;
2) nitrate input: the idea was to study the fate of nitrate starting from the water table, so units should be chosen where nitrate could be expected to reach the water table at first; therefore, the local production of manure was considered;
3) the shallow aquifer should at least locally serve as an important resource;
4) a reasonable repartition over the Flemish territory was aimed at, with one test site for each Flemish province.

The following five hydrogeologically homogeneous units and zones, with their related test sites, have been selected:

- unit 20: Pleistocene valley, with thick Quaternary cover, with test site Adegem in zone 21 (Flemish Valley);
- unit 30: Palaeogene clay layers with thin Quaternary cover, with test site Torhout in zone 32 (Ypresian clay);

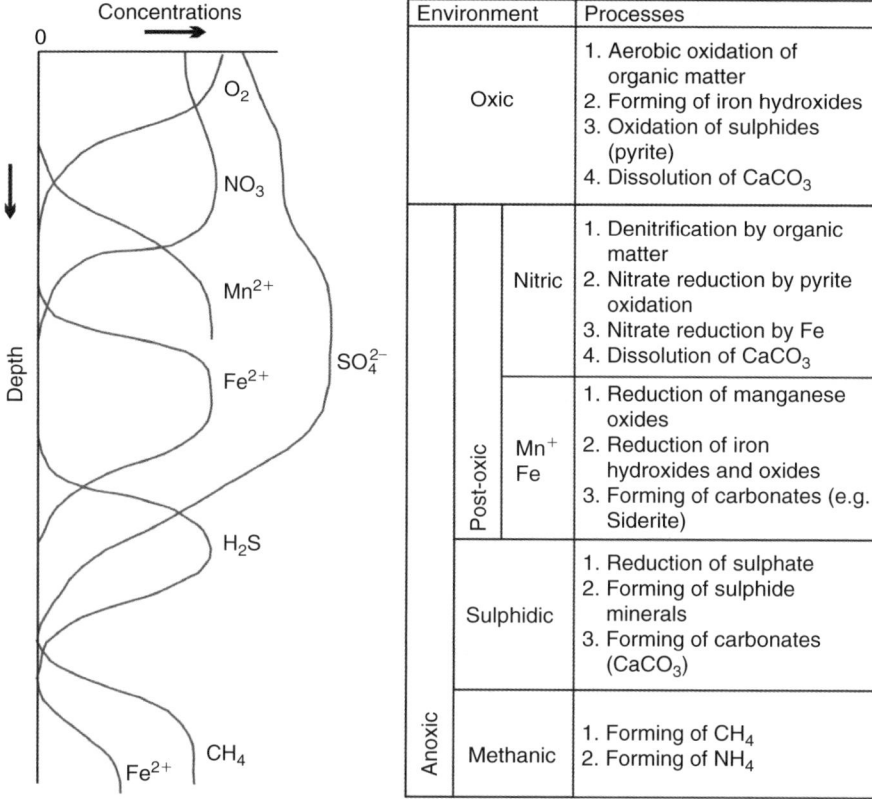

Figure 3. Redox zones in groundwater (modified after Berner 1981).

- unit 50: Plio-Pleistocene sands with thin Quaternary cover, with test site Mol in zone 52 (Formation of Mol);
- unit 60: Neogene sands with thin Quaternary cover, with test site Peer in zone 62 (Formation of Kasterlee);
- unit 70: Palaeogene sands with thin Quaternary cover, with test site Kampenhout in zone 71 (Brusselian sands).

REDOX ZONES

When investigating nitrate distribution in groundwater, it is important to consider the redox zones as a function of depth. According to this concept, with increasing depth, the following redox zones are found subsequently (Figure 3; Berner 1981):

I)   Oxic zone: where O_2 is still present

II) Anoxic zone: after O_2 has been consumed (by the oxidation of organic matter and pyrite), the following reactions can take place consecutively, each one characterizing a further zone in the redox zonation (with conditions gradually turning more strongly reducing):

1)   nitrate reduction: $NO_3^- \rightarrow N_2$

2)   Mn (IV) oxides reduction $\rightarrow$ Mn^{2+}
3)   Fe (III) oxides and hydroxides reduction $\rightarrow$ Fe^{2+}
4)   sulphate reduction $\rightarrow$ sulphides
5)   methanogenesis.

From this redox zoning, important conclusions can be drawn with respect to nitrate pollution and monitoring:

- as long as oxygen is still present, no nitrate reduction can occur;
- from monitoring wells in too strongly reducing conditions (and thus at too large a depth, for example, from the redox zone where Fe^{2+} exists in solution), no conclusions can be drawn as to whether or not nitrate pollution exists here near the water table; thus only wells in the oxic zone allow to distinguish between presence/absence of nitrate pollution. This has important consequences for the characteristics of nitrate monitoring wells. They should have their screen in the oxic zone, except for the purpose of confirming a deeper redox zone, and to monitor a possible shift of redox zones towards depth with time: indeed, the sediment's reduction capacity is depleted as redox reactions proceed and especially in lowly reactive aquifers to which highly nitrate-loaded infiltration waters are recharged, noticeable shifts with depth may be observed over the years.

PRELIMINARY FLEMISH NITRATE MONITORING NETWORK

The preliminary Flemish monitoring network for nitrate pollution of groundwater consisted of 260 pumping wells on farms, with a depth <20 m. For these wells, the redox state of the sampled groundwater was determined, mainly based on the groundwater analyses. It was established that only 97 wells (37% of the total) indeed have their screens in the oxic zone. The other wells could not be used for conclusions about the presence or absence of nitrate pollution.

Out of the 97 suitable wells, 70% contain $>50\,mg\,l^{-1}$ nitrates, and only 11% have $<25\,mg\,l^{-1}$ nitrates. This confirms the assumption that nitrate pollution of shallow groundwater in Flanders is widespread. On the other hand, the number of suitable monitoring wells is much too low to draw definite conclusions. Therefore, the Flemish Community (Administration for Environment, Nature, Land and Water 'AMINAL') is now implementing a new monitoring network of about 2,000 wells.

NITRATE SENSITIVITY OF HYDROGEOLOGICALLY HOMOGENEOUS ZONES

The nitrate sensitivity of hydrogeologically homogeneous zones was based on the depth at which nitrates can penetrate into the soils. It was postulated that the deeper nitrate is able to penetrate, the more sensitive is the hydrogeologically homogeneous zone. Indeed, as nitrates can penetrate deeper:

- the larger the vertical extent of nitrate pollution is;
- the longer the remediation time is;
- the further the lateral flow distance of nitrates may be, as deeper flow pathways, can move further laterally.

In areas where nitrates penetrate only up to a limited depth, this can be due to:

1) the large total reduction capacity of the sediments (high amounts of organic matter or sulphides), such that nitrates are quickly reduced; in this case, nitrates are removed from the environment;

2) the limited vertical downward flow, for example, due to the occurrence of a clay layer; in this case, nitrates persist in the environment, and are transported by groundwater flow to outflow areas, for example, rivers, where they can contribute to surface water pollution.

Three classes of nitrate sensitivity of groundwater were defined, based on the possible penetration depth of nitrates:

<10 m depth: less sensitive
$10-20$ m depth: moderately sensitive
>20 m depth: very sensitive.

This depth dependence is determined by the characteristics of the shallow groundwater reservoir, which are typical for each hydrogeologically homogeneous zone. The considered characteristics were: hydraulic conductivity, hydraulic gradient, degree of oxidation during deposition, thickness of unsaturated zone, thickness of water-saturated zone, absence of effective reduction capacity. A score ranging from 0 (not relevant) to 3 (very strong contribution of this characteristic to nitrate sensitivity of the zone) was attributed to each characteristic for every hydrogeologically homogeneous zone (Table 1). The sum of scores for all characteristics for a given hydrogeologically homogeneous zone gives the total score for this zone, which is classified into one of three classes:

Score: <9: less sensitive
$9-13$: moderately sensitive
>13: highly sensitive.

The following hydrogeologically homogeneous zones (HHZ) were in this way determined to be highly sensitive:

HHZ22: Meuse-Rhine deposits
HHZ23: High Terrace deposits
HHZ76: Landenian
HHZ77: Heersian
HHZ82: Cretaceous deposits.

Other zones are very sensitive in the hilly areas:
HHZ63: Formation of Diest
HHZ71: Brusselian sands
HHZ73: Ledo-Paniselian sands
HHZ74: Egem sands
HHZ75: Mons-en-Pévèle sands.

CONCLUSION

A methodology was developed for classifying the sensitivity of hydrogeologically homogeneous zones for nitrate pollution in terms of depth penetration potential. The most sensitive zones will require special consideration when delimiting nitrate-vulnerable areas as meant by the EC Nitrate Directive.

Table 1. Contribution of hydrogeological characteristics to nitrate sensitivity of hydrogeologically homogeneous zones (HHZ).

HHZ	Hydraulic conductivity	Hydraulic gradient	Oxidation during deposition	Thickness unsaturated zone	Thickness saturated oxic zone	Absence effective reduction capacity	Total score
00-Polders	1	1	1	0	0	0	3
10-Dune areas	3	3	3	2	1	2	14
21-Flemish Valley	3	2	2	1	2	1	11
22-Meuse-Rhine deposits	3	1	3	1	3	3	14
23-High Terrace deposits	3	2	3	2	3	3	16
32-Quaternary on Ypresian clay	2	1	2	1	0	1	7
33-Quaternary on Paniselian clay	2	1	2	1	0	1	7
34-Quaternary on Bartonian clay	2	1	2	1	0	1	7
35-Quaternary on Boom clay	2	1	2	1	0	1	7
40-Campine Complex	2	2	2	1	2	2	11
51-Formation of Brasschaat	2	1	2	1	1	1	8
52-Formation of Mol	3	2	3	1	2	1	12
61-Formation of Lillo-Poederlee	2	1	1	1	1	1	7
62-Formation of Kasterlee	2	2	3	1	3	3	14
62-Formation of Kattendijk	2	1	2	1	1	1	8
63-Formation of Diest	3	1	2	1	2	2	11
63-Formation of Diest in hilly areas	3	3	2	3	3	3	17
64-Formation of Berchem (and Bolderberg)	3	1	1	1	1	1	8

Formation							Total score
64-Formation of Bolderberg in hilly areas	2	3	2	3	2	1	13
71-Formation of Brussels	2	2	1	2	2	1	10
71-Formation of Brussels in hilly areas	3	3	1	3	3	3	16
72-Lower-Oligocene	2	2	2	3	2	1	12
73-Ledo-Paniselian sands	2	1	1	1	2	2	9
73-Ledo-Paniselian sands in hilly areas	2	3	1	3	3	3	15
74-Egem sands (Mont-Panisel)	2	1	1	1	2	2	9
74-Egem sands in hilly areas	2	3	1	3	3	3	15
75-Mons-en-Pevèle sands	2	1	1	1	2	2	9
75-Mons-en-Pevèle sands in hilly areas	2	3	1	3	3	3	15
76-Landenian	2	3	2	3	3	2	15
77-Heersian	2	3	1	3	3	2	14
78-Formation of Eigenbilzen	2	3	1	2	2	1	11
82-Cretaceous deposits	3	3	1	3	3	3	16
90-Brabant Massif	3	3	1	1	2	2	10

Parameter score: 0 – negligible; 1 – little/small; 2 – moderate; 3 – important/strong.
Total score: <9 less nitrate sensitive; 9–13 moderately nitrate sensitive; >13 highly nitrate sensitive.

The results of the preliminary monitoring network for nitrate contamination of groundwater were evaluated in the framework of redox zones. Recommendations were formulated for the implementation of a new monitoring network.

REFERENCES

Berner, R.A. 1981: A new geochemical classification of sedimentary environments. *Journal of Sedimentary Petrology* 51: 359—365.

De Breuck, W., De Moor, G., Maréchal, R. and Tavernier, R. 1974. *Depth of the fresh-salt water interface in the unconfined aquifer of the Belgian coastal area (1963—73).* Map 1:100.000. Military Geographical Institute, Brussels.

De Moor, G. & Heyse, I. 1978. The morphological evolution of the Flemish Valley (in Dutch). *De Aardrijkskunde* 4: 343—375.

Eppinger, R., Van Camp, M. and Walraevens, K. 2002: *Investigation of the spread of nitrates in groundwater in Flanders* (in Dutch). 219 pages. Study financed by Ministry of the Flemish Community. Report TGO 99/18. Ghent University.

Wouters, L. and Vandenberghe, N. 1994: *Geology of the Campine area* (in Dutch). 208 pages. NIRAS, Brussels.

Monitoring and protection of groundwater from nitrate pollution

CHAPTER 22

Agricultural nitrate contamination in groundwater in England and Wales: an overview

I.R. Davey
National Groundwater Regulations Process, Environment Agency, United Kingdom

T.J. Besien
Thames Region, Environment Agency, United Kingdom

S. Evers and R. Ward
National Groundwater and Contaminated Land Centre, Environment Agency, United Kingdom

ABSTRACT: This paper considers the susceptibility of groundwater in England and Wales to nitrate pollution from agriculture. It considers the state of contamination, methods for assessment and means for protection. Voluntary controls for farmers are set out in the 'Code of Good Agricultural Practice' and mandatory measures apply under the Nitrate Directive Action Plan. Restrictions for previously established 'Nitrate Sensitive Areas' have improved groundwater quality in some cases. Subsequently, 68 Nitrate Vulnerable Zones (NVZ) were established under the Directive. The extent of Zones is being reviewed, which may result in the designation of vulnerable zones across most of England in order to protect both surface water and groundwater resources. A comprehensive review of groundwater monitoring is also underway, since current assessment of groundwater quality in England and Wales relies heavily on measurements at public supply boreholes. Geostatistical methods are also being developed to help assess the spread of nitrate contamination.

INTRODUCTION

Groundwater is an important source of drinking water in the UK, particularly in south-east England where it provides up to 80% of drinking water supply (Figure 1). In rural parts of the country groundwater is also used by tens of thousands of households for their private domestic supply.

Baseflow to rivers, of particular importance during the summer months, supports abstractions and helps maintain ecological health and amenity value. It is essential that groundwater be protected from pollution. This paper considers nitrate pollution of groundwater particularly related to agriculture.

IMPLICATIONS OF THE DRINKING WATER DIRECTIVE

The EC Drinking Water Directive (80/778/EEC) introduced a maximum admissible nitrate concentration in drinking water of $50\,\text{mg}\,\text{l}^{-1}$. Prior to this there was an advisory

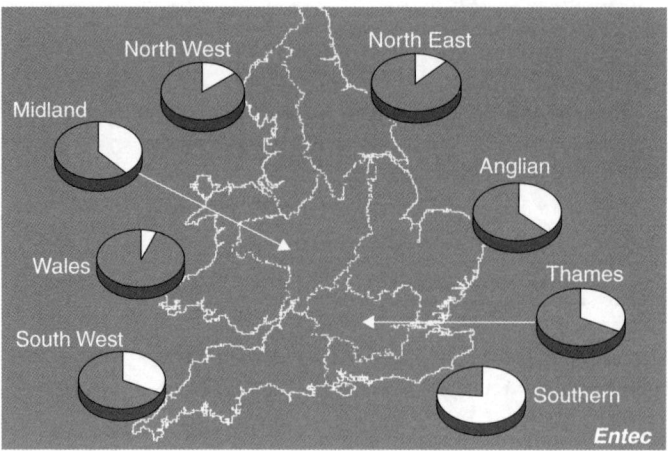

Figure 1. Groundwater use for public supply in England and Wales.

limit of $100\,\mathrm{mg}\,\mathrm{l}^{-1}$ in the UK. When the new limit was introduced, in many cases water suppliers had to blend high and low nitrate water to meet the standard or, in some cases, had to close down water sources used for supply.

Doubts were expressed over the contribution of agriculture to nitrate contamination of water. Many considered that measures for reducing contamination should not be aimed solely at agriculture. The Fertiliser Manufacturers' Association, the National Farmers Union and water suppliers objected to the new standard and maintained that the previous level was acceptable, although it was accepted that excessive concentrations could be harmful.

SOURCES OF NITRATE CONTAMINATION

Several Government studies took place in the 1980s and 1990s, to improve the understanding of nitrate leaching and to identify practical measures to control this. Table 1 summarizes the results of various research (cited in Environment Agency 1996) and shows loadings of N from various sources and the concentration in groundwater that could be related to these loadings.

Atmospheric deposition of N exceeds $30\,\mathrm{kg}\,\mathrm{ha}^{-1}\,\mathrm{a}^{-1}$ in some industrial areas of South Wales and the north-west of England. However, in most of the rural areas where NVZs were initially identified, this figure is below $20\,\mathrm{kg}\,\mathrm{ha}^{-1}\,\mathrm{a}^{-1}$ and in many parts less than $15\,\mathrm{kg}\,\mathrm{ha}^{-1}\mathrm{a}^{-1}$. There is therefore only a limited effect of atmospheric N deposition in these areas.

A simple spreadsheet model was developed as part of this study to estimate the impacts on groundwater from leaching of N for different scenarios. This considered the nitrate loading for each scenario, in particular the amounts that would be expected to leach into groundwater. It assumed an abstraction rate equal to the annual recharge. No account was taken of travel times and equilibrium conditions were assumed.

This shows the contributions of NO_3-N in the abstracted water. It assumes a catchment of $8\,\mathrm{km}^2$ and recharge rates of $200\,\mathrm{mm}\,\mathrm{a}^{-1}$ (abstraction rate of 4.4 ml/day), typical of the

Table 1. Sources of N loading and impacts on groundwater.

Nitrogen source	Loading or application rate (kg N ha^{-1}a^{-1})	N concentration in effluent or leachate*	Groundwater NO$_3$-N concentration (mg l^{-1})
Atmospheric deposition	5–35		
Sewage sludge to land	150–700	40–80 mg l^{-1}	
Losing rivers		7 mg l^{-1}	7 mg l^{-1}
Septic tanks		20–100 mg l^{-1}	8–40 mg l^{-1}
Leaking sewers		32–132 mg l^{-1}	10–100 mg l^{-1}
Highway drainage	3–5	0.2–3 mg l^{-1}	1–3 mg l^{-1}
Landfills		30–115 mg l^{-1}	0–2 mg l^{-1}
Recreational grounds	25–480	0–18.9 mg l^{-1}	
Intensive livestock	420		
Slurry storage		37–68 mg l^{-1}	
Effluent recharge		10–20 mg l^{-1}	2–5 mg l^{-1}

* NO$_3$ or NH$_4$ as N

Table 2. Estimate of impacts on groundwater from N leaching.

Source	N contribution (mg l^{-1})	
	Recharge (200 mm/year)	Recharge (350 mm/year)
Urban areas	0.59	0.40
Highway drainage	0.25	0.25
Recreational grassland	1.25	0.71
Atmospheric deposition	0.5	0.29
Livestock units	0.98	0.56
Landfills	2.5	1.51
Airfield de-icing	3.12	1.78
Losing rivers	2.32	1.70
Septic tanks	0.12	0.07
Winter barley	2.5	
Salad crops	7.5	

drier eastern parts of England, and 350 mm a^{-1} (abstraction rate of 7.7 ml/day), more representative of the wetter western parts.

There are clearly a number of sources that contribute to nitrate contamination of groundwater, although in many cases these are point sources and impacts will tend to be localized. However, the ploughing up of long-term grassland in the second half of the twentieth century, together with the increase in agricultural fertilizers is now accepted as the major cause of the widespread elevated concentrations seen in the aquifers in the UK. It has been estimated that around 80% of nitrate in rural areas in the UK is from agriculture (Rosso Grossman 2000).

The dramatic increase in the use of nitrate fertilizers that took place in the twentieth century is illustrated in Figure 2 (adapted from Environment Agency[1]).

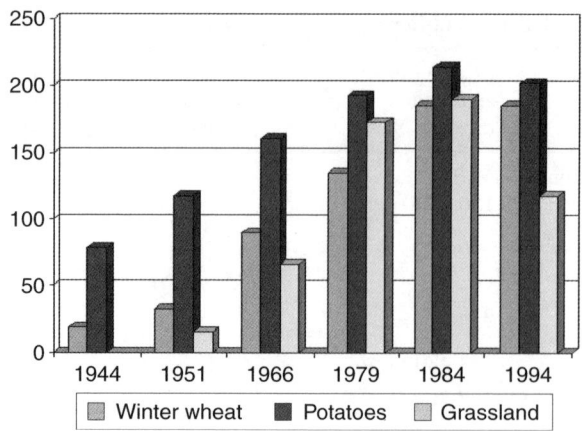

Figure 2. Average amounts of N applied to crops 1944–1994 $(kg\,ha^{-1}\,a^{-1})$.

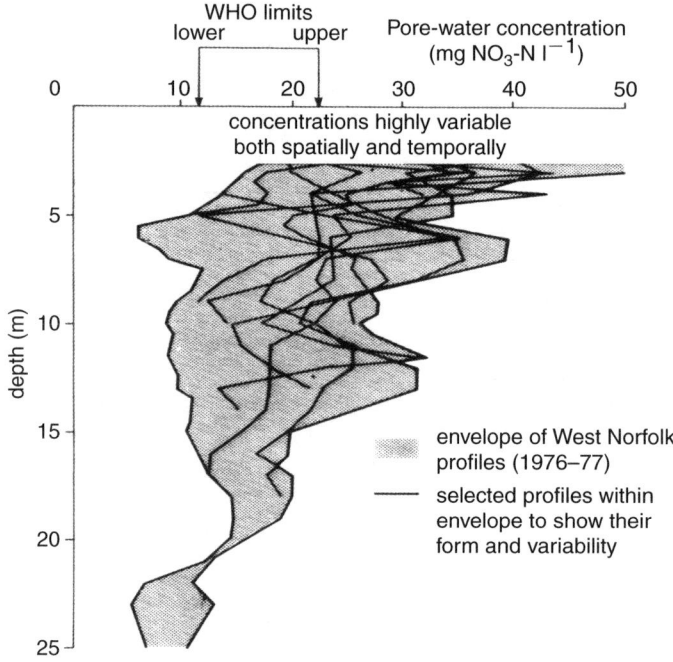

Figure 3. Unsaturated zone profiles from beneath long-standing arable land in the Chalk of eastern England (BGS 1986).

Numerous investigations have been carried out on agricultural land that show how the variation in pore water concentrations of nitrate can be related to changes in the amount of fertilizer used over the years. This is illustrated in Figure 3. These changes are not considered to be due to denitrification.

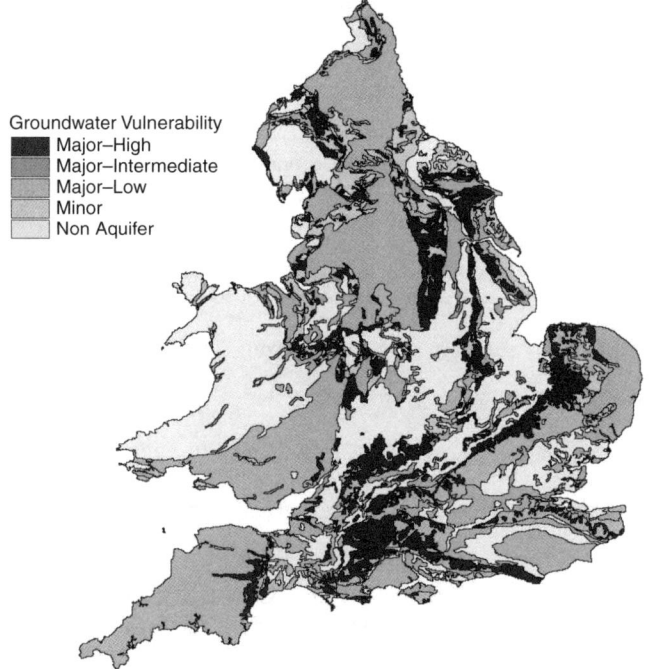

Figure 4. Groundwater vulnerability in England and Wales.

GROUNDWATER VULNERABILITY AND MONITORING

The Environment Agency has published vulnerability maps for England and Wales, which were designed initially to consider the susceptibility of groundwater to contamination by nitrate. The surface geology is considered as major, minor or non-aquifers depending on the ability to yield water for abstraction, which is a reflection of flow mechanisms and rates. This is considered together with the leaching characteristics of the overlying soils. Figure 4 shows that the most important aquifers in the south and east of England are also the most vulnerable.

The most vulnerable areas also coincide with the areas of least effective rainfall, which are also the areas used most intensively for arable farming, with the associated high use of fertilizers.

NITRATE SENSITIVE AREAS (NSA) SCHEME

This pilot scheme was introduced on a trial basis in catchments initially to ten public supply boreholes in 1989 (MAFF 1993) and was expanded to include a further 22 areas in 1994. Catchments were selected on the basis of existing high nitrate concentrations and their rapid response to rainfall. It could therefore be expected that reductions in N fertilizer use would be followed rapidly by reductions in N concentrations in the abstracted water.

Table 3. Costs for Nitrate Sensitive Areas Scheme.

Option	Requirement	Payment ($\text{€ ha}^{-1}\text{a}^{-1}$)	Cost of NO_3-N reduction ($\text{€ mg}^{-1}\text{l}^{-1}\text{ha}^{-1}\text{a}^{-1}$)
Basic	• Reduce N fertilizer use below economic optimum • Introduce cover crops over winter • Apply $<175\,\text{kg a}^{-1}$ N in organic manure • Not to apply poultry manure July to October	88–152	1.76
Premium A	Not to apply fertilizer or manure and not to graze grassland	448–608	4.14
Premium B	Not to apply fertilizer or manure; grazing permitted	400–560	3.92
Premium C	Not to apply more than $150\,\text{kg N ha}^{-1}$; grazing permitted	272–432	3.15
Premium D	No fertilizer; tree planting (Farm Woodland Scheme)	288–448	3.82

Farmers were compensated financially for voluntarily restricting N fertilizer use, the amount of compensation depending on how much of the farm was included in the scheme and what level of reduction in N use was achieved. The options were given in Table 3.

Table 3 also shows an estimate of the cost of reducing nitrate concentrations in water abstracted at boreholes. It assumes average NSA payments per ha, and long term effects arising from 250 mm drainage each year (Manteuffel Szoege, Crabtee and Edwards 1996).

Considerable monitoring of groundwater and also of pore waters has been carried out in some of these Nitrate Sensitive Areas. Results have varied because of variable rates of participation in the voluntary scheme, as well as the differing hydrogeological and climatic conditions. Besien, McCubbine and Chalmers (2001) have described results in one of these Areas, a limestone catchment, where there has been a significant decline in nitrate concentrations (Figure 5).

The reversal of the rising trend in nitrate concentration demonstrates that major long-term changes in farming practices can be effective in improving groundwater quality. However, there have been more mixed results in other catchments.

NITRATE DIRECTIVE

Under the Nitrate Directive, 68 Nitrate Vulnerable Zones (NVZs) were identified in 1993, taking into account both surface water and groundwater considerations. Groundwater catchments were included on the basis of existing or likely breach of the drinking water standard at public supply boreholes. This was assessed by simple trend analysis in the first instance.

Catchments to boreholes can vary considerably with abstraction rates or seasonal changes, an example of which is shown in Figure 6. This needs to be considered when defining the field areas on the ground within which pollution prevention measures need to apply under the Directive.

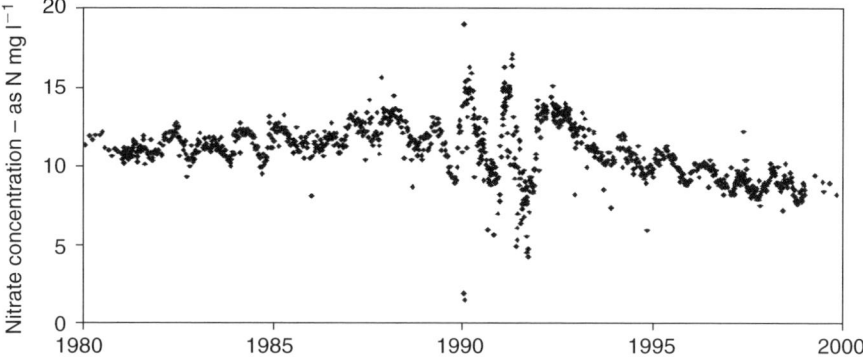

Figure 5. Old Chalford public water supply spring.

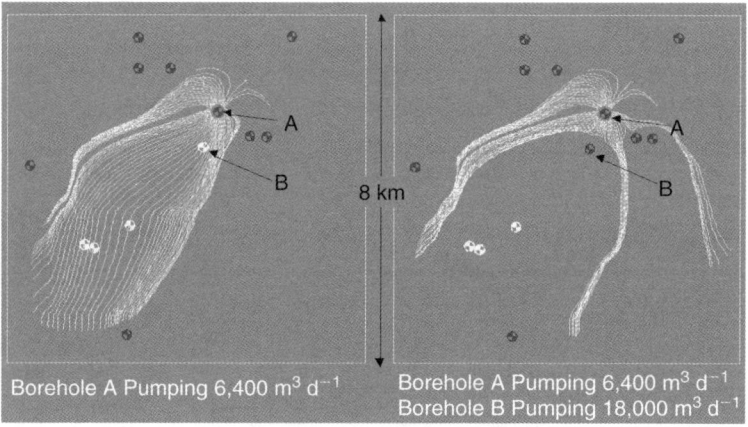

Figure 6. Changes in protection zones with abstraction rate.

To accommodate this, a 'zone of confidence' (ZOC) was identified, which *always* supplies the borehole with water, and a 'best estimate zone' (BEZ), which *usually* supplies the borehole with water. For enforcement of pollution prevention programmes, fields within the ZOC are included in the NVZ unless more than half the field falls outside the BEZ.

Figure 7 shows the distribution of NVZs in England and Wales, the dots showing the groundwater zones (MAFF 1994), which were formally designated in 1996. (The diagram also shows the Ministry of Agriculture Farming and Fisheries regional boundaries.)

PREVENTING NITRATE POLLUTION BY GOOD AGRICULTURAL PRACTICE

A Code of Good Agricultural Practice was published in 1991 for England and Wales to encourage voluntary good practice. Although this included guidance on reducing nitrate

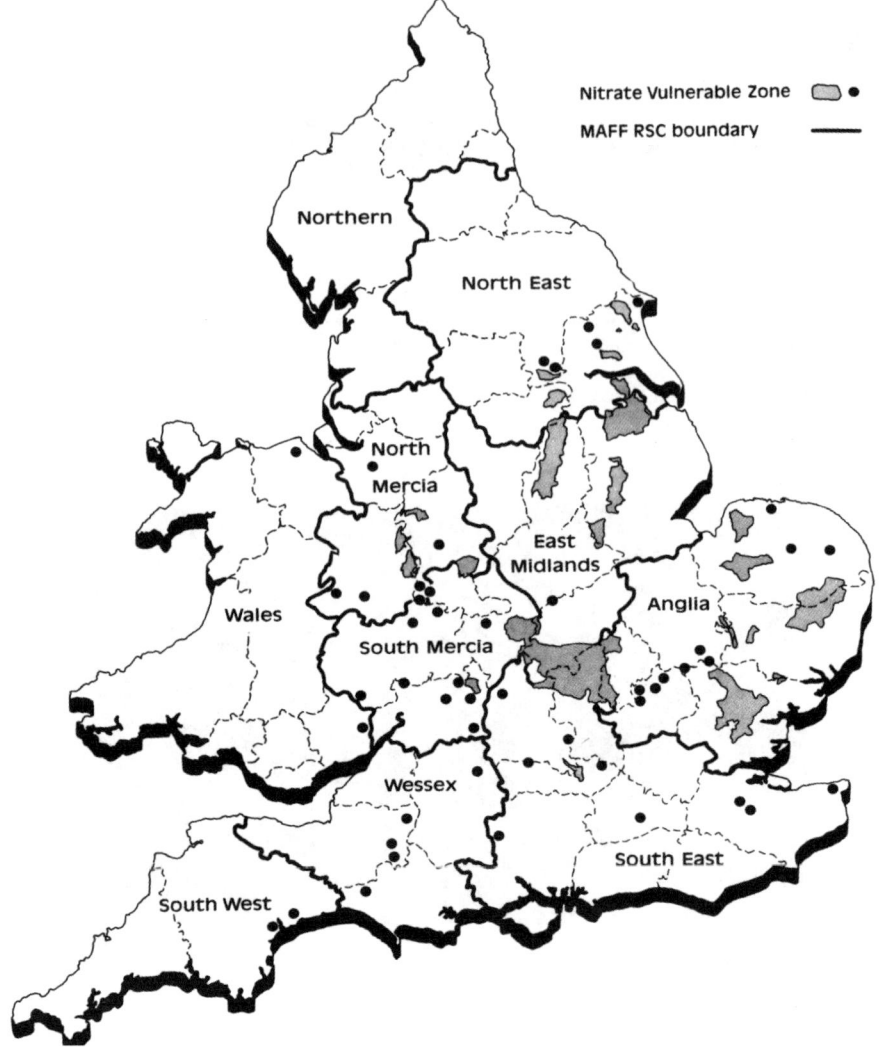

Figure 7. Nitrate Vulnerable Zones in England and Wales.

leaching, it was not specifically directed at the Nitrate Directive. A revised version of the Water Code was published in 1998 (MAFF 1998) specifically for the purposes of the Nitrate Directive. Farmers in NVZs must comply with the mandatory measures in the Water Code, as well as rules that apply specifically to NVZs (MAFF 1998). For other farmers compliance with the code is voluntary.

In summary, farmers within NVZs must:

- not apply fertilizers or organic manures within autumn and winter closed periods
- not apply more than $250\,kg\,N\,ha^{-1}$ from organic manures each year
- only apply N when it can be used by the crop
- not apply N to flooded, frozen or steeply sloping fields

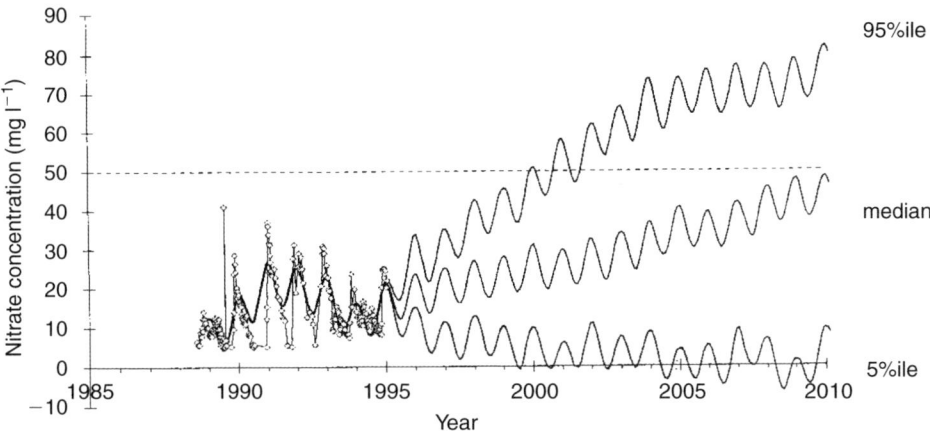

Figure 8. Probabilistic extrapolation of nitrate data.

- not apply N fertilizers or organic manures in a way that contaminates watercourses
- ensure sufficient slurry storage for the autumn closed period
- keep adequate records of all N use

The Environment Agency is responsible for enforcement of the mandatory measures in NVZs.

RECENT MONITORING AND ASSESSMENT

A review of NVZs was begun in the mid-1990s and more sophisticated modelling assessments than previously used were considered. Piecewise linear regression was trialled initially, which took into account fluctuations in nitrate concentrations because of climatic or other variations evident in the historical record. The method was used to extrapolate the often limited data sets by 'statistically mimicking' the underlying but continuously changing trend, to estimate the probability that concentrations would exceed the $50\,\text{mg}\,\text{l}^{-1}$ limit in future years. Figure 8 shows the results of simulations for the median, 5th percentile and 95th percentile paths.

Before the review was established fully, a more comprehensive assessment of the implementation of the Directive was instigated by the UK Government. This was a response to the European Court of Justice Judgement, which considered that the UK's reliance on monitoring data only from public supply abstractions was insufficient for the purposes of the Directive. It was considered that this would not give a proper view of the impact of agricultural nitrate on the overall groundwater resource.

As a consequence, the nitrate-monitoring network has been expanded to about 3,770 boreholes (Figure 9), of which 33% are non-public supply sources, although this is still considered by the UK to be insufficient.

A comprehensive review of the groundwater monitoring strategy is currently underway, to improve the density of distribution of monitoring boreholes in order to satisfy the needs of both the Nitrate Directive and the Water Framework Directive. However, groundwater lies at significant depth in many of the aquifers in England and Wales. Availability of

Figure 9. Nitrate monitoring network.

funds will limit the numbers of new monitoring boreholes that can be installed and there
will be continued reliance on existing boreholes, wells and springs.

In the meantime, other techniques, such as geostatistical methods, have been developed
to support assessment of the spread of nitrate contamination (Evers, Fletcher, Ward *et al.*
2001). Data from the nitrate-monitoring network were used to support this assessment.
Gaps in the network were filled by extrapolating between points, taking account of
groundwater vulnerability and using kriging techniques.

REVIEW OF NVZS: GOVERNMENT CONSULTATION

From a consideration of groundwater and surface water issues, the total area assessed as
being appropriate for designation as NVZs in England would be about 80%. The
Government has recently undertaken a national consultation on whether parts of the
country identified should be designated in this way, or whether all of England should be
designated. (Following devolution of some powers, Scotland, Wales and Northern Ireland
are making separate proposals). Figure 10 indicates the approximate areas that may be
designated on the basis of existing or likely nitrate contamination of groundwater (Evers

Figure 10. Potential NVZs from groundwater considerations.

et al. 2001). Figure 11 shows the potential combined areas for on the basis of groundwater and surface water considerations (DEFRA 2002).

CONCLUDING POINTS

The UK has conducted many studies and trials to understand nitrate leaching and identify suitable approaches to reduce contamination from agricultural sources. The costs of prevention are considerable. The UK Government has estimated that the costs which will arise to the agricultural community for designating the all of England would be about 50 million Euros and about 36 million Euros for designating 80% of the country. However this needs to be considered in relation to the costs for continued treatment of drinking water to remove nitrate, or, indeed, the replacement of sources.

Treatment for nitrate contamination is already undertaken for many public supply abstractions from boreholes. However under the Nitrate Directive, this is not acceptable as a substitute for prevention of pollution at source. It is hoped that measures being introduced within NVZs will alleviate nitrate contamination from agriculture. Undoubtedly though, treatment facilities will need to be installed at further sites to

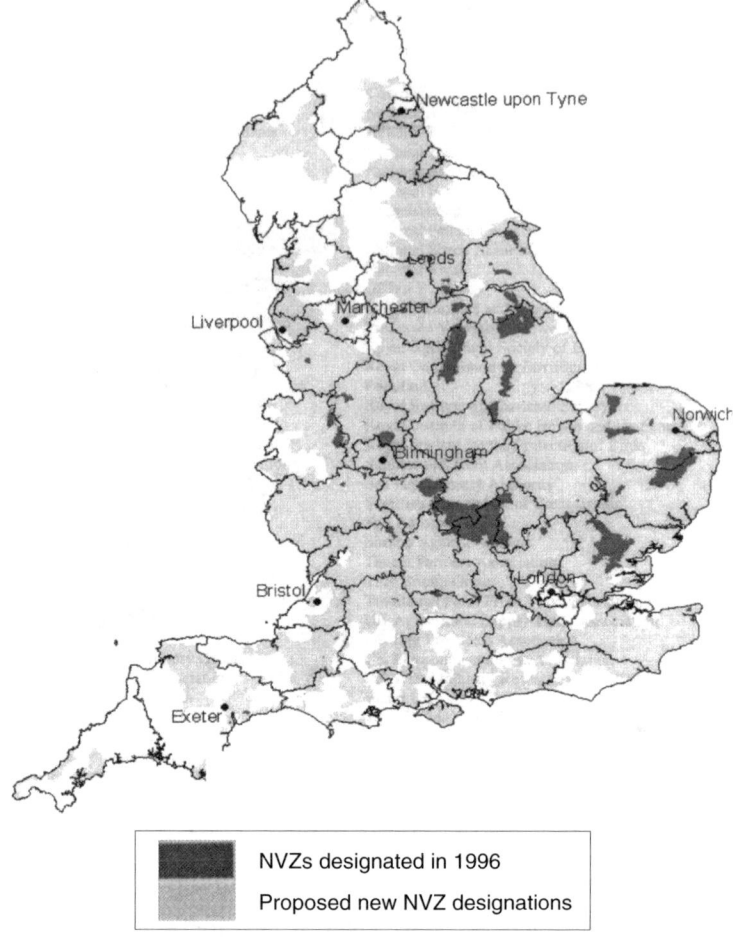

Figure 11. Potential NVZs from groundwater and surface water considerations (England only).

tackle not only current contamination but also to deal with rising trends before prevention measures become effective.

REFERENCES

80/778/EEC Council directive relating to the quality of water intended for human consumption.
91/676/EEC Council directive concerning the protection of waters against pollution caused by nitrates from agricultural sources.
Besien, T., McCubbine, S. and Chalmers, A. 2001: *The effects of the nitrate sensitive areas scheme on groundwater nitrate concentrations at Old Chalford, West Oxfordshire, England.* In: *Future groundwater at risk*, IAH Conference Proc June 2001, Lisbon.
BGS 1986: The groundwater nitrate problem: A summary of research on the impact of agricultural land-use practices on groundwater quality between 1976 and 1985. *Hydrogeol. Rep. Br. Geol. Surv.*, No 86/2.

DEFRA 2002: How should England implement the 1991 Nitrates Directive? A consultation paper. Department of Environment, Food and Rural Affairs, UK.

DOE 1994: UK review group on the effects of nitrogen deposition on terrestrial ecosystems. Department of Environment, UK.

Environment Agency[1] 1999: Understanding Rural Land Use, Environment Agency of England and Wales, UK.

Environment Agency 1996: R&D Technical Report P32: Identification and quantification of groundwater nitrate pollution from non-agricultural sources, Environment Agency of England and Wales, UK.

Evers, S., Fletcher, S., Ward, R. and Harris, R. 2001: National Groundwater and Contaminated Land Centre Report NC/7079NVZ: Development of a new groundwater NVZ definition methodology, Environment Agency of England and Wales, UK.

MAFF 1993: Solving the nitrate problem. MAFF Publications.

MAFF 1994: Consultation Document: Designation of Nitrate Vulnerable Zones under the EC Nitrate Directive (91/676). MAFF Publications.

MAFF 1998: Guidelines for farmers in NVZs. MAFF Publications.

MAFF 1999: The Water Code. MAFF Publications.

Manteuffel Szoege, M., Crabtee, R. and Edwards, A. 1996: Policy cost effectiveness for reducing non-point agricultural groundwater pollution in the UK. *Journal of Environmental Planning and Management* 39(2): 205–222.

Rosso Grossman, M. 2000: Nitrates from agriculture in Europe: the EC Nitrates Directive and its implementation in England. In *Boston College Environmental Affairs Student Publications*, v27 No.4 pp 567–630.

UK Groundwater Forum 1998: Groundwater: our hidden asset, *Br. Geol. Surv.* UK.

CHAPTER 23

Can we manage the nitrate concentration in the Ljubljana Field aquifer (Slovenia)?

B. Bračič-Železnik
Vodovod Kanalizacija Public Utility, Ljubljana, Slovenia

ABSTRACT: Ljubljana Field is an aquifer with intergranular porosity and a dynamic capacity of $3.5\,m^3\,s^{-1}$, making it one of the biggest drinking water resources in Slovenia. It is a flat area where the interests of agriculture, urbanization, traffic, industry and public water supply all intersect. Influences of all the activities present in the area are reflected in the quality of the groundwater. Groundwater monitoring has been performed for several decades. Nitrate is one parameter that has varied over time and is regularly increasing. There is 1,990 ha of agricultural land with intensive vegetable production inside drinking water protection zones. Whilst the quality and quantity of vegetable production depend on the use of fertilizers and plant protection measures, farmers often exaggerate the amount of agrochemicals. Despite legal acts limiting the input of nitrogen fertilizers inside protection zones, nitrate concentrations in the groundwater have continued to increase. However, agriculture is not the only source of groundwater nitrate pollution. The old and leaky sewer system and areas lacking sewerage contribute significantly to nitrate concentrations. Some activities have been started in order to lower nitrate concentrations in the groundwater. The results of restoration and control mechanisms, such as groundwater quality monitoring and farmer education indicate that it is possible to restore and manage the nitrate concentration in the groundwater.

INTRODUCTION

Due to both its central position in the state and its geographical location, the Ljubljana Field gravel plain is situated where the interests of urbanization, industry and agriculture intersect. Ljubljana, the capital of Slovenia, has developed over the last hundred years from a town of 20,000 inhabitants to a city of 300,000 inhabitants. The small town situated at the foot of the Castle hill has expanded over the great part of the Ljubljana Field and merged with villages in the vicinity.

The intergranular porous aquifer of Ljubljana Field is the most important source of the drinking water for the city. The public water supply dates from the year 1890, when the first water field, Kleče, began operation. Today the drinking water supply is sourced from five water fields, four of them situated on the Ljubljana Field (Kleče, Hrastje, Šentvid and Jarški Brod). In the past the water fields were situated far away from the town, but nowadays are surrounded with the intensive urban areas and industrial zones. Today the 41 million m³/year of groundwater pumped still has suitable sanitary quality. Increasing pumping has enlarged the catchment zones and today the city is situated over its source of drinking water, but in consequence arrangements have been implemented for protection and for unexpected pollution affecting the groundwater quality.

Nitrate concentration in the groundwater of Ljubljana Field is still within allowed limits. Groundwater quality monitoring indicates nitrate concentrations exceeding recommended levels in the Hrastje and Kleče water fields. These both have intensive agriculture and settlements without sewerage systems in their catchment zones.

1,990 hectares of agriculture land are situated on the second water protection zones and define the limitations and prohibitions. The city, as a major consumer of fresh vegetables, fruit and flowers, stimulates the intensive local production. Agricultural loading of the environment, as the main suspect for nitrate and pesticide contamination of the groundwater, has been the subject of many researches. Agricultural impacts can be divided into two groups: point sources of pollution (stables with storehouses for the manure, silage, greenhouses etcetera) and diffuse sources of pollution (agricultural land with leaching of fertilizers and pesticides spread over the surface).

LJUBLJANA FIELD

The groundwater body of Ljubljana Field is built up from of an intergranular aquifer that extends over an area of 95 km^2. The groundwater body is 18 km long, 8 km wide and 35 to 100 m thick. It is one of the biggest and most important aquifers in Slovenia.

The area of Ljubljana Field is a bowl-shaped tectonic sink consisting of river sediments reaching to a thickness of more than 100 m in the deepest part. The impermeable base of permocarbonic shaley mudstone and sandstone began to sink in the Quaternary period. In the last million years, the River Sava filled up the basin, thus forming a field. The lower part of the aquifer is composed of Pleistocene gravel and sand whilst in the upper part there are Holocene sediments of sand and gravel. Amongst the alluvium of sand and gravel on the Ljubljana Field there are several layers of conglomerate lenses. Above the lenses, there are many clay deposits which, together with conglomerates, represent a barely permeable complex. However, on the other hand, the conglomerate in which caverns are formed due to dissolution of carbonate gravels represents a medium with high horizontal hydraulic conductivity.

The flat area of Ljubljana Field is surrounded by hills built up from low resistance permocarbonic schist's and sandstone. In a few places on the west side, Triassic limestone and dolomite with high horizontal permeability is present.

Hydrogeological characteristics

The hydraulic conductivity of Ljubljana Field is very high, from 10^{-2} m s^{-1} in the central part to $3-7 \times 10^{-3}$ m s^{-1} on the borders of the field and $1-5 \times 10^{-4}$ m s^{-1} at the foot of the hills. Groundwater flow velocities cover a wide range of values according to the infiltration of precipitation and the River Sava, and reach several metres to several tens of metres per day. The River Sava flows by the north side of the Ljubljana Field. It has an alpine river character and is an important hydrodynamic element of the area. Its transversal recharge along the Brod-Roje and Tomačevo-Jarše infiltration areas is very important in the time of high water level. Along the River Sava the groundwater is shallow below the surface, from 4 to 8 m in depth. Groundwater is the deepest on the Pleistocene terrace in the west part of Ljubljana Field, situated at depths of 20−30 m.

Hydrometeorological characteristics

Ljubljana Field has a south alpine climate with temperate continental characteristics. Abundant annual precipitation, temperate hot summers and temperate cold winters are prominent. The mean annual temperature is 11.5°C, the annual range of air temperature is 21°C, and the daily temperature range is between 9.5° and 11°C. Mean annual precipitation for the period 1981–1990 was 1,351 (mm/year). Average monthly precipitation is between 80–155 mm in June.

Precipitation is lowest in the winter and early spring, and highest from June to November. In the summer the majority of precipitation falls as heavy rainfall. From the view of protecting the groundwater from nitrate leaching, large quantities and spatial dispersion of precipitation are very inconvenient, as we know that on the protection zones the soil is mostly very permeable. Precipitation is the most important element influencing the quality and quantity of the leaching of pollution from the surface to the groundwater.

Pedology

There are two terraces on the Ljubljana Field: the lower terrace situated near the River Sava and the upper Pleistocene terrace. On the upper terrace soil has developed on the carbonate gravel sandy alluvium. On the lower terrace, carbonate, sandy clayey riverside soil has developed. On the upper terrace, redzina (marly soils) of up to 40 cm depth have developed. These are very permeable and sensitive to drought. Although saturated with bases, especially with calcium, it contains a considerable amount of humus, giving the soil A stable structure that binds fertilizers very easily.

On the Šentvid water field area, medium to deep carbonate riverside soil is present. This soil is appealing for agriculture because of the high groundwater level and satisfactory water regime, even in the dry season.

On the Kleče water field area redzinas of 40–45 cm depth and eutric brown soils of 50–80 cm depth are present. On the Hrastje water field area, medium to deep carbonate riverside soils and eutric cambisol are present.

Groundwater pollution depends on the soil type (texture, thickness, organic substance and clay mineral content) as well as the precipitation regime. Leaching will occur if the percolated water exceeds the field capacity.

Aquifer vulnerability

Ljubljana Field aquifer vulnerability is defined by its natural characteristics:
- The soil above the aquifer is sandy and very permeable.
- The unsaturated zone is thick (3 m–20 m), very well aired and permeable.
- Ljubljana Field is an intergranular aquifer with good hydraulic conductivity.
- The River Sava infiltrates to the groundwater, contributing 50% of it's dynamic stock.

Risk assessment for the Ljubljana Field groundwater depends on anthropogenic influences and interventions into groundwater body, such as:
- Diffuse sources of contamination (fertilizers, irrigation, traffic).
- Point sources of contamination (discharging cesspools, cesspits, greenhouses).
- Pumping.

- Interventions into groundwater body or catchments area that influence the natural aquifer sensitivity.

With regards to recorded facts the Ljubljana Field aquifer has a high vulnerability and is assessed as high risk.

THE ORIGIN OF NITRATE

The biochemical cycle of nitrogen in the hydrosphere is closely linked with the hydrological cycle. Interaction of nitrogen ions with water, acting as a transport medium in the hydrologic cycle, influences the anthropological input of nitrogen (from urban, industrial, agricultural; disperse and point sources). For effective drinking water source management in urbanized areas with developed industry and agriculture, knowledge of the anthropological sources of nitrogen is necessary. Transformation of nitrogen in the hydrological cycle is influenced by numerous processes from atmospheric deposition, outflow, leaching and infiltration, to evaporation, transpiration, percolation and microbiological and exchange reactions along the groundwater flow. Notably, it is influenced by processes from both planned and unplanned human interventions. Agriculture (manure and fertilizers) and urbanization (uncontrollable outflow from sewage system and cesspools) are quantitatively the most important sources of nitrogen in the groundwater of Ljubljana Field aquifer. The soil is the most dynamic environment, where the transformations of nitrogen ions are very fast. Together with physical flow transport, the most important processes influencing nitrogen transfer inside the system are denitrification, nitrification, volatilization, assimilation and, to a minor extent, nitrogen fixation.

Monitoring on state, community and company levels has measured nitrate, nitrite and ammonium concentrations in the groundwater. Due to the high expense, isotopic analyses of nitrate ions that could be used to define more precisely the origin and the source of the pollution have not been carried out.

Natural isotopes of $\delta^{15}N$ and $\delta^{18}O$ in dissolved nitrates may be used to define the nitrogen origin regardless of transformation, compounds and oxidation conditions. Knowing the origin and source (artificial and natural fertilizers, sewage system, contamination from the surface water, traffic and atmosphere) of nitrates is important for sanitation planning. The use of isotope ($\delta^{15}N/\delta^{14}N$), microbiological and chemical markers has provided clear evidence of sewage inputs.

Basic statistical data analysis suggests an estimated natural background nitrate level of 1.7 to 5.6 mg l^{-1} NO$_3$. Pollution is considered to begin at 7.2 mg l^{-1} NO$_3$.

THE NITRATE CONCENTRATION IN THE LJUBLJANA FIELD AQUIFER

The nitrate concentration in the groundwater is changing over time, so we must continuously monitor the groundwater quality and try to determine any relation with other observations (geological, hydrogeological, hydrological and meteorological) in order to investigate and understand the reciprocal effects. The water distributed through the water supply system must meet the regulations on health requirements for drinking water. Therefore, it is necessary that we monitor the groundwater quality in the catchment areas in order to follow pollution trends.

The groundwater quality varies spatially with the position of sampling points. To the north of the Kleče water field, where the infiltration of the River Sava to the groundwater is intensive, the concentration of measuring sites is low in comparison with the sampling points on the southern part of Hrastje water field, where part of the groundwater is coming from beneath the city. The concentrations of ions such as calcium, magnesium, sodium and potassium in the groundwater are increasing proportionally along the flow path through the aquifer and consecutively with the greater part of precipitation. Their origin is partly anthropogenic. Elevated concentrations of nitrates, sulphates, organic compounds and microelements in the groundwater are undoubtedly of anthropogenic origin. The concentration of any particular parameter is related to the water level and precipitation. Precipitation washes pollution from the surface to the deeper levels. Changes in parameter concentrations in the groundwater depend on the quantity of pollution present at or close to the surface (leakage of waste water from pervious sewage system), as well as the intensity and duration of precipitation. In the dry season, when virtually no vertical pollution inflow is present, the groundwater level decreases, causing increased concentration of pollutants within the saturated zone. The extent of the rise depends on the level of pollution present in the aquifer.

Groundwater monitoring has been established here since 1987. Maximum nitrate concentrations have not been exceeded in any samples to date, even on the particular well fields where increasing nitrate concentration has been noticed. Analyses of nitrate concentrations have become part of regular drinking water tests since 1992. Today three complementary groundwater-monitoring bodies are established. Monitoring on the state level has been carried out twice a year at nine sample sites since 1992. Groundwater quality has also been monitored on the community level at the same sites since 1997. This extended monitoring includes some elementary physical–chemical and bacteriological analyses. Parameters have been selected taking into account past analyses. Samples have been taken 6–24 times per year. The sampling frequency depends on the importance of the sampling site for the public water supply and past appearance of pollutants. Monitoring financed by the water supply company is carried out twice a year on sample sites that are not incorporated in the state monitoring, helping complete the picture of groundwater quality.

To date, nitrite and ammonia concentrations have always been under the limit of detection. Elevated concentrations of nitrate have been detected in the area of water field Hrastje, where the additional analyses of groundwater quality indicate that rising nitrate concentrations are not just a consequence of intensive agriculture in the vicinity of the water field, but are also due to the leaky sewage system and use of cesspools in the villages upstream from the Hrastje water field. **Even the regular groundwater quality observation in the single wells indicates that according to the present state, the permitted concentrations in groundwater will be exceeded within a few years**.

Whilst the concentrations of other types of pollutants (VOC, atrazine, Cr^{+6}) in the aquifer on the south side of the water field Hrastje are comparable with the concentrations in the pumped water, the nitrate concentration in the south side is lower than that in the pumped water. In comparison to the other sampling sites, nitrate concentrations on the west part of Kleče water field are higher, but are not rising. The reason for this could be that the west part of the water field is recharged mostly by precipitation.

Nitrate concentration varies between individual wells in Kleče water field. The highest nitrate concentration was detected in 1987 in well XI (45.2 mg NO_3/l). This well has the highest concentration among the wells that are occasionally monitored. The groundwater

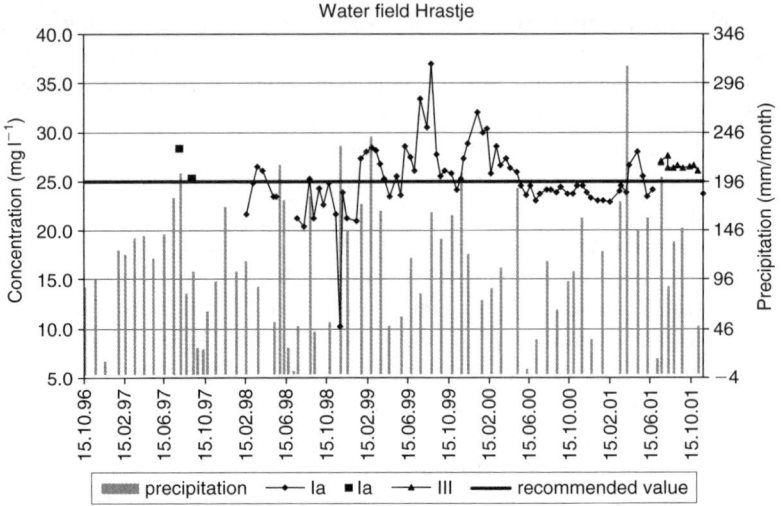

Figure 1. Nitrate concentration in the groundwater pumped from well Ia and well III in Hrastje water field.

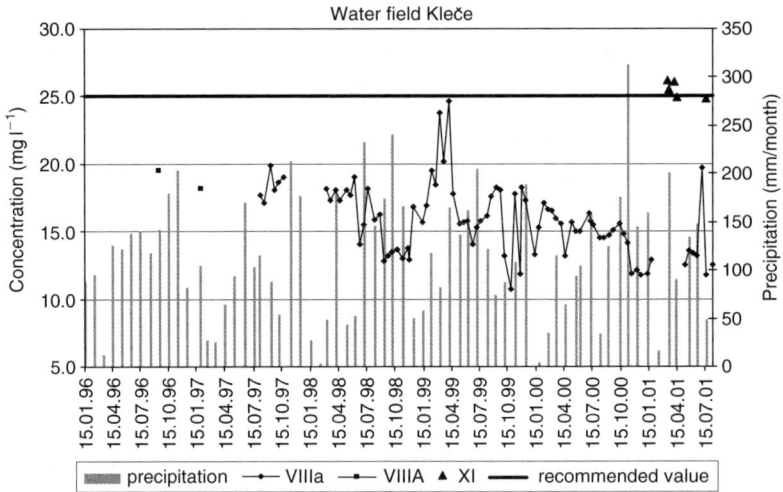

Figure 2. Nitrate concentration in the groundwater pumped from well VIIIa and well XI in Kleče water field.

pumped in Šentvid water field never has over 25 mg NO_3/l. Chemical analysis indicates stability over recent years.

MONITORING OF PLANT NUTRIENTS IN THE SOIL

The beginnings of plant nutrient monitoring on the second water protection zone of the Kleče, Hrastje, Jarški Brod and Šentvid water fields date back to 1996 and 1997. Sampling

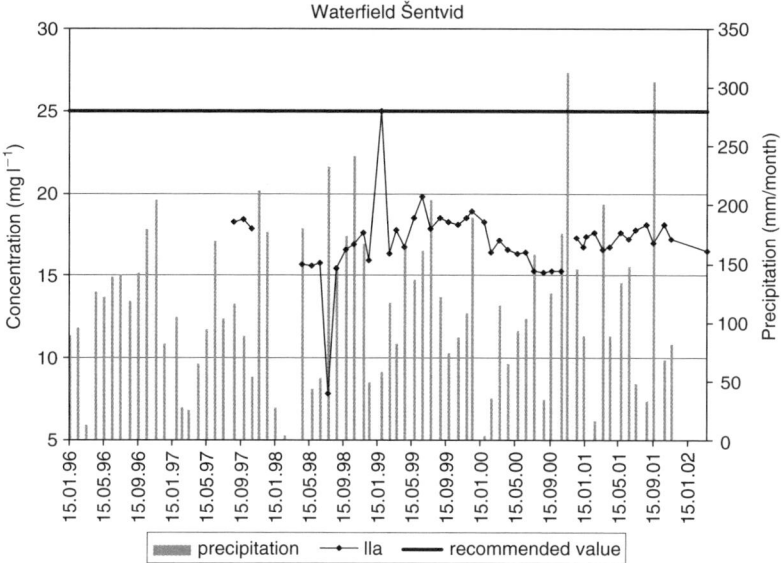

Figure 3. Nitrate concentration in the groundwater pumped from well IIa in Šentvid water field.

was performed on 50 sites in autumn and springtime. The samples were taken from depths of 0–30 cm, 30–60 cm and 60–90 cm. The content of mineral nitrogen was determined. The abundance of nitrates in the soil increased before the beginning of winter on one separate field only. The mineral nitrogen content was observed to decrease in less than half the samples during the winter period.

On the Hrastje waterfield area, where the highest concentrations of nitrates in the groundwater were detected, the content of mineral nitrogen has decreased in one sample only. The decreased content of the nitrate in soils in the time period from December 1996 to February 1997 indicates that during the winter period some nitrate was washed from the upper soil profile into the deeper layers of the gravel base. It is expected that it reached the groundwater level. Results also indicate that for the high concentration present in the groundwater, sources other than just the leaking sewage system may also be responsible.

To prevent the excess intake of fertilizers into the soil, 173 soil analyses were performed in 1999 and 216 in 2000. In the year 2000 fertilizing tests were carried out on four locations on the water protection zones, studying the optimal dosage of nitrogen in the planting period and additional fertilizer application during growth. Additionally, equipment for rapid determination of soil mineral nitrogen content was acquired, enabling determination of the actual immediate local nitrogen requirements of the plants. This method allows the dosage to be kept within the recommended quantity that can be used by plants, therefore preventing increases in the groundwater.

In 2001 monitoring of plant nutrition in soils of the second water protection zone of four water fields was introduced. The aim is to establish a permanent unified control system that monitors the abundance of plant nutrition substances. Results will be used for consultant services that will give information on the correct use of fertilizers. This monitoring system will be introduced step by step in 240 parcels. Besides mineral nitrogen, pH and phosphates, potassium and dry organic matter will also be measured. The

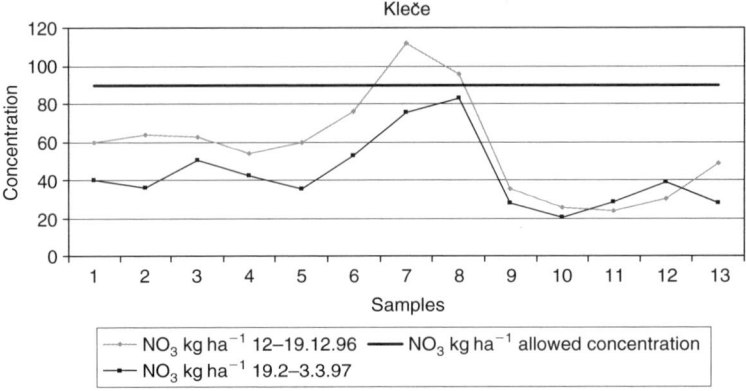

Figure 4. Nitrate concentration in the soil on the second protection zone for Kleče water field.

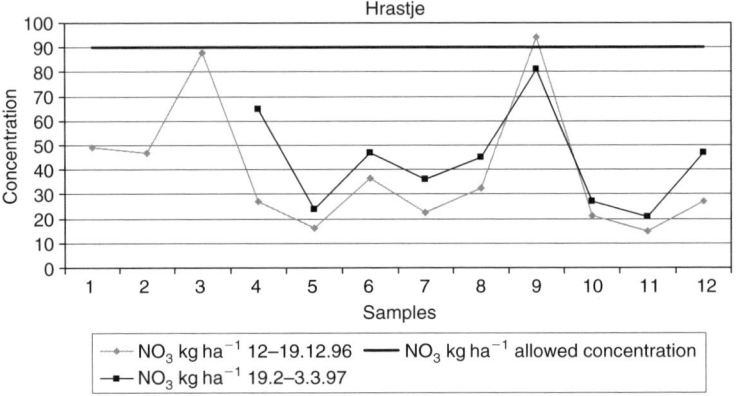

Figure 5. Nitrate concentration in the soil on the second protection zone for Hrastje water field.

Table 1. Status of Ljubljana Field groundwater body in view of WFD.

Year	1992	1993	1994	1995	1996	1997	1998	1999	2000
AM	21,629	16,128	18,363	18,674	18,274	15,43	17,832	17,925	16,494
CL(AM)	25,855	19,564	20,923	21,042	20,903	18,342	21,283	21,13	19,857

register will be assessed once a year by a consultant who is responsible for warning farmers if activities are showing deviations from the plan.

EUROWATERNET

The Ljubljana Field aquifer was included in the EEA 'EUROWATERNET-groundwater' project. The pilot study focused on the nitrogen status of groundwaters in Europe. In order to improve the assessment of nitrogen problems in groundwaters the study recommends not to focus only on nitrate, but to also pay attention to nitrite, ammonium and dissolved oxygen.

It is necessary to develop EUROWATERNET-groundwater to meet the requirements for groundwater monitoring and reporting laid down in the proposed EC Water Framework Directive. Furthermore, it is necessary to develop an update mechanism for EUROWATERNET-groundwater to be used as a tool for day-to-day work and informing groundwater information users.

The results of analyses from 1991 to 1997 were included in the research. The results of the study show that the Ljubljana Field aquifer is not overloaded with nitrates and has good status.

LEGISLATION

Although Slovenia is not a member of the EU, the legislation on the groundwater and drinking water field is harmonized with EU directives. On the state level the drinking water quality is regulated by Regulations on Health Requirements for drinking water. It prescribes a maximum allowed nitrate concentration of 50 mg NO_3/l and a recommended value of 25 mg NO_3/l. Until 1997 the Regulation on hygienic irrevocability was in force and 44.27 mg NO_3/l was the maximum allowed concentration in the groundwater.

The groundwater quality status is regulated by the Ordinance on groundwater quality that was prepared in accordance with Water Framework Directive (WFD 2000/60/EC). The directive's guidelines have been used for bodies of groundwater where the average daily groundwater pumping volume for public water supply is more than $100 \, m^3$.

The quality of water in the body of groundwater is estimated on the basis of two parameters:
- Average of pollutant concentration in all sampling points (AM)
- 95% upper confidence limit CL(AM).

The groundwater body has a good chemical state if the annual average value of either of the parameters does not exceed the maximum allowed concentration.

When the increased nitrate concentration was detected in the groundwater, agriculture was the first on the list of suspects for nitrate pollution. There are 1,990 ha of agricultural land with vegetable production on the protection zones. This activity is connected with plant nutrient and pesticide use.

On the local community level, The Decree on the Protection of Drinking Water Sources defines the appropriate activities and arrangements on water protection zones. Agriculture is a permitted activity but only under the condition that it is not intensive. Intensive agriculture is connected with the use of fertilizers, plant protection agents and irrigation systems that make it possible for agricultural production to be independent from weather conditions. The agricultural influence on groundwater quality is indicated in pesticides exceeding 500% of the maximum allowed concentration.

Nitrate concentration in the groundwater is increasing even though the Ordinance on input of hazardous substances and plant nutrients to soil is valid and has limited the nitrogen input on water protection zones. Fertilization with manure and liquid manure on agricultural areas is forbidden between 15 November and 15 February.

The Ordinance also manages the storage of animal manure, which may only be collected in the proper cesspools and dunghills. These reservoirs must be made from concrete and be water tight with no outflow to surface or groundwater. Storage reservoirs must have capacities appropriate to the period over which fertilization is forbidden.

The results of plant nutrient monitoring will be used to prepare the annual fertilization plan. The plan will include estimation of the common volume of manure, the time period when fertilization is allowed and the measurement of manure for all available land.

As agriculture is still one of the biggest sources of groundwater nitrate pollution, the Slovene agricultural environmental programme was prepared. The purpose is to popularize agricultural produce that suits consumers whilst protecting their health, assuring the sustainable use of natural sources and preserving the biodiversity and characteristics of the Slovene landscape. There is a stress on environmental components and the programme is divided into three elementary groups that define the nature and contents of arrangements:

Group I. Reducing negative influences of agriculture on the environment.

Group II. Preserving natural conditions, biodiversity, fertility of the soil and traditional countryside.

Group III. Protection of protection areas.

The villages on the fringe of the town are partly without sewage systems. Technical guidance from 1985 on dunghills and cesspool settlement is out of date and inspections are unsuitable. Consequently, numerous point sources pollute groundwater with nitrate.

The sewage system in Ljubljana was constructed after acceptance of the first urbanization plan in 1899. The sewage system is mostly mixed, with waste and rainwater collected in 505.3 km of channels. 196.9 km of channels collect only sewage, 161 km of channels only rainfall water. The sewage system is constructed under an area of 5,532 ha. 20,299 units are connected to the sewage system, one third less than to the water supply system. Wastewater from other units is accumulated in the cesspools, which, in many cases are permeable and pollute the areas that are protected due to drinking water sources.

CONCLUSIONS

Nitrates are present in the Ljubljana Field aquifer. Agriculture is still the most important source of nitrate pollution, but the urban source of nitrates is too often forgotten. Despite negative influences, groundwater quality still meets the requirements for drinking water. Through realizing actions of:

- fertilizer schemes
- cesspools and dunghills sanitation
- construction of sewage systems in the protection zones

it may be expected that the complex control over emissions of nitrate to the groundwater and the state of groundwater quality will improve over the next few years.

REFERENCES

Brečko-Grubar, V., Kušar, S. and Plut, D. 2000: Regionalna vloga in pokrajinska obremenjenost talne vode Ljubljanskega polja, Geografija mesta, Ljubljana, Založba ZRC.

EEA 2000, EUROWATERNET-Pilot implementation EUROWATERNET-groundwater in selected groundwater bodies with reference to nitrogen, Technical report No. 39.

Jamnik, B. and Urbanc, J. 2000: Izvor in kakovost podzemne vode Ljubljanskega polja, Ljubljana, RMZ, let. 47, št.2.

Jamnik, B., Bračič-Železnik, B., Loose, A. and Jankovi, M. 2001: Groundwater as a source of drinking water and its management in the city of Ljubljana, Phare Workshop in Maribor, Environmental protection agency, Maribor.

Poročilo o stanju in ogroženosti podtalnice, ki je vir pitne vode na območju mestne občine Ljubljana za obdobje 1992−1998, http://www.mesto-LJ.si/onesna/vode.html.

Predanič, M. 1993: Ocena kvalitete podtalnice Ljubljanskega polja glede na fizikalno-kemijske parametre, Ljubljana, arhiv JP Vo-Ka.

Predanič, M. and Jamnik, B. 1998: Poročilo o kakovosti podtalnice Ljubljanskega polja na podlagi rezultatov republiškega monitoringa in strokovnih podlag za zavarovanje pomembnih vodnih virov in vodnih zalog, Ljubljana, arhiv JP Vo-Ka.

Rejec-Brancelj, I. 2000: Okoljski učinki intenzivnega kmetovanja v rastlinjakih, Geografija mesta, Ljubljana, Založba ZRC.

CHAPTER 24

Nitrate monitoring and CFC-age dating of shallow groundwaters — an attempt to check the effect of restricted use of fertilizers

T. Laier

Geological Survey of Denmark and Greenland (GEUS) Øster Voldgade 10, 1350 Copenhagen

ABSTRACT: Age dating of groundwater from over 1,000 monitoring wells using tritium and CFC showed that approximately 20% of the groundwater monitored pre-dated 1940. Only 10% of the monitored groundwater was recharged after the late 1980s, when restriction on the use of fertilizers was decided by the Danish parliament. No clear decreasing trend in groundwater nitrate concentration, reflecting the reduced consumption of fertilizers, can be observed for the most recent groundwater due to a fairly large scatter. The scatter in groundwater nitrate concentration for some shallow wells is due to the fact that water belonging to different flow lines is being sampled over relatively short time intervals as documented by repeated Chlorofluorocarbon (CFC)-age dating. Changes in groundwater flow lines being sampled from a particular monitoring well may occur with changes in the groundwater level caused by changes in net-precipitation.

INTRODUCTION

In order to reduce the concentration of nutrients in surface waters as well as in groundwater, it was decided by the Danish parliament to include restrictions on the use of fertilizers in agriculture, along with a number of other measures taken. It was also decided to set up a monitoring programme on groundwater quality in order to be able to check the effect of various measures taken to protect groundwater against potential pollutants. The monitoring programme has been running for 12 years. However, none of the 600 borings — including 1,100 individual sampling levels — have shown any continuous decrease in nitrate concentration during that period. During 1997–1998, CFC-age dating was performed on waters from all borings (ca 800) that had tritium above background level — determined by previous measurements in the early 1990s. As expected, a general decrease in nitrate concentrations with groundwater age was observed, although the data were scattered considerably depending on the differences in geology and of land use in the various areas (Stockmarr and Nyegaard 2004).

In order to make a meaningful comparison between the change in consumption of fertilizer with the change in groundwater nitrate concentration, the latter have been plotted versus date of groundwater recharge, that is date of sampling minus CFC-model age of the groundwater determined in 1997–1998. As denitrification may obscure any relation between groundwater nitrate concentration and the amount of fertilizer being applied, only data for oxygen-containing groundwater (oxygen being measured in the field) have been included in Figure 1.

Figure 1 indicates that the cause for the increase in groundwater nitrate concentration is the increase in fertilizer consumption since the 1950s, although the range in nitrate concentration varies greatly among the different monitoring wells (Stockmarr and Nyegaard 2004). It is more difficult, however, to show a positive effect of the reduced fertilizer consumption since 1990 on groundwater nitrate concentration due to the limited amount of data. Only 10% of the monitoring wells produced water that had been recharged within the last 10 years, when a decrease in fertilizer consumption has been noted (GEUS 1999).

The median nitrate concentration of the relatively few wells representing the youngest groundwater may not reflect the general decrease in fertilizer consumption, as it is likely to be more sensitive to local variations as well as change in nitrate concentration due to variations in net-precipitation. Also the CFC-age dating of groundwater has to be very precise in order to correctly determine the year of recharge within the relatively narrow time interval of interest.

No clear decrease in nitrate concentration with time was observed for any of the monitoring wells producing young water. Some of the wells even showed an increase in nitrate concentration with time for waters recharged after 1988 (GEUS 1999). There may be a number or reasons for this: 1) The amount of nitrate transferred from the root zone depends not only on the amount of fertilizer used, but will also depend on net precipitation, soil temperature during the winter period and agricultural practice. In fact large variations in nitrate concentrations were observed over periods with large variations in net precipitation (GEUS 1999); 2) Regular sampling from a particular level may not represent a smooth change in groundwater age due to fluctuations in groundwater flow lines; 3) Denitrification may take place even in groundwater containing oxygen, thereby obscuring the change in the amount of nitrate initially added.

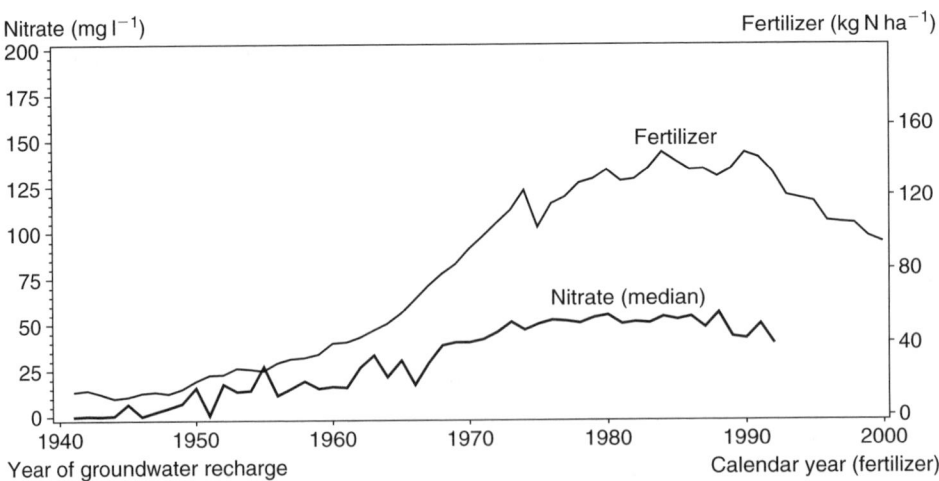

Figure 1. Nitrate (median) concentration in groundwater versus date of recharge obtained from the Danish groundwater quality-monitoring programme. Only samples containing oxygen with nitrate concentrations above $1 \, mg \, l^{-1}$ are included in the data set. Data on the fertilizer consumption each year were obtained from the Danish Plant Directorate.

This paper examines how factors, other than changes in fertilizer consumption, may affect the groundwater nitrate concentration observed in each monitoring well, by including regular CFC-age dating in the monitoring programme for three selected wells. The paper will also describe the principle of the CFC-age dating method, and the testing of it against tritium data in one monitoring programme locality. In addition, an example will be presented that illustrates how young groundwater of similar age flowing in glacial deposits shows large differences in nitrate concentration with time within relatively short distances. Fractured chalk deposits form important groundwater reservoirs in Denmark, therefore an example of how fracture flow may influence nitrate monitoring data will also be given.

CFC-AGE DATING OF GROUNDWATER

Chlorofluorocarbons (CFC), also called 'freons', have accumulated in the earth's atmosphere due to the inert nature of these compounds. The increase in CFC concentration since the mid-1970s has been fairly well documented, due to the potential threat freons constitute to the ozone layer. By calculating CFC concentrations prior to 1975 using production numbers and half-lives of the compounds, a complete set of atmospheric CFC values since the start of CFC production in the late 1930s have been obtained (Figure 2).

Using this information, Busenberg and Plummer (1992) were able to determine the age of groundwater fairly accurately, ±2 years under optimal conditions, given the equilibrium of CFCs between air and groundwater (Warner and Weiss 1985). The CFC-age dating technique applied by GEUS is similar to that described by Busenberg and Plummer (1992). First, samples of groundwater are collected in flame-sealed 60 ml boron silicate ampoules, which have been flushed with pure nitrogen prior to sampling. Then,

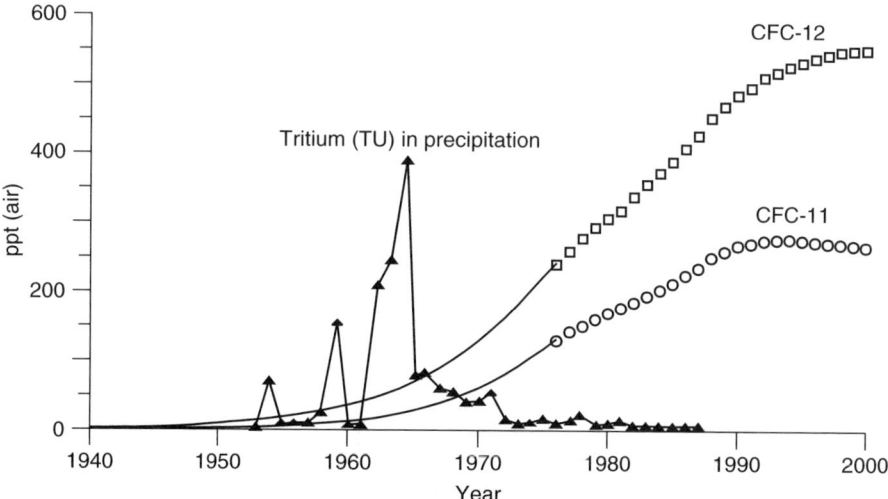

Figure 2. Concentration of CFC in the atmosphere and tritium in precipitation since 1940. Units are parts per trillion (ppt) and tritium units (TU). Maximum tritium in precipitation, measured in 1963, due to hydrogen bomb testing.

approximately 30 ml of the sample is transferred to a purge and trap system and finally the gases are swept into a gas chromatograph equipped with an electron capture detector (ECD) to quantify the amount of CFC gases dissolved in the water. Using the equilibrium constants determined by Warner and Weiss (1985), the CFC concentration in the atmosphere at the time of recharge, and thus the age of the groundwater, can be calculated using the information given in Figure 2.

Checking the CFC-method versus tritium

In order to test the accuracy of the CFC-method in Denmark, CFC-age dating was first performed in a fairly homogenous sandy aquifer at Rabis Creek (Postma, Boesen, Kristiansen *et al.* 1991). Groundwater flow lines in this aquifer had been modelled using tritium analyses on samples from several multilevel monitoring wells (Engesgaard, Jensen, Molson *et al.* 1996). Strictly speaking, the year of recharge determined by the CFC-method refers to the time of transition from the unsaturated zone to that of groundwater, and not to the year of precipitation. The residence time of water in the unsaturated zone of the area determined by Andersen and Sevel (1974) using tritium is, however, similar to that of air diffusion through the zone determined by CFC-analyses on pore-space gases (Engesgaard, Højberg, Hinsby *et al.* in press). Thus, the age of the groundwater determined by the CFC-method should be comparable to that modelled using tritium analyses (Figure 3).

CFC-age dating results using CFC-12 (CCl_2F_2) agree with groundwater age modelled by the tritium method, whereas CFC-11 (CCl_3F) results indicate much greater ages at greater depths. The large increase in CFC-11 age (decrease in CFC-11 concentration)

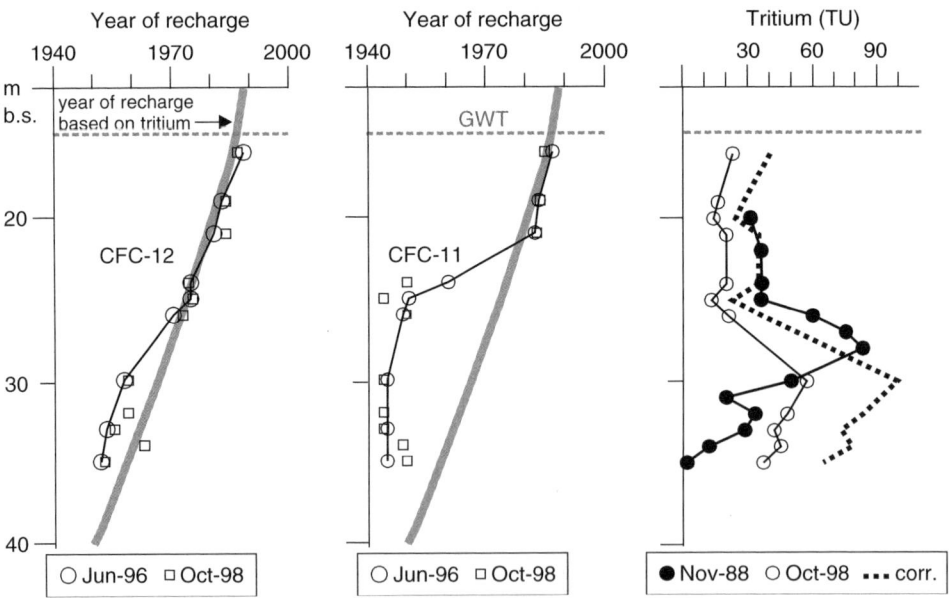

Figure 3. Year of recharge for groundwater in the Rabis Creek area, location A on Figure 4, determined by CFC-analyses on samples from the multilevel well T1. The recharge curve was modelled from tritium analyses of several multilevel wells including T1 (Engesgaard and Molson 1998). GWT = ground water table. Tritium results for October 1998 have been corrected for 10 years of decay ($t_{1/2} = 12.43$ y).

occurs in a zone of intense denitrification evidenced by fairly large concentrations of N_2O at this level, and is probably due to degradation of CFC-11. CFCs, and particularly CFC-11, has been shown to degrade under anoxic conditions in laboratory experiments (Lovely and Woodward, 1992). Oster, Sonntag and Münnich (1996) estimate that CFC-11 was being degraded ten times faster than CFC-12 under anoxic conditions. However, Shapiro, Schlosser, Smethie *et al.* (1997) did not observe any measurable degradation of CFC-12 under methanogenic conditions in a natural system, comparing CFC-12 with tritium and tritium/helium-3. Plummer, Busenberg, McDonald *et al.* (1998), also found that CFC-12 persists even under methanogenic conditions in nature.

The somewhat greater age indicated by CFC-12 compared to tritium at greater depth (Figure 3) may thus be explained by degradation of CFC-12. However, the difference between CFC-12 and tritium may also reflect heterogeneities in the Rabis aquifer, as pumping indicated lower yields for some of the deeper levels. Also the material obtained from drilling the T1 well is somewhat finer at greater depths (Hansen and Gravesen 1990). In order to model the age of water using tritium, Engesgaard and Molson (1998) had to assume that the Rabis Creek aquifer was completely homogeneous for that section containing the multilevel wells, which is probably not realistic for this glacial melt water deposit (location A on Figure 4.)

Whether or not degradation of CFCs takes place probably depends on the actual microbial activity in the subsurface, for example, no degradation of CFC-11 relative to CFC-12 was observed under anoxic conditions (nitrate below detection) in a multilevel well located in a nearby forest area. The low nitrate in groundwater in this area was not due to denitrification, as was the case for the T1 well at Rabis Creek (adjacent to arable land), but due to low nitrate in water infiltrating from the forest.

Comparing CFC-age dating with tritium dates, it may be concluded that the CFC-age dating method works just as well in Denmark as in the USA (Busenberg and Plummer 1992). Whether or not CFC-12 is being degraded to a measurable degree under anoxic conditions in nature is still open to question. However, for the examples presented below, this is not important as they all deal with oxic waters.

INCREASE IN GROUNDWATER NITRATE WITH TIME IN THE CHALK AQUIFER AT DRASTRUP

Very high nitrate concentrations in groundwater ($>100 \, \text{mg} \, l^{-1}$) have become a serious problem for the fresh water supply to the city of Aalborg (160,000 inhabitants). Most water is pumped from a fractured chalk reservoir covered by thin Quaternary deposits, which offer little reducing capacity toward nitrate infiltrating from the farmland around the city. The chalk itself holds very little reducing material, therefore oxic groundwater may be found at relatively great depth (60 m) in this area. The municipal water works of Aalborg, therefore, anticipated a decrease in groundwater nitrate concentration following the decisions made by the parliament in 1987 to reduce the consumption of fertilizers, and to forbid the spreading of manure during the non-growing season. However, groundwater nitrate concentration continued to increase at shallow levels during the first eight years of the groundwater quality-monitoring programme. In the upper level of the four-level well at Drastrup, location B (Figure 4), just south of the city, the nitrate concentration increased from 100 to 120 $\text{mg} \, l^{-1}$ (Figure 5).

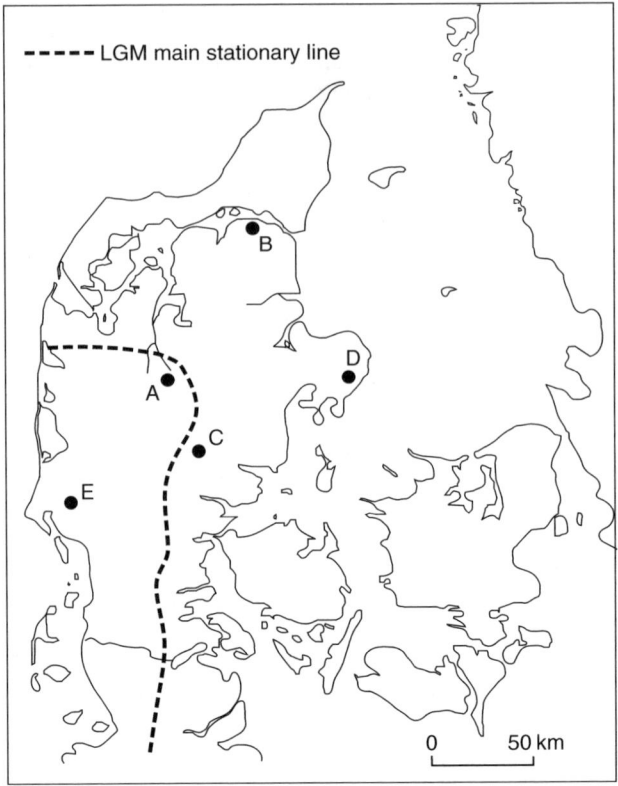

Figure 4. Location map – the dashed line indicates the extension of glaciers during the late glacial maximum. No Weichselian till has been observed south and west of this line.

The management of the water works therefore asked GEUS to undertake age-dating of the groundwater at all four levels of the well. The age-dating of the Drastrup water in May 1997 was the first to be performed on the monitoring network after having tested the CFC- method in the Rabis Creek aquifer.

The age of the groundwater from the upper level, 9 metres below surface, appeared to be 20 years old, so should not have shown any effect due to a decrease in fertilizer consumption, see Figure 2. The age of water from the lower three levels of the well increased from 25 years at 21 metres to 45 years at 61 metres.

Dissolved oxygen was present at all four levels so it is assumed that denitrification has not occurred. Therefore Figure 5A shows the true change in nitrate concentration with time over a 40-year period since the late 1940s. However, the co-variance in nitrate concentration with sampling time for the upper three levels (Figure 5B) indicate that nitrate concentrations are all being affected by some common feature from time to time. This may be due to the fact that groundwater produced from each of the 0.5 m filter sections mainly came from fractures in the chalk, as the matrix-permeability of chalk itself is very low. Therefore, the water sampled at each level may be composed of mixtures from different levels, some of which may be quite shallow, giving rise to short term variations in nitrate concentrations.

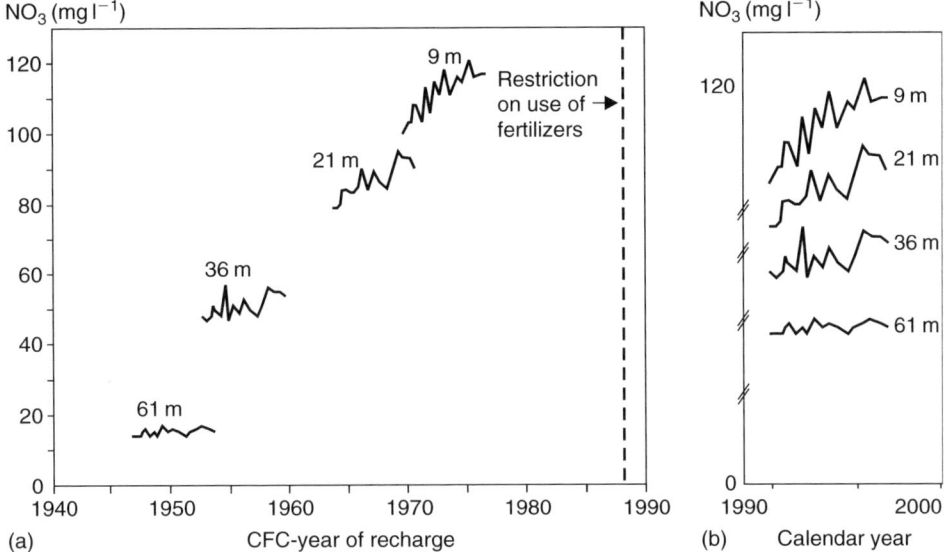

Figure 5. Increase in groundwater nitrate concentration with time in the chalk aquifer at Drastrup, location B on Figure 4. (a) Groundwater age has been subtracted from the time of sampling in May 1997 to obtain the real change in nitrate with time. (b) Nitrate concentrations at sampling dates for the four levels.

When CFC-analyses are performed on water samples consisting of mixtures of groundwater from different levels, the term 'CFC-age of groundwater' has no real meaning. It is probably not correct either, to use the term 'average-age' since the increase in CFC in the atmosphere was not linear through time (Figure 2). From a practical point of view, it may be argued that this is not very important since the proportion of young water with a high nitrate concentration is roughly reflected by the CFC content in the water, and so 'CFC-age' dating may still be considered useful even for fractured rocks. This may be true for the case of the long-term pollution with nitrate. However, for short-term pollution, for example, with pesticides, the 'average CFC-age' may indicate a date of groundwater recharge several years prior to the first use of a particular pesticide, which may have been found in the water. Such incidents have occurred in the groundwater quality-monitoring programme in Denmark.

LARGE DIFFERENCES IN NITRATE CONCENTRATION IN WATERS OF SIMILAR AGE

Young oxic waters of the Thyregod area, location C on Figure 4, show very different patterns with respect to nitrate concentration with time within distances of a few kilometres (Figure 6). The geology of the area is typically that of many areas east of the late glacial maximum line (Figure 4) consisting of melt water sand and clay as well as glacial till (Figure 6). The whole Thyregod area consists mainly of farmland with no major cities. Thus, one might expect to observe similar patterns in nitrate concentration with time for waters of similar age, if no denitrification took place. However, the presence of

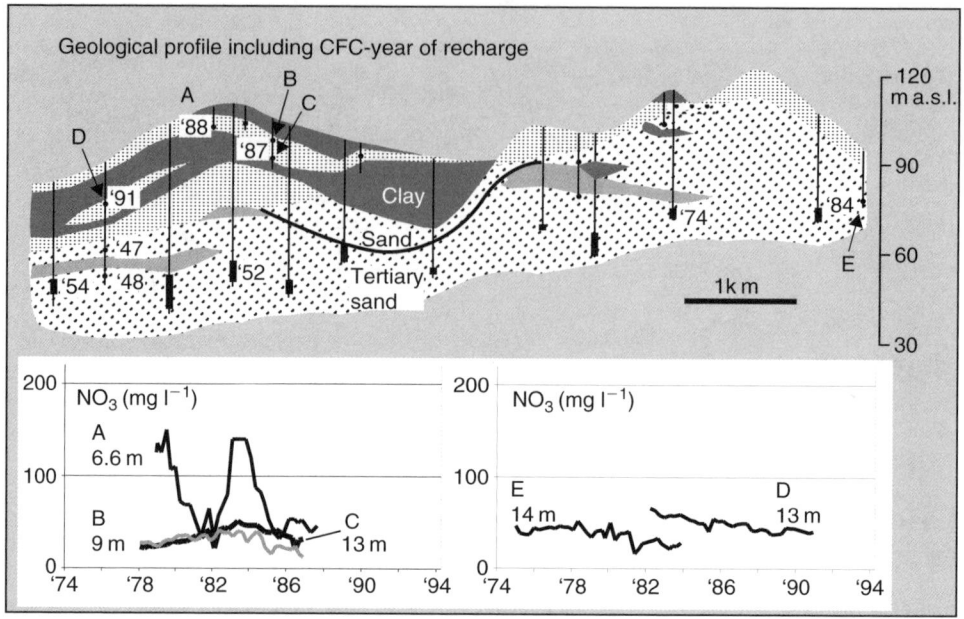

Figure 6. Geological profile of the Thyregod locality including CFC-year of recharge of groundwater. Nitrate has been plotted versus time of recharge for wells producing the youngest waters.

nitrous oxide (N_2O) indicates that denitrification did take place even though dissolved oxygen was measured in water during sampling.

The geological profile shown in Figure 6 was constructed from existing wells in the area and may not be sufficiently detailed to track groundwater flow paths from surface to monitoring wells. Increasing the sampling density would probably reveal an even more heterogeneous pattern of the area with respect to lithology, as has been found from previous detailed mapping of such areas. Therefore, it is not possible decide whether denitrification took place in clays, which are known to reduce the nitrate content of groundwater, or in sands which occasionally contain lignite and pyrite.

The very large variations in nitrate concentration from $140 \, \text{mg} \, l^{-1}$ to $30 \, \text{mg} \, l^{-1}$ and up to $130 \, \text{mg} \, l^{-1}$ again over relatively short periods of time at well 'A' (Figure 6) was probably not just due to a change in denitrification rate. Significant changes were also observed for some wells in other areas in the same period of recharge. During that period, marked variation in net precipitation occurred which may have been responsible for the large variations in groundwater nitrate concentrations (GEUS 1999). The large variations may be due to different amounts of nitrate being washed from the root zone in response to differences in net precipitation, but fluctuation in groundwater may also occur which change the age of groundwater at a particular sampling level.

FLUCTUATION OF GROUNDWATER FLOW LINES

Two examples of this will be given below, one at Homaa location 'D' east of the main stationary line (Figure 4) and one at Forumlund location 'E' west of the line.

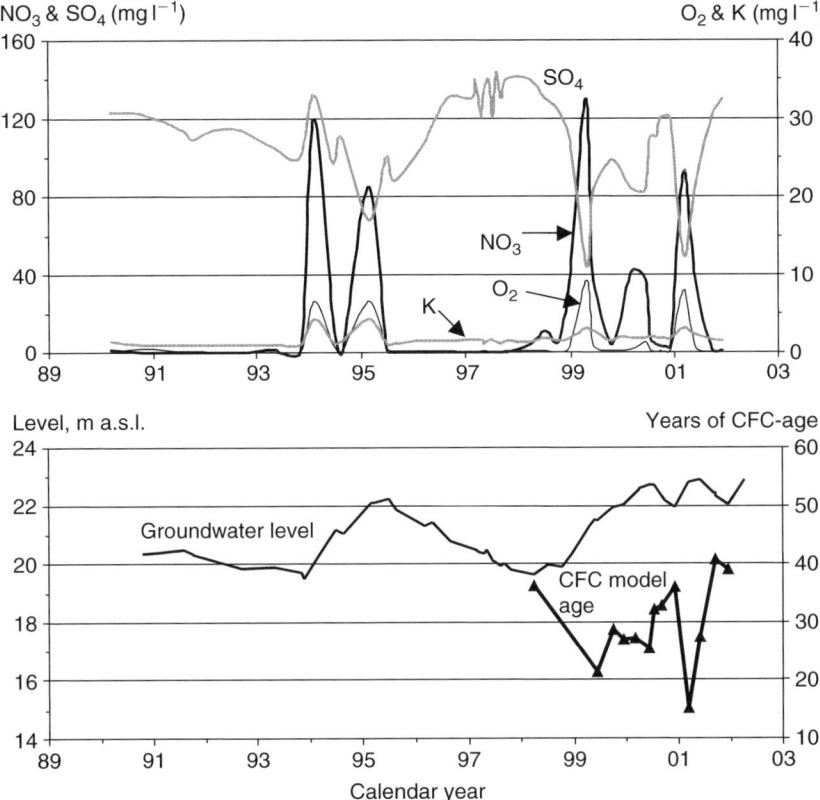

Figure 7. Variation in groundwater chemistry with time at Homaa, location D. Relatively lower CFC-ages are seen for periods with high nitrate.

Homaa example

Regular CFC-age dating of groundwater has been performed on samples from one well, which showed remarkable changes in groundwater chemistry over short time periods. The well is located in a sandy aquifer with several thin layers of clays. High to very high nitrate concentrations were observed in five short periods between 1990 and 2002. Other than these, nitrate was below detection limit (Figure 7). CFC-age dating covered three of the five periods with nitrate peaks and a relative decrease in groundwater age could be documented in these periods (Figure 7). The age of the groundwater varied as much as 25 years within the year of 2001 (Figure 7).

In four of the periods with high nitrate, a significant decrease in sulphate concentration was observed. Also a minor, but distinct, increase in potassium concentration was observed in the periods with high nitrate. Thus, there can be little doubt that different types of groundwater have been sampled from time to time at this particular level, 20–22 m below surface. The reason for the change in groundwater type is most likely fluctuation in groundwater flow lines with the changing level of water which can be as much as 3.3 m at this location, meaning that the distance between water level and sampling level varies from 5 to 8 m. However, no simple relation between groundwater level and groundwater age can be deduced from the data shown in Figure 7.

Forumlund example

In a sandy aquifer at Forumlund, location E (Figure 4), different patterns in groundwater nitrate concentration with date were observed at two levels separated by 5 metres (Figure 8).

CFC-age dating of the groundwater on 26 samples from each level since 1998 did not show a smooth increase in time since recharge, as might have been expected. Instead, fairly large fluctuations in groundwater age, up to 10 years, was observed over short time intervals, particularly for the deeper level 22–23 m below surface (Figure 8). For the shallow level 17–18 m below surface, the variation in age was less than five years for the whole period. Furthermore, the age of the groundwater at the shallow level appears to be younger when the groundwater level is low and the distance between water level and sampling level is short. The age increased from 14–15 years in 1998–99, when this distance was less than 1.5 m, to 20 years in 2000 when the distance was 4 m. For the deeper level, the situation appears to be that, increases in water level, irrespective of the absolute level, create pulses of younger water which are only observed at this level and

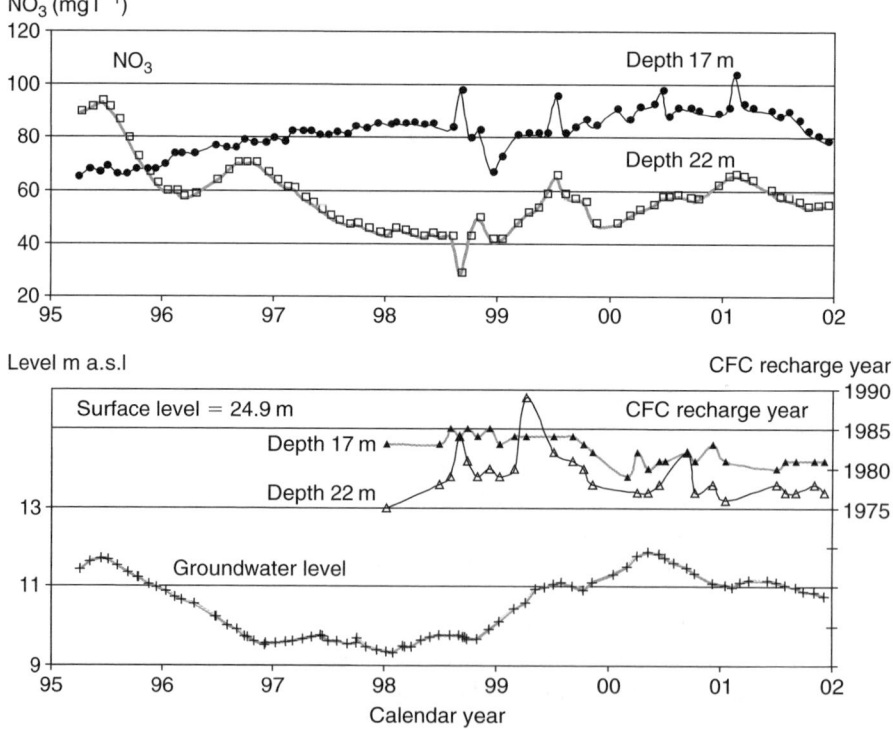

Figure 8. Variation in groundwater nitrate concentrations with date of sampling at two levels at Forumlund location E. Larger variation in nitrate as well as in groundwater age is seen for the lower level compared to the upper one.

not the upper one (Figure 8). Interestingly, the change in water level measured at both levels in the multilevel well was the same within a few centimetres, indicating good hydraulic contact among different layers of the aquifer.

Differences in recharge temperature will affect the amount of CFC being dissolved in the water, and thus the estimate of the CFC-age (Warner and Weis 1985). Therefore, one has to be certain of the recharge temperature in order to make an accurate estimate of the age of the water. Determination of recharge temperature may be undertaken by analysis of the content of inert gases dissolved in the groundwater (Busenberg and Plummer 1992). This has not generally been done by GEUS. However, for the examples presented in this paper, the water level was located 10–15 m below surface, so it is probably safe to conclude that recharge temperature at groundwater level was constant, close to the average temperature in Denmark, that is 8°C.

CONCLUSIONS

CFC-age dating of groundwater from 800 monitoring wells has made it possible to compare the change in nitrate concentration with that of fertilizer application. The results clearly document that the increase in fertilizer consumption from 1950 to 1990 is reflected in the nitrate content of the groundwater recharging in this period.

The effect of the decrease in fertilizer application since 1990 on groundwater nitrate concentration cannot yet be documented because less than 10% of the monitoring wells produce water that is less than 10 years old. Furthermore, fluctuation of groundwater flow lines at shallow depths due to change in net precipitation, documented by repeated CFC-age dating, showed that a longer period of monitoring is required to document such an effect.

REFERENCES

Andersen, L.J. and Sevel, T. 1974: Six years' environmental tritium profiles in the unsaturated and saturated zones, Grønhøj, Denmark. In: Isotope Techniques in Groundwater Hydrology, Vol. 1. IAEA Proceedings Series No. STI/PUB/373. IAEA: Vienna; 3–20.

Busenberg, E. and Plummer, L.N. 1992: Use of chloroflurocarbons (CCl3F and CCl2F2) as hydrologic tracers and age-dating tools: The alluvium and terrace system of central Oklahoma. *Water Resources Research* 28: 2257–2283.

Engesgaard, P. and Molson, J. 1998: Direct simulation of groundwater age in the Rabis Creek aquifer, Denmark. *Groundwater* 36: 577–582.

Engesgaard, P., Jensen, K.H., Molson, J., Frind, E.O. and Olsen, H. 1996: Large-scale dispersion in a sandy aquifer: Simulation of subsurface transport of environmental tritium. *Water Resources Research* 32: 3253–3266.

GEUS 1999: Groundwater quality monitoring 1999. GEUS special edition, 115 pp. (In Danish with an English summary.)

Hansen, M. and Gravesen, P 1990: Geologiske forhold i oplande til Karup Å og Rabis Bæk, Jylland. NPO-projekt 2.1. *Danmarks Geologiske Undersøgelse*, Intern Rapport Nr. 35.

Engesgaard, P., Højbjerg, A., Hinsby, K., Jensen, K.H., Laier, T., Larsen, F., Busenberg, E. and Plummer, L.N. (in press): Transport and time lag of chlorofluorocarbons (CFC) gases in the unsaturated zone, Rabis Creek, Denmark. *Vadoze Zone Journal*.

Lovely, D.L. and Woodward, J.C. 1992: Consumption of Freon CFC-11 and CFC-12 by anaerobic sediments and soils. *Environmental Science and Technology* 26: 925–929.

Oster, H., Sonntag, C. and Münnich, K.O. 1996: Groundwater age dating with chlorofluorocarbons. *Water Resources Research* 32: 2989–3001.

Plummer, L.N., Busenberg, E., McDonald, J.B., Drenkard, S., Schlosser, P. and Michel, R.L. 1998: Flow of river water into a Karstic limestone aquifer.1. Tracing the young fraction in groundwater mixtures in the Upper Floridan Aquifer near Valdosta, Georgia. *Applied Geochemistry* 13: 999–1015.

Postma, D., Boesen, C., Kristiansen, H. and Larsen, F. 1991: Nitrate reduction in an unconfined sandy aquifer: Water chemistry, reduction processes, and geochemical modeling. *Water Resources Research* 27: 2027–2045.

Shapiro, S.D., Schlosser, P., Smethie, W.M. and Stute, M. 1997: The use of 3H and triogenic 3He to determine CFC degradation and vertical mixing rates in Framvaren Fjord, Norway. *Marine Chemistry* 59: 141–157.

Stockmarr, J. and Nyegaard, P. 2004: Nitrate in Danish groundwater (this volume).

Warner, M.J. and Weiss, R.F. 1985: Solubilities of chlorofluorocarbons 11 and 12 in water and seawater. *Deep-Sea Research* 32: 1485–1497.

CHAPTER 25

Reducing nitrate leaching from arable agriculture: preliminary results from the Netherlands

J.W.A. Langeveld and A.L. Smit
Plant Research International Wageningen University and Research Centre,
P.O. Box 14, NL-6700 AA Wageningen, The Netherlands

B. Fraters
National Institute for Public Health and the Environment Bilthoven, The Netherlands

ABSTRACT: Nutrient losses to surface- and groundwater are among the most persistent environmental problems in agriculture. Reduction of (nitrate) emissions in the Netherlands is pursued applying monitoring programmes, research projects and policy instruments (mainly MINeral Accounting System or 'MINAS'). Research projects combine nitrate measurements with measurement of so-called nutrient efficiency indicators. These indicators (nitrogen surplus, residual mineral soil nitrogen) are cheaper, are more easily measured (calculated), can be more easily linked to fertilization practices and have a stronger appeal to farmers. Preliminary project results link highest nitrate concentrations to high nitrogen applications, low groundwater levels and potato, scorzonera and vegetable cultivation. Average farm nitrate concentration is almost twice the EU objective (50 mg nitrate/l) which therefore probably only can be realized on arable farms lacking intensive vegetable cultivation. The paper discusses the potential of the given approach to reduce nitrate emissions in other countries.

INTRODUCTION

The large impact of Dutch agriculture on the environment is associated with high application rates of (in)organic fertilizers and chemicals. Despite recent cut-downs, consumption of fertilizers and agro-chemicals in the Netherlands remains high in comparison to other industrialized countries. As a result, agriculture significantly contributes to major environmental problems, causing 40% of all acidifying emissions and (over) two thirds of the nutrient loads to land and surface waters (Table 1).

The effects of these emissions are considerable, and in order to reduce both their extent and their effect, application of chemical and organic fertilizers has been subjected to increasingly stringent regulation. Dutch legislation is based on restrictions in the allowed farm-level nutrient surplus. In the so-called MINeral Accounting System (MINAS), farm level nutrient surpluses are calculated using real input data (fertilizers, manure, nitrogen fixation, additional nutrient sources) and reference nutrient removal figures (165 kg of nitrogen and 65 kg of phosphate), allowing farmers limited surpluses of nitrogen (100 kg of nitrogen per ha of standard soils and 60 kg on leaching-sensitive soils in 2003) and

Table 1. Role of agriculture in major environmental problems in the Netherlands.

Theme Sector	Emissions			$\%^1$	$\%^2$
	1990	1995	2000		
Climate change					
CO_2-equivalents (Tg)					
All	217	226	222		
Agriculture	27	27	24	11	−11
Acidification (billion acid-equivalents)					
All	31	25	21		
Agriculture	13	11	9	43	−31
Nutrient load to soil (Gg)					
Nitrogen					
All	408	462	340		
Agriculture	406	460	338	99	−17
Phosphorus					
All	73	63	48		
Agriculture	73	63	48	100	−34
Nutrient load to surface water (Gg)					
Nitrogen					
All	142	136			
Agriculture		91	92	68	+1
Phosphorus					
All	14	10			
Agriculture		5	6	60	+23

[1]% of total; [2]% of change since 1990.
Source: calculated from RIVM (2001).

phosphate (20 kg per ha). Additional policy measures relate to a maximum nitrogen input through manure application, levies on nutrient surpluses and compulsory manure sales contracts for intensive animal husbandry farms.

In addition to these measures, extensive monitoring schemes have been implemented in order to assess the impact of agriculture on water quality. The most important scheme is the National Monitoring programme for effectiveness of the Minerals Policy (LMM), which was initiated for sand regions (45% of the agricultural acreage) in 1987. Continuous threats to valuable groundwater in sand regions triggered further research and monitoring, and in 1992 a programme was started to monitor the effectiveness of the Dutch emission policy in agriculture. In 1993, LMM was extended to the clay regions (covering 40% of the agricultural area). This was followed by a monitoring programme in the winter of 1996–1997. In the peat region (15% of the acreage) research and monitoring started in winter 1995–1996. On-farm monitoring includes both agricultural practices and water quality (Fraters, Van Eerdt, De Hoop *et al.* 2000). Since 1992, LMM has monitored some 100 farms annually.

LMM results (Figure 1) show that while nitrate levels in the sand regions have decreased over the last decade they still remain too high, averaging around 100 mg l^{-1}. In the clay regions nitrate concentrations in tile drain water are on average slightly above 50 mg l^{-1}. In the peat regions nitrate concentrations in the upper groundwater and ditch water are low (<5 mg l^{-1}). Figure 2 shows that considerable differences exist between

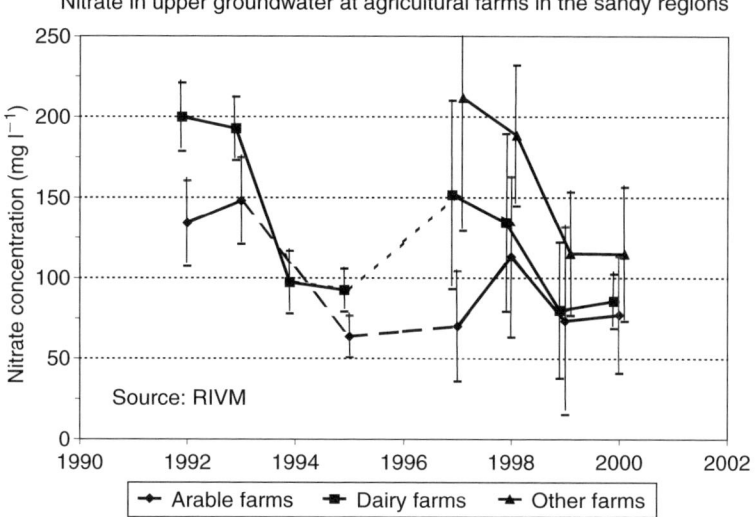

Figure 1. Nitrate concentration for several sectors in the LMM.

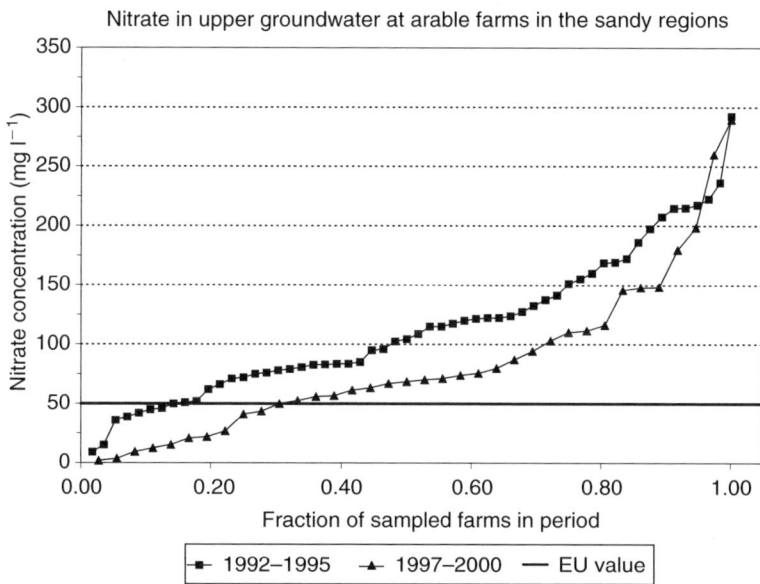

Figure 2. Frequency distribution of nitrate concentration found on arable farms in the LMM.

individual farms. In the period 1997–2000 about 30% of arable farms in the sand regions participating in LMM have complied with the EU directive limit of 50 mg l^{-1}, the highest concentrations being more than 10 times higher than the lowest.

Table 1 showed that, since 1990, agriculture-related acidifying emissions have fallen by 30%, while nitrogen and phosphorus loads to soils were reduced by 17% and 34% respectively. Although this is promising, the speed of emission reductions recently has been slowing down and much remains to be done. Water quality in the Netherlands, for example, is still far from the desired standards (as given in Table 2). Moreover, on specific issues the country has difficulties in complying with international regulations, such as the EU Nitrate directive. Therefore, towards the end of the 1990s, the Dutch government has been taking additional initiatives to assist the agricultural sector to speed up the reduction of nutrient emissions. For this, additional financial resources have been made available.

Activities supported by these funds include regional and national research and extension projects that are mainly related to arable and dairy farming. In the current paper we restrict ourselves to two national projects related to arable farming, *Telen met toekomst* ('Farming with a future') and *Sturen op nitraat* ('Steering on nitrate'). 'Farming with a future' designs and tests environmentally friendly arable farming systems, while 'Steering on nitrate' focuses on the description of the relation between agricultural practices and measured nitrate concentrations. Both projects mainly focus on farms in the sandy regions, as sandy soils are most susceptible to nitrate leaching. They combine nitrate measurements with intensive monitoring of agricultural practices, sharing the philosophy that nitrate measurements are essential during the research phase, where more sustainable farming systems are designed and tested. Once these systems have been developed, however, intensive nitrate measurements are no longer needed and water quality can best be monitored using so called nutrient efficiency indicators.

The design and testing of these indicators, the major focus of 'Steering on nitrate', is discussed in Ten Berge (2002) and Schröder, Neeteson, Oenema *et al.* (2000). The major

Table 2. Quality standards on nitrogen and phosphorus in the Netherlands.

	Source	Value	Unit	Remarks
Air quality				
NO$_2$ concentration	EU air quality standards	40	µg m^{-3}	to be realized in 2010
NH$_3$ emission	International protocol	100	kt	id. (86 kt from agriculture)
Water quality				
NO$_3$ groundwater	EU nitrate directive	50.0	mg NO$_3$/l	
N surface water	International protocol	2.2	mg N/l	
		−50%	kt	as compared to 1985
P groundwater	Committee advice	3	mg P/l	(clay, peat soils)
		0.4	mg P/l	(sandy soils)
P surface water	International protocol	0.15	mg P/l	
		−50%	kt	as compared to 1985
Soil quality				
Mineral N in	Committee advice	45	kg N/ha	(dry sandy soils)
autumn		70	kg N/ha	(all other soils)
Farm nutrient budget				
N surplus	National Legislation	100	kg N/ha	realized in 2003 (all soils)
		60	kg N/ha	id. (leaching-sensitive soils)
P surplus	National Legislation	20	kg P$_2$O$_5$/ha	to be realized in 2003
N input through manure	EU legislation	170	kg N/ha	

Source: Langeveld, Van Keulen, De Haan *et al.* (in press).

reason for using indicators instead of nitrate measurements is found in the enormous and potentially prohibitive costs of nitrate measurements. Additional arguments for using indicators are found in the fact that establishment of the relation between incidental exceedence of pre-defined thresholds and fertilization actions of individual farmers is often cumbersome, and the fact that water quality standards are not easily translated into day-to-day farmer decision-making, which makes it even more difficult to convince them to change fertilization practices.

The use of nutrient efficiency indicators has two major advantages: they are easier (quicker) to measure (and hence cheaper), and are more appealing to the end-users. The advantage of such indicators depends however on their reliability and the quality of their relation with nitrate concentration. Indicators that have been considered so far include nitrogen surplus (calculated with nutrient mass balance or with MINAS), and amount of residual mineral nitrogen measured in the soil profile. Soils in the Netherlands generally are deep. The amount of mineral nitrogen present in the top 90 cm of the soil at the beginning of the winter is an indicator of the amount of nitrogen that will probably be leached to groundwater. In winter, evapotranspiration is negligible and most of the annual leaching occurs. In summer, potential evapo-transpiration usually exceeds precipitation and leaching is therefore negligible.

In the remainder of this paper the set-up and results of 'Farming with a future' and 'Steering on nitrate' will be discussed. The main focus will be on the measurements of both nitrate and nutrient efficiency indicators.

APPROACH FOLLOWED

'Farming with a future' designs and tests environmentally friendly arable farming systems, paying particular attention to water quality and agro-chemical applications. Research on experimental farms is combined with on-farm testing of preliminary results on pilot farms (see Booij, Van Dijk, Smit *et al.* 2001, and Langeveld *et al.*, in press, for details). Observations include direct nitrate (and other water quality) measurements as well as determination of several nutrient efficiency indicators. Nitrate concentrations are determined in a laboratory and averaged at farm level figure (Table 3). Information on the variation over plots and crops can be derived from 'Nitracheck' measurements that are

Table 3. Measurements of nitrate and potential indicators in 'Farming with a future'.

Parameter	Years	Within the year	Plots
Nitrate[1]	All project years	Once (Apr.−Sep.)	All permanent plots[2,3]
Soil quality	Every second year	November	All permanent plots[2]
Mineral nitrogen at harvest	All project years	When applicable	All cultivated plots[4]
Mineral nitrogen in autumn			
(0−30, 30−60 cm)	All project years	November	All cultivated plots[4]
(60−90 cm)	All project years	November	Selection of plots[5]

Source: Booij *et al.* (2001).
[1] Measured in groundwater, surface water and drainwater (if applicable); [2] Permanent plots are owned or permanently rented; [3] Four composite sub-samples taken from 16 (for clay soils) or 48 (for sandy soils) samples taken on whole farm area; [4] Including annually rented plots; [5] Four plots randomly selected for each farm.

Table 4. Mean, minimum (min) and maximum (max) amounts of residual mineral nitrogen (kg N/ha) in soil profile (0–90 cm) in autumn by region in 2000 and 2001.

Region	2000			2001			Objective
	Mean	Max	Min	Mean	Max	Min	
Arable farming north-east	62	78	45	53	79	37	45
Arable farming south-east	79	97	48	82	91	75	45
Arable farming south-west	132	175	86	77	113	49	70
Vegetable farming Brabant	147	239	86	118	256	59	45
Vegetable farming Limburg	163	267	61	126	184	75	45

done for all observation points of the farm. Other measurements include residual mineral nitrogen in autumn and at harvest. Additionally, nitrogen surpluses are calculated using nutrient mass balance and MINAS approaches.

'Steering on nitrate' aims at the selection of nutrient efficiency indicators that can be used for evaluation of water quality in day-to-day practice. Activities of this project include a desk study on potential indicators, as well as an intensive measurement programme on a limited number of farms distributed over the country (Ten Berge 2002). Measurements include both direct nitrate concentrations as well as a number of potential indicators such as mineral nitrogen and nitrogen surplus. Results are used in the selection of a proper indicator.

RESULTS

Preliminary results of 'Steering on nitrate' confirm earlier findings that in the Netherlands, fertilization practices and groundwater levels equally influence nitrate concentrations, high nitrogen application levels and low groundwater levels leading to increased nitrate concentrations in the upper groundwater. Crops associated with high nitrate concentrations are potato, scorzonera and double-cropped vegetables. Not surprisingly, cereal plots show low concentrations. These findings are confirmed by preliminary results from 'Farming with a future', which further suggest an average nitrate concentration which is almost double the EU objective of $50 \, \text{mg} \, \text{l}^{-1}$. This is mainly due to high values in vegetable cultivation.

A review of potential indicators in 'Steering on nitrate' suggests that residual mineral nitrogen in autumn and nitrogen surplus can be valuable indicators for nitrate concentrations of groundwater. The theoretical relation between nitrate and potential indicators is explained in Figure 3 which depicts residual mineral nitrogen, nitrogen surplus and nitrate figures of individual farms in a three-quadrant figure. Ideally, a farm complies with all objectives that are set: nitrate, nitrogen surplus and residual mineral nitrogen (Farm 1). In practice, however, farms may comply with only a limited number of objectives, for example, for nitrogen surplus (Farm 2), residual mineral nitrogen (Farm 3), or nitrogen surplus and residual mineral nitrogen (Farm 4). Some farms will not comply with any objective (Farm 5).

Nitrogen surpluses were calculated for arable farms in 'Farming with a future' using a nutrient mass balance approach (Figure 4). They show that large differences exist between the regions represented in the project. A surplus of 80–90 kg N/ha is found on sandy soils

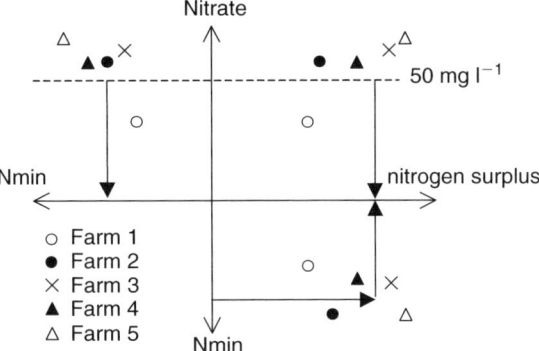

Figure 3. Theoretical relation between nitrate measurements (nitrate axis), residual mineral nitrogen and nitrogen surplus (Nmin and nitrogen surplus axes respectively). Figure design by H. ten Berge (pers. comm.).

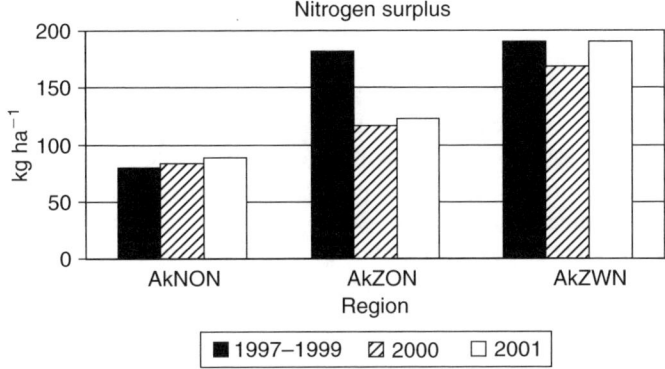

Figure 4. Nitrogen surplus calculated with nutrient mass balances at farm level on arable farms of 'Farming with a future'. Data for the average of the period 1997–1999, and to 2000 and 2001. (AkNON = arable farming in north-east of the country; AkZON = arable farming in the south-east; AkZWN = arable farming in the south-west.) Source: Anon. (2002).

in the north-east of the country. Surpluses of over 160 kg are found in the clay area of the south-west, while the sandy region in the south-east takes an intermediate position. Data refer to the two project years that have been analysed so far (2000 and 2001). These differences are in line with earlier experiences. Traditionally, crop rotation and soil distribution on farms in the north-east allow them to realize higher nutrient efficient use. In the south-east, where different crop patterns are found, most farms combine arable farming with animal production, so a manure surplus has triggered high to very high fertilization, while autumn application of manure in the south-west reduces nitrogen use efficiency by crops grown in this region.

In addition, average figures for a three-year period (1997–1999) prior to the start of the project are given for reference. Figures for 2001 exceed those for 2000, which seems odd. It is explained by a number of incidents (changes in farm structure, adoption of new fertilization strategy, hiring of heavily fertilized plots). Most incidents are not

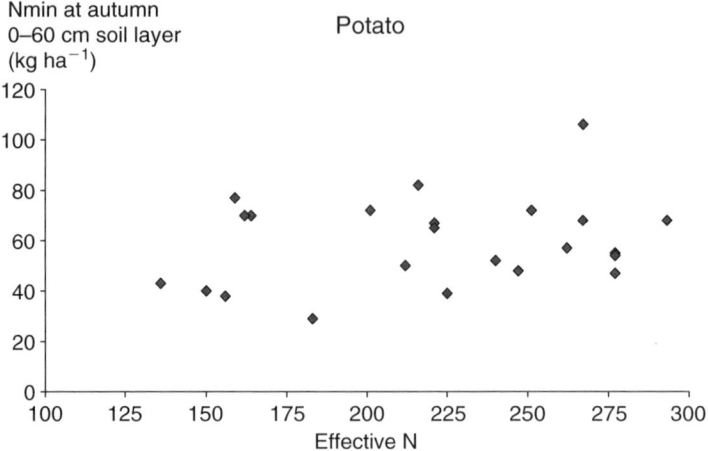

Figure 5. Mineral nitrogen in autumn resulting from nitrogenuous fertilization of potato plots in 'Farming with a future'. Data refer to 2001. (See text for explanations.)

expected to occur again. Over time, important progress has been made in the south-east and south-west. In the project, a threshold value for the nitrogen surplus of $90 \, \text{kg ha}^{-1}$ is applied. Comparison of the results presented in Figure 4 shows that this objective can be met in the north-east, while it may be within reach for arable farms in the south-east.

Measured residual mineral nitrogen is presented in Table 4. Data refer to project years only. On average, mineral nitrogen in 2001 is below the 2000 level. Comparison to objective value ($45 \, \text{kg ha}^{-1}$ on sandy soils, $70 \, \text{kg ha}^{-1}$ on clay soils) shows that objectives are within reach for farms in two of the three regions.

For comparison, figures for vegetable cultivation have been added. They indicate that mineral nitrogen levels in vegetable cultivation exceed those of arable farming. This is due to intensive cultivation practices including multiple sequential cropping (lettuce, endives) which combine high nitrogen applications with low efficiency of nitrogen use. The latter, of course, is caused by the fact that vegetable crops are harvested in a growth phase (where nitrogen uptake still is high) whereas arable crops generally are harvested during or after a ripening phase (where nutrients can be harvested leaving relatively low mineral nitrogen levels in the profile).

Results presented are used in the evaluation of farming practices. Starting in 2001, the effect of fertilization on nitrate leaching is determined both before and after the growing season. Results are presented to the pilot farmers and new, alternative schemes are discussed. In this way, farmers are continuously stimulated to improve their fertilization practices. Notwithstanding, large differences are found between farms and even within individual farms. This is demonstrated by Figure 5, where mineral nitrogen in autumn is depicted against the amount of effective nitrogen (nitrogen available for the crop in the year of application: initial soil mineral nitrogen, applied chemical fertilizers and both mineral nitrogen and easy-degradable organic nitrogen from applied organic manure) that is applied. A wide range of mineral nitrogen values is found, most of which are between 45 and $70 \, \text{kg ha}^{-1}$, while there is no clear relation between mineral nitrogen in autumn and applied effective nitrogen.

Final results of nitrate measurements (not yet available) will be used to depict the relation between nitrate on the one side and nitrogen surplus and residual mineral nitrogen on the other side.

DISCUSSION AND CONCLUSION

As nutrient emissions from agriculture are a serious problem in the Netherlands, considerable efforts are made to assist farmers in increasing fertilization efficiency and reducing emissions, such as nitrate emissions, to groundwater. It is too early to determine whether the approach followed is successful. So far, large differences were found in nitrate concentrations between crops and between farms. Alternative fertilization practices currently are being explored. The set-up of the two research projects discussed here combines direct nitrate measurements with measurement and calculation of so called nutrient efficiency indicators. As soon as sufficient nitrate measurement results become available, the relation between nitrate concentrations and potential indicators can be determined. Preliminary calculations suggest that nitrogen surplus and residual mineral nitrogen can serve as indicators satisfactorily, but threshold values corresponding to the EU nitrate concentration in groundwater ($50 \, mg \, l^{-1}$) have not yet been determined. Results further suggest that there is plenty scope for a significant reduction of nitrogen (nitrate) emissions. It is not clear, however, whether the EU nitrate objective can be realized. Problems are expected especially for double cropping in vegetable production.

Literature on the situation in other European countries (for example, Sapek 2000) suggests that nitrate and, more generally, nitrogen emission problems there also could benefit from the approach applied in the Netherlands. In Poland, for example, nitrate leaching is the second major nutrient emission problem. Additional efforts are required to assist farmers to reduce their nutrient emissions. Calculations for Polish agriculture (Langeveld 2000) suggest that at least 40% of nitrogen input is being lost. One third of this (being 15% of all input) is lost through nitrate leaching. As a result, Poland is the single most important contributor of nitrate to the Baltic Sea. Thus there is need for improvement. As the economic situation of Polish agriculture is difficult, the major focus should be on increasing nitrogen use efficiency as a way to save expensive fertilizer inputs. It is thought that the use of nutrient efficiency indicators that refer directly to farmers' perception of good farming practices (such as nutrient balances and residual mineral nitrogen) can be especially beneficial here. This will probably also be the case in other countries.

ACKNOWLEDGEMENTS

Support for 'Steering on nitrate' and 'Farming with a future' by the Ministry of Agriculture, Nature Management and Fisheries and the Ministry of Housing, Spatial Planning and the Environment is gratefully acknowledged. The authors would like to thank Peter Uithol for his excellent work on figures and tables.

REFERENCES

Anon. 2002: *Telen met toekomst: kansen en knelpunten in zicht. Jaaroverzicht 2001.* Plant Research International, Wageningen.

Booij, R., Van Dijk, W., Smit, B., Wijnands, F., Langeveld, H., De Haan, J., Pronk, A., Schröder, J., Proost, J., Brinks, H., Dekker, P. and Ehlert, P. 2001: *Detaillering projectplan 'Telen met toekomst'.* Telen met toekomst Publicatie no. 3. Praktijkonderzoek Plant en Omgeving, Lelystad.

Fraters, B., Van Eerdt, M.M., De Hoop, D.W., Latour, P., Oltshoorn, C.S.M., Swertz, O.C., Verstraten, F. and Willems, W.J. 2000: Landbouwpraktijk en waterkwaliteit in Nederland. Achtergrond-informatie periode 1992–1997 voor de landenrapportage EU-nitraatrichtlijn. Report 718201003. National Institute of Public Health and Environmental Protection, Bilthoven.

Langeveld, J.W.A., Van Keulen, H., De Haan, J.J., Kroonen-Backbier, B.M.A. and Oenema, J. in press: The nucleus and pilot farm research approach: examples from the Netherlands. Submitted to *Agric. Systems.*

Langeveld, J.W.A. 2000: *Application of nutrient balance sheets in analysis and design of agricultural systems; examples from the Netherlands and Poland.* In: A. Sapek (ed.): Scientific basis to mitigate the nutrient dispersion into the environment. Proceedings of a conference held in Falenty, 13–14 December 1999. (IMUZ) Institute for Land Reclamation and Grassland Farming, Falenty, pp. 78–89.

RIVM 2001: Nationale Milieubalans. National Institute of Public Health and Environmental Protection, Bilthoven.

Sapek, A. (ed.) 2000: *Scientific basis to mitigate the nutrient dispersion into the environment.* Proceedings of a conference held in Falenty, 13–14 December 1999. (IMUZ) Institute for Land Reclamation and Grassland Farming, Falenty.

Schröder, J.J., Neeteson, J.J., Oenema, O. and Struik, P.C. 2000: Does the crop indicate how to save nitrogen in maize production? Reviewing the state of the art. *Field Crops Research 66*: 151–164.

Ten Berge, H.F.M. (ed.) 2002: *A review of potential indicators for nitrate loss from cropping and farming systems in the Netherlands.* Reeks Sturen op nitraat 2. Plant Research International, Wageningen, pp. 119–126.

CHAPTER 26

Water and agriculture management: environmental problems in the Fucino Plain (central Italy)

M. Petitta and A. Del Bon
Dipartimento Scienze della Terra, Università 'La Sapienza' – P. Aldo Moro 5 – 00185 Roma, Italy

E. Burri and G. Pannunzio
Dipartimento Scienze Ambientali, Università dell'Aquila, Via Vetoio, Località Coppito, 67100 l'Aquila, Italy

ABSTRACT: The reclamation of Lake Fucino, an endorheic basin lacking natural outlets and the third largest lake in Italy ($200\,km^2$), was carried out in the 1800s through the construction of an artificial underground canal. This led to the complete and irreversible disappearance of the lake.

The initial agricultural crops were progressively replaced by much more profitable vegetable crops. This new tendency, which has been more marked in recent years, was accompanied by high water requirements, exacerbated by the practice of two/three harvests per year. As a consequence, many wells were drilled in the area, generally in the surrounding carbonate structures, in order to exploit groundwater.

The paper reports the results of a research programme that was started in 1998 with a view to: i) monitor some wells; ii) make an initial evaluation of water withdrawal; iii) identify type and quantity of water-intensive crops; and iv) assess nitrates, if any, in streams and groundwater.

The results of nitrate studies, started in 2001, showed the presence of nitrates and ammonia both in springs and streams. Nitrates proved to have their maximum spatial distribution in the western and northern portions of the Fucino Plain and to be more concentrated in streams than in springs. The association of nitrates with the ammonium ion made it possible to discriminate between nitrates of agricultural origin and those of geological origin, given the presence of clay deposits of Neogene age representing a potential natural source of nitrogen compounds. The distribution of nitrates over time, which was very variable, enabled the assessment of the actual impact of agriculture on groundwater, as well as the natural purification rate of the outcropping deposits.

Using the final results of the study, a groundwater management plan, taking into account the Fucino Plain agricultural needs and aquifer vulnerability, will be formulated.

INTRODUCTION

The Fucino Plain, over $200\,km^2$ wide, once hosted the largest lake in central Italy, which was reclaimed for agricultural use in the 1800s. As part of research activities, nitrate concentrations in streams and groundwater in different seasons were monitored. The paper reports the results of a water resource monitoring programme, which was conducted on the basis of the local hydrogeological and agricultural setting.

The final aim of the programme is to develop a plan for better management of water resources in the Fucino Plain, taking into account the characteristics of its streams, groundwater, irrigation needs and types of agricultural crops. This planning activity should be co-ordinated by all the authorities operating in the Plain (Drinking Water Authority, Regional Agency for Agricultural Development Services – ARSSA, Irrigation Authority and Industrial Development Authority).

The current situation has many imbalances, which are increasing and causing environmental problems, such as groundwater lowering, spring exhaustion, decreased canal discharge in summer and, of course, surface and groundwater pollution risk connected with agricultural fertilizer and pesticide use.

HYDROGEOLOGICAL AND AGRICULTURAL SETTING

The Fucino basin extends for about $900 \, \text{km}^2$ over the carbonate ridges of the Latium-Abruzzo Apennines and is morphologically dominated by an alluvial plain. The depression is bounded by extensional or thrust-belt tectonic features. Furthermore, it may be rapidly filled with terrigenous syn-orogenic sediments (flysch) or with detrital and lacustrine alluvial deposits, which may reach $1,000 \, \text{m}$ thick (Giraudi 1994; Cavinato, Carusi, Dall Asta *et al.* 2002). The fractured and karstified carbonate aquifers encircling the plain are drained at their boundary by high-discharge springs (Figure 1), which ensure regular discharges even during the dry season (Boni, Bono and Capelli 1986; Celico 1983). The

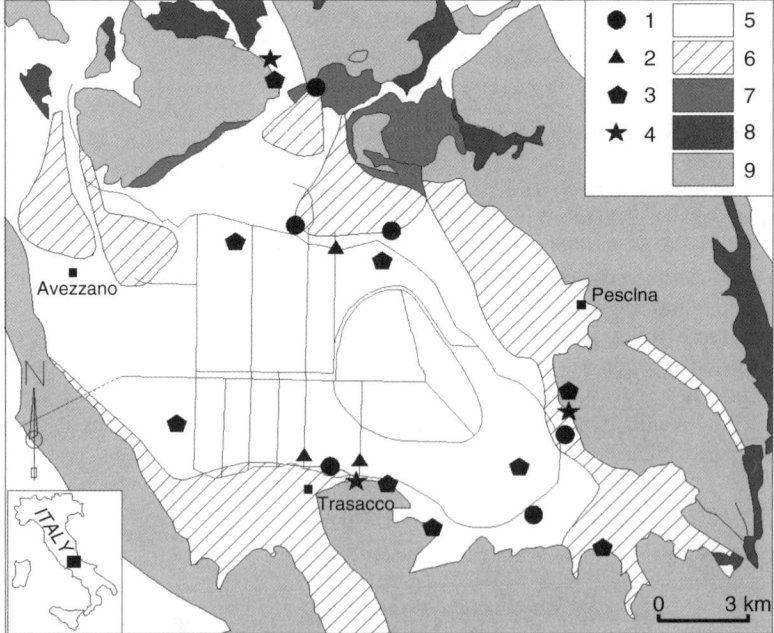

Figure 1. Hydrogeological map of the Fucino basin. 1) Main springs; 2) main streambed springs; 3) irrigation water well fields; 4) drinking water well fields; 5) alluvial and lacustrine deposits; 6) detrital and fan deposits; 7) ancient alluvial deposits; 8) siliclastic deposits (aquiclude); 9) carbonate deposits (aquifer).

aquifer of the plain is supplied in part by groundwater seepage from the carbonate ridges and in part by direct infiltration. The permeability of the porous aquifer is extremely variable, depending on the grain size of the alluvial deposits. The Fucino watershed consists of multiple man-made canals converging into a 'main canal', which collects the water into the tunnelled outlet.

Climatic data show that evapotranspiration is equal to 450 mm/year and that average water surpluses amount to as little as 240 mm/year (Petitta and Capelli 1994). Water surpluses only occur in the November–March period. From April through October, rainfall does not contribute to the water balance of the basin. Consequently, in summer, large amounts of water are used to irrigate farm crops.

The drainage of Lake Fucino, completed in 1876, represents one of the most extensive and radical environmental changes of historical times. The project had been conceived and partly implemented by the Romans 19 centuries earlier (Burri 1990). The reclamation was completed in 1875 by Prince Torlonia and the necessary infrastructure was built: 210 km of roads, over 100 km of canals, 618 km of drainage ditches, in addition to farmhouses, store-houses and stables (Brisse and De Rotrou 1876).

In 1951, the Torlonia estate was expropriated, the land was transferred to the local farmers, and the 'Ente Fucino' (Fucino Basin Authority, now ARSSA) was established. This Authority had the task of assigning the land and of launching and sustaining a new model of farmland management. Moreover, the growing of three main crops (wheat, potatoes and sugar beets) on the basis of a three-year rotation was encouraged. This system remained practically unaltered until 1962. These crops required minimal amounts of water and the demand was satisfied by the natural flows in the canals. However, in the past few years, farmers gradually switched to vegetable crops (Figure 2), such as carrots, salad greens, fennel, celery, etc. Vegetable crops are very widespread, thanks to favourable circumstances, including the option of repeating two to three growing cycles on the same field, provided that adequate water resources are available. At the same time, fertilizer and pesticide use increased: in the 1990s, the amount of pesticides used in the Fucino Plain exceeded 35 kg ha^{-1}, in comparison with an Italian national average of 12.8 kg ha^{-1} and of only 9.9 kg ha^{-1} in the rest of the Abruzzi region (Burri and Petitta 1998).

Water requirements for agricultural and other uses have increased since the 1950s, leading to the digging of more than 200 public wells. Since 1989, ARSSA has been directly managing the most important wells. There are another three water management authorities

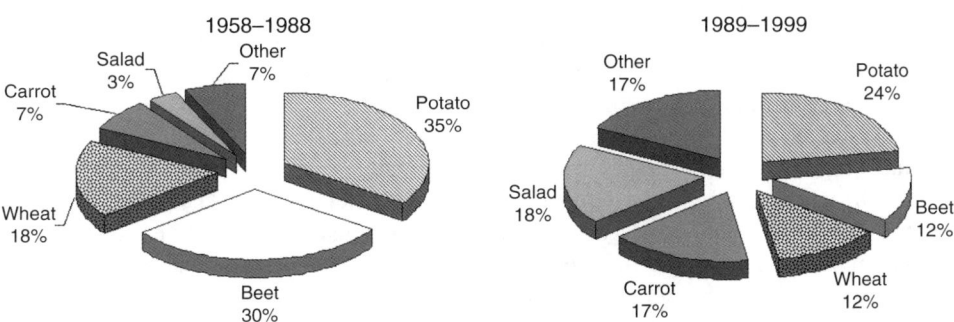

Figure 2. Evolution of farming in the Fucino Plain – comparison of crop distribution in the 1958–1988 and the 1989–1999 periods.

in the Fucino Plain: Drinking Water Authority, Reclamation Authority and Industrial Development Authority. A map of the current main well fields is shown in Figure 1.

Water consumption for farming is concentrated in the May–September period: water is generally withdrawn from surface canals via mobile pumps connected to tractor engines.

Both ARSSA and the Reclamation Consortium tap groundwater from the well fields and discharge it into the canal network, where it is exploited by individual users. The local consumption of water has never been accurately determined, owing mainly to the diversification of uses (different users, different patterns and periods of use).

As a consequence of reclamation works, but especially of the recent patterns of utilization of the local water resources, the hydrogeological setting underwent major changes. In the last few decades, these changes created high-risk factors, related to depletion and pollution of groundwater resources, both in the plain and in the surrounding karst aquifers (Burri and Petitta 1998, 2002).

WATER AVAILABILITY

In the last decade, pumping from the wells of the Fucino area, both in the plain and on the edges of the surrounding carbonate aquifers, gradually increased in order to cover the growing needs of agricultural crops, still watered by wasteful sprinkler irrigation methods.

At the same time, both the output of the springs and the natural outflow from the tunnelled outlet decreased. The mean outflow from the Fucino basin over the last 30 years was $6 \, m^3 \, s^{-1}$.

This value dropped significantly in the past few decades, from a yearly average of $7.6 \, m^3 \, s^{-1}$ from 1968 to 1987, to $3.3 \, m^3 \, s^{-1}$ in the decade 1988–1998 (minimum of $1.5 \, m^3 \, s^{-1}$ in 1990). Outflow values in summer (June–September) declined from $4.4 \, m^3 \, s^{-1}$ in the 1968–1987 period to $0.9 \, m^3 \, s^{-1}$ in the last decade (Burri and Petitta 1999). Today, water discharge into the Fucino Basin in summer is fed by groundwater. Water withdrawal from wells for irrigation uses reached a value of $1 \, m^3 \, s^{-1}$ in 1999–2001.

The latest measurements indicated that the flow of springs, reported in the literature as 6–$7 \, m^3 \, s^{-1}$ in the 1970s (Boni *et al.* 1986; Celico 1983), is currently no greater than $2 \, m^3 \, s^{-1}$ and that the springs dry up during the irrigation season. This fact resulted in the disappearance of many of the main natural springs and significantly reduced flows in all other springs (Burri and Petitta, 2004).

In this context, a research programme was started in 1998 in order to assess the local hydrological balance, monitor groundwater (Burri and Petitta 2002), evaluate the distribution of crops and estimate their water requirements.

The research programme, whose data are stored in a dedicated database, includes: groundwater level monitoring; census of springs and wells (over 200 water points); collection of precipitation and temperature data from ten stations; discharge measurements along streams and canals with hydrometric data; analysis of crop distribution, plot by plot, in the southern half of the Fucino area (about 6,000 ha), with determination of the variations from 1989 until today and with particular reference to the practices of repeated harvests in the same year (second harvest and, although rarely recorded, third harvest).

On the basis of appropriate conversion tables, the potential water requirements of the crops were determined to lie in the range of 10 million m^3 per season for the whole plain.

This demand obviously increased in recent years (up to 12 million m³ per season). In summer, it is roughly distributed as follows: 5% in June, 35% in July, 45% in August and 15% in September. Springtime demand, easily satisfied by precipitation, was disregarded.

Since actual natural outflows in summer do not reach 9 million m³ on average, use is made of groundwater; as mentioned previously, groundwater exploitation in recent years exceeded 10 millions of m³ during the irrigation period. In principle, groundwater withdrawal should coincide with water deficits for irrigation. In practice, however, the distribution system does not allow optimal yields and even favours waste, caused mainly by sprinkler irrigation methods. This fact, in turn, increases evapotranspiration, which is potentially very high in summer months. The adoption of sprinkler irrigation systems, coupled with the increase in groundwater exploitation and water requirements by intensive crop growing, cause a risk of pollution of the Fucino Plain groundwater, a risk that may extend to the surrounding carbonate aquifers (Burri and Petitta 1998), given the small piezometric difference between the two types of aquifers.

NITRATE DATA MONITORING

To obtain a preliminary indication of the agricultural impact on groundwater in the Fucino Plain, a water-monitoring programme was conducted from 2001 to 2002, in order to evaluate nitrate releases and concentrations in surface water and groundwater.

Water samples were collected in October 2001 (52 water points, mainly spring), in February 2002 (89 water points, mainly canals) and in May 2002 (96 water points, springs and canals). The third sampling survey was focused on many of the previous sampling sites, with a view to evaluate seasonal variations. Figures 3–4–5 and 6 show the sampling sites, including surface water (canals) and groundwater (springs).

Chemical parameters were lab-determined with standard procedures based on Spectroquant Kits (Merck): i) ammonia nitrogen as NH_4^+ (mg l^{-1}); ii) nitric nitrogen as NO_3^- (mg l^{-1}); iii) phosphoric ions PO_4^{3-} (mg l^{-1}).

During the first survey, only NH_4^+ and NO_3^- were determined. From the second survey, PO_4^{3-} was also analysed.

The results showed a maximum concentration in NO_3^- of 96 mg l^{-1}, with several samples having minimum values (below 1 mg l^{-1}) during the three surveys. The parameter NH_4^+ ranged from <0.1 to a maximum of 35 mg l^{-1}. The ion PO_4^{3-} ranged from <1 to 20 mg l^{-1}. More details are shown in Table 1. The spatial distribution of the parameters, processed with the kriging method, was determined on the basis of the winter surveys (October 2001 + February 2002, Figures 3 and 5) and of the spring survey (May 2002, Figures 4 and 6).

The results of analyses suggest the following:
- Surface water samples have a higher concentration of nitrates and ammonia ions than groundwater samples; this fact demonstrates the low permeability of the outcropping deposits in the plain, which favours runoff into the canals;
- the remarkable seasonal variations in nitrates and ammonia are connected to agricultural activities and to the dilution effect due to precipitation; higher values are observed in winter surveys, when fertilizer use is very common; otherwise, during springtime, the abundance of surface waters lowers nitrate concentration;

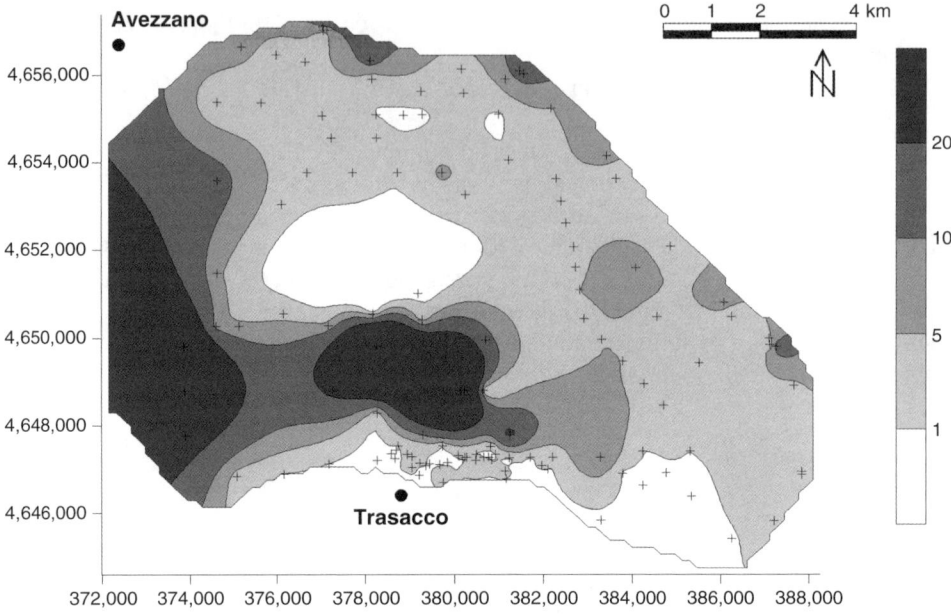

Figure 3. Map of NO_3^- concentration ($mg\,l^{-1}$) in streams and groundwater, during the winter period. The crosses identify the sampling sites.

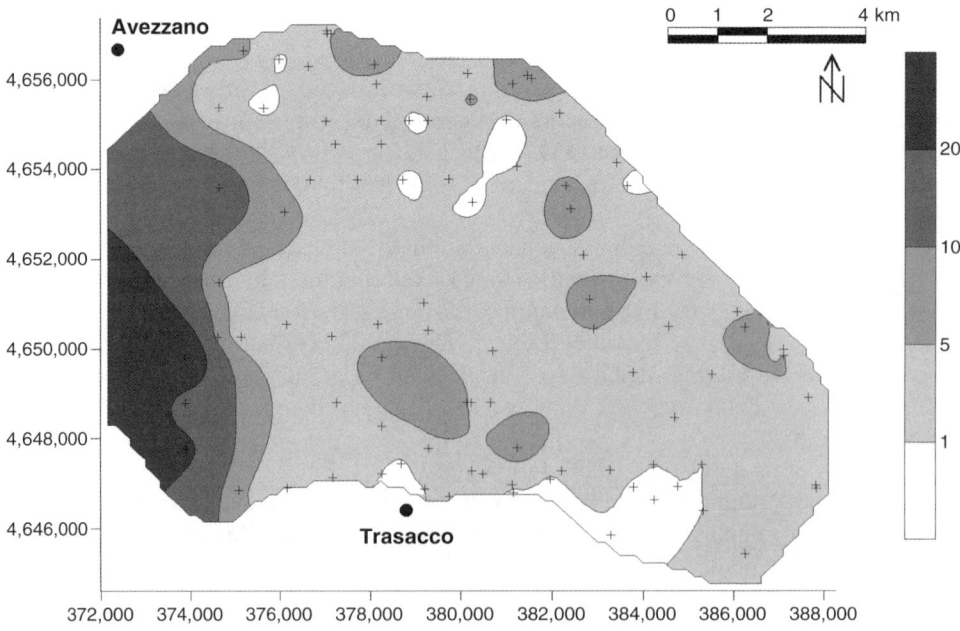

Figure 4. Map of NO_3^- concentration ($mg\,l^{-1}$) in streams and groundwater, during the spring period. The crosses identify the sampling sites.

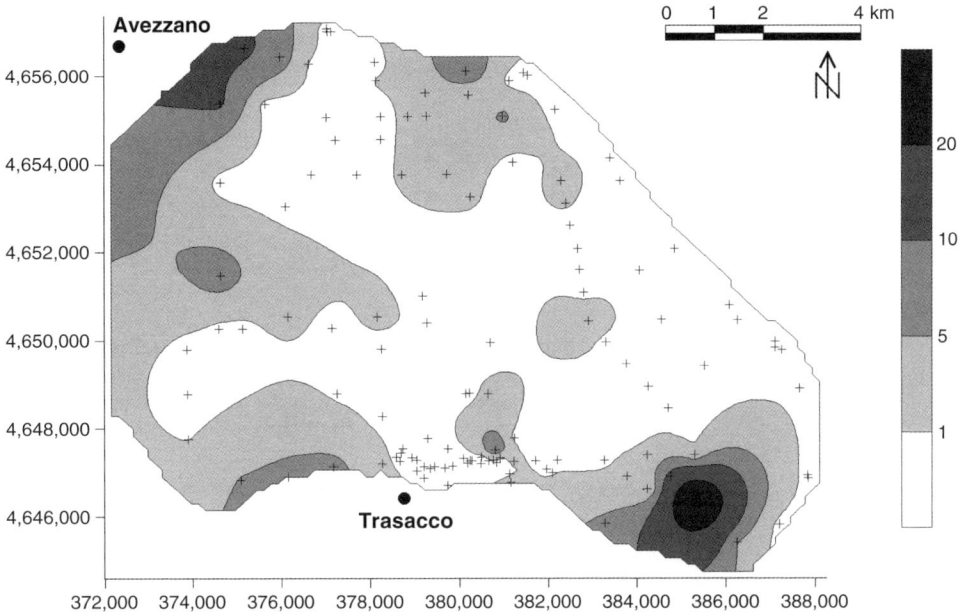

Figure 5. Map of NH_4^+ concentration (mg l^{-1}) in streams and groundwater, during the winter period. The crosses identify the sampling sites.

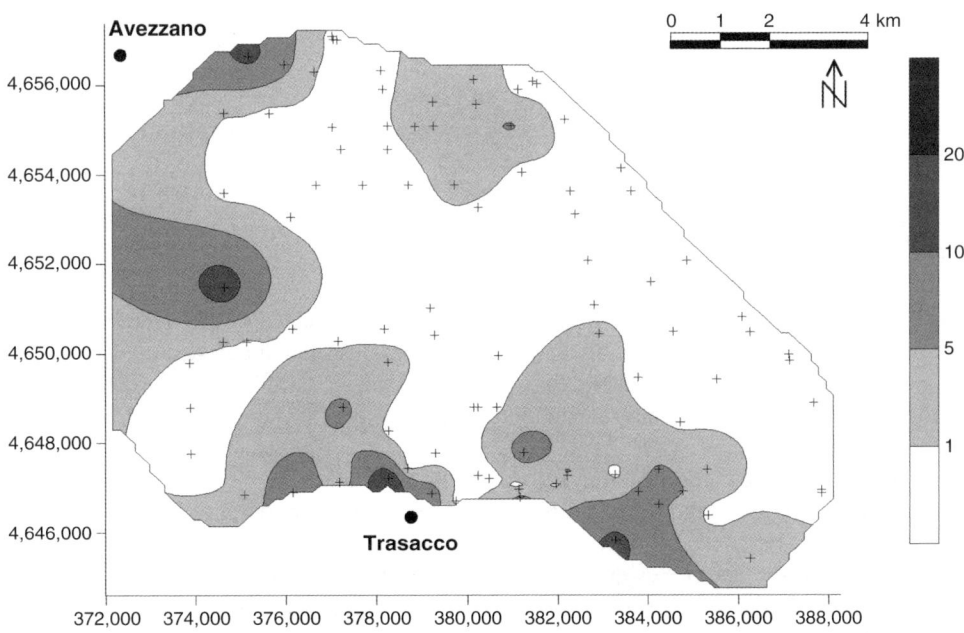

Figure 6. Map of NH_4^+ concentration (mg l^{-1}) in streams and groundwater, during the spring period. The crosses identify the sampling sites.

Table 1. Measured values of ion concentration in the waters of the Fucino Plain during the three surveys 2001–2002 (total: 237 samples).

Parameter	Value ($mg\,l^{-1}$)	1st survey (October 2001)	2nd survey (February 2002)	3rd survey (May 2002)
NO_3^-	Max	13	96	35
	Average	0.5	7.2	2.8
	Min	<1	<1	<1
NH_4^+	Max	11.2	35	15.5
	Average	0.6	2.0	3.9
	Min	<0.1	<0.1	<0.1
PO_4^{3-}	Max	Not measured	20	10
	Average	Not measured	2.0	2.0
	Min	Not measured	<1	<1

- the origin of ammonia in the groundwater is also to be ascribed to the presence of organic deposits in the lacustrine rock sequence; its contribution to the nitrogen cycle should be evaluated before evaluating pollution;
- the spatial distribution of nitrates depends also on the local hydrogeological setting; in fact, in areas where the aquifer is shallow and perched on a low-permeability bed (clays), nitrate concentration is higher (on the western and the northern side); in areas where the aquifer is directly connected to the karst aquifer, nitrates have lower values, due to the dilution effect of seepage (on the southern and eastern side, near the main springs).

Today, pollution due to agricultural activity in the Plain, as shown by nitrates, is moderate, with its highest values at the end of the dry season, when fertiliser use is at its maximum and the dilution effect of precipitation and groundwater is minimum. The low permeability of the Plain deposits prevents and slows down the migration of nitrates into groundwater. At the same time, this situation causes an increase in the pollution of canals, where low discharge worsens the situation.

CONCLUSIONS

The research carried out in the Fucino Plain shows the influence of the hydrogeological setting on nitrate distribution in soil and groundwater. Despite the decrease in discharges, groundwater pollution is mild. Nitrates have higher concentrations in surface waters, whose low discharge, especially at the end of the dry period (summer–autumn), causes environmental problems.

In detail, the distribution of nitrates shows higher values at the western and south-western border of the Plain, for two different reasons: along the northern border, there is a perched water table, extremely vulnerable to nitrates; agricultural activities are concentrated in the south-western area and nitrate dilution is very difficult, due to the lack of groundwater. Only near the main springs of Trasacco the nitrate values decrease very quickly. The ammonia distribution is probably connected with the presence of organic deposits in the lacustrine sequence; at the north-western border, the city of Avezzano, with its sewage, might represent a pollution source.

Nevertheless, the observed groundwater pollution, albeit low, might be a first indicator of future involvement of groundwater in the pollution process.

Hence, the need arises not only for conducting hydrogeological studies, but also for changing methods of irrigation and use of pesticides and fertilizers, in order to minimize the risks of over-abstraction from, and pollution of, the local streams and groundwater. Only a modern strategy of integrated water management can guarantee the future survival of the man–water–agriculture system of the Fucino Plain.

ACKNOWLEDGEMENTS

We thank ARSSA (Franco Ciofani and Vittorio Di Giamberardino) for its financial and logistic support and Valentina Marinelli for field sample collection. The comments made by an anonymous referee were highly appreciated.

REFERENCES

Boni, C., Bono, P. and Capelli, G. 1986: Schema Idrogeologico dell'Italia Centrale. *Memorie Società Geologica Italiana 36*: 991−1012.

Brisse, A. and De Rotrou, L. 1876: Desséchément du Lac Fucino executé par S.E. le Prince Alexandre Torlonia. Precis historique et techniques. Imprimerie de la Propagande, Rome.

Burri, E. 1990: Storia di un lago. Il Fucino in Abruzzo. *Terra 3* (10): 42−52.

Burri, E. and Petitta, M. 1998: Groundwater lowering in karstic aquifers due to agricultural activity in the Fucino Plain (Abruzzi, Central Italy). *Acta Carsologica 27* (1−2): 27−46.

Burri, E. and Petitta, M. 1999: *Farming and water management in the Fucino Plain (Central Italy) in the last century.* In: Proceedings of the XVII° International Congress ICID, Granada, 1D, pp. 257−268.

Burri, E. and Petitta, M. 2002: *Groundwater monitoring as a management tool for irrigation in the Fucino Plain (Abruzzo, Central Italy).* In: Luis Ribeiro (ed.): Future Groundwater Resources at Risk; Proceedings of the 3rd International Congress, Lisbona, June 2001. Fomento Grafico, Lisbona, pp. 425−432.

Burri, E. and Petitta, M. 2004: *Agricultural changes affecting water availability: from abundant to scarcity conditions (Fucino Plain, Central Italy).* Irrigation and Drainage, 53: DOI: 10.1002/ird.119.

Cavinato, G.P., Carusi, C., Dall'Asta, M., Miccadei, E. and Piacentini, T. 2002: Sedimentary and tectonic evolution of Plio-Pleistocene alluvial and lacustrine deposits of Fucino Basin (central Italy). *Sedimentary Geology 148*: 29−59.

Celico, P. 1983: Idrogeologia dei massicci carbonatici, delle piane quaternarie e delle aree vulcaniche dell'Italia centro-meridionale: Progetti speciali per gli schemi idrici nel Mezzogiorno. *Quaderni Cassa Mezzogiorno* 4/2:1−225, Rome.

Giraudi, C. 1994: *Origine ed evoluzione geologica recente del bacino del Fucino.* In: E. Burri (ed.): Il lago Fucino e il suo emissario. Pescara, Carsa, pp. 14−34.

Petitta, M. and Capelli, G. 1994: *Inquadramento idrologico del bacino del Fucino.* In: E. Burri (ed.): Il lago Fucino e il suo emissario. Pescara, Carsa, pp. 46−61.

CHAPTER 27

Nitrate monitoring in groundwater of Dutch provinces

M. Van der Aa, B. Van Der Grift, R. Busink and H.P. Broers
Netherlands Institute of Applied Geoscience TNO-NITG, Schoemaker straat 97, P.O. Box 6012, 2600 JA Delft, The Netherlands

ABSTRACT: In the Netherlands, eight out of twelve provinces have both soil and groundwater quality monitoring networks. The effects of (changing) land use practices are evaluated by monitoring temporal and spatial trends in soil and groundwater quality. The soil quality monitoring networks focus on the upper metre of the soil and groundwater. The groundwater quality monitoring networks focus on groundwater at depths from about 5 to 30 metres. In this research, data of both soil- and groundwater monitoring networks are made comparable and evaluated together. In this way it is possible to present detailed concentration-depth profiles for areas with common characteristics. The results show that the nitrate concentration–depth profile is related to the combination of soil type, land use and hydrological situation. Nitrate is present in low concentrations in shallow groundwater of clayey and peaty soils, while it is absent in the deeper groundwater of these areas. In shallow groundwater of sandy soils with agricultural land use, nitrate concentrations are high and often exceed the EU nitrate directive of $50 \, mg \, l^{-1}$. However, the deeper groundwater of sandy soils with agricultural land use in recharge areas (dry anthrosols and podzols) shows much higher nitrate concentrations than the deeper groundwater in groundwater discharge areas (wet anthrosols and podzols). This is related to the older age of the deeper groundwater in discharge areas, which mostly is above 50 years. Around 1950 agricultural practices were rather different and manure loads in the recharging groundwater were much lower. The results also show that two provinces with similar site characteristics but different manure loads have a similar nitrate concentration–depth profile. This is related to regional differences in geochemical characteristics of the deeper aquifers, which cause nitrate reduction in one province, but not in the other.

INTRODUCTION

History of soil and groundwater quality monitoring in Dutch provinces

In the Netherlands, most of the 12 provinces have their own soil and groundwater quality monitoring networks. The effects of (changing) land use practices are evaluated by monitoring temporal and spatial trends in soil and groundwater quality. While some qualitative measurements of ground water have been performed since the beginning of the century, true groundwater quality monitoring only started in 1978 when the national government issued it to the National Institute of Public Health and Environmental Protection (RIVM) (Mol 2002). The national monitoring network for groundwater quality was completed by 1984 (van Drecht, Reijenders, Boumans *et al.* 1996; Dufour 1998). The main goal is to follow the quality of the compartment as it changes over time, and to relate these changes to either natural or anthropogenic causes. Provinces only started monitoring groundwater when, as a result of the Groundwater Act (1981), they became responsible for the groundwater quality. Most provinces considered the sampling density of the National Monitoring Network insufficient for their needs and extended the network with provincial sampling points. All these extensions were made between 1988 and 1993. The goal of these provincial monitoring networks is identical to that of the national network, but many provinces do some additional work, such as the installation of shallow filters and the measurement of additional parameters. This depends on local natural phenomena and sometimes on the preferences of the people in charge, resulting from differences in background and specific expertise (Mol 2002). Currently the national and provincial groundwater quality monitoring networks together consist of around 850 sampling sites.

Regarding soil monitoring, the National Institute of Public Health and Environmental Protection (RIVM) set up a pilot soil-monitoring network on a national scale in 1988/1989. After evaluation in 1993 a design was chosen that focuses on the relation between agricultural activities (considered an important source) and soil contamination. Since 1991 some provinces also set up provincial soil monitoring networks. The provincial soil monitoring networks address three themes: contaminant spread, eutrofication and acidification. However, they show slight variation in their designs depending on problems related to the specific soil and geomorphological properties of the various provinces. Other factors influencing the details of the networks are the personal preferences of the provincial employees and the actual consultancy that was selected for the project management (Mol 2002). Eight out of the twelve Dutch provinces currently have operational networks, consisting of over 1,600 sampling sites (Busink and Postma 2000).

Monitoring goals

The soil and groundwater monitoring networks of the Dutch provinces share similar goals. They both aim to detect temporal and spatial trends in the quality of soil/groundwater, and relate these to anthropogenic (agricultural) influences. The provincial groundwater quality monitoring networks are also used for the evaluation of policy measures for sustainable protection of drinking water sources.

The design criteria for these two types of monitoring networks are rather different, since monitoring of soil and nutrient leaching to groundwater is strongly influenced by short-term seasonal climatic changes and local site characteristics, while the deeper groundwater generally shows more gradual quality changes representative of a much larger recharge area. In the soil quality monitoring networks soil, soil moisture and the upper metre of the groundwater (shallow groundwater) are sampled. In most cases sampling is performed by taking the average of four sub samples per parcel with homogeneous soil and land use. The groundwater quality monitoring networks focus on deeper groundwater between about 5 to 30 metre depth. In this monitoring network the groundwater is sampled at fixed sampling locations (permanent wells).

Towards a more integrated analysis of soil- and groundwater quality monitoring

Because the current soil- and groundwater quality monitoring networks focus on different compartments, there is no integral system concept and thus no insight into relationships between the load on the soil and the quality of the shallow and the deeper groundwater at regional level. There is, however, an increasing need for a more integrated examination.

An important bottleneck is the fact that the data from the two monitoring networks cannot or can hardly be compared as a result of differences in the various design and realization options. Now that the policy is moving towards a more integral approach, the relevant questions are: what are the relationships between compartments, which ones can be measured and what advantages can be obtained by attuning them?

In this research, seven out of the eight Dutch provinces with both soil- and groundwater quality monitoring networks joined efforts to investigate the possibilities for a more integrated evaluation of the monitoring networks. The provinces involved are Friesland, Groningen, Drenthe, Gelderland, Utrecht, Noord-Brabant and Limburg. Figure 1 shows

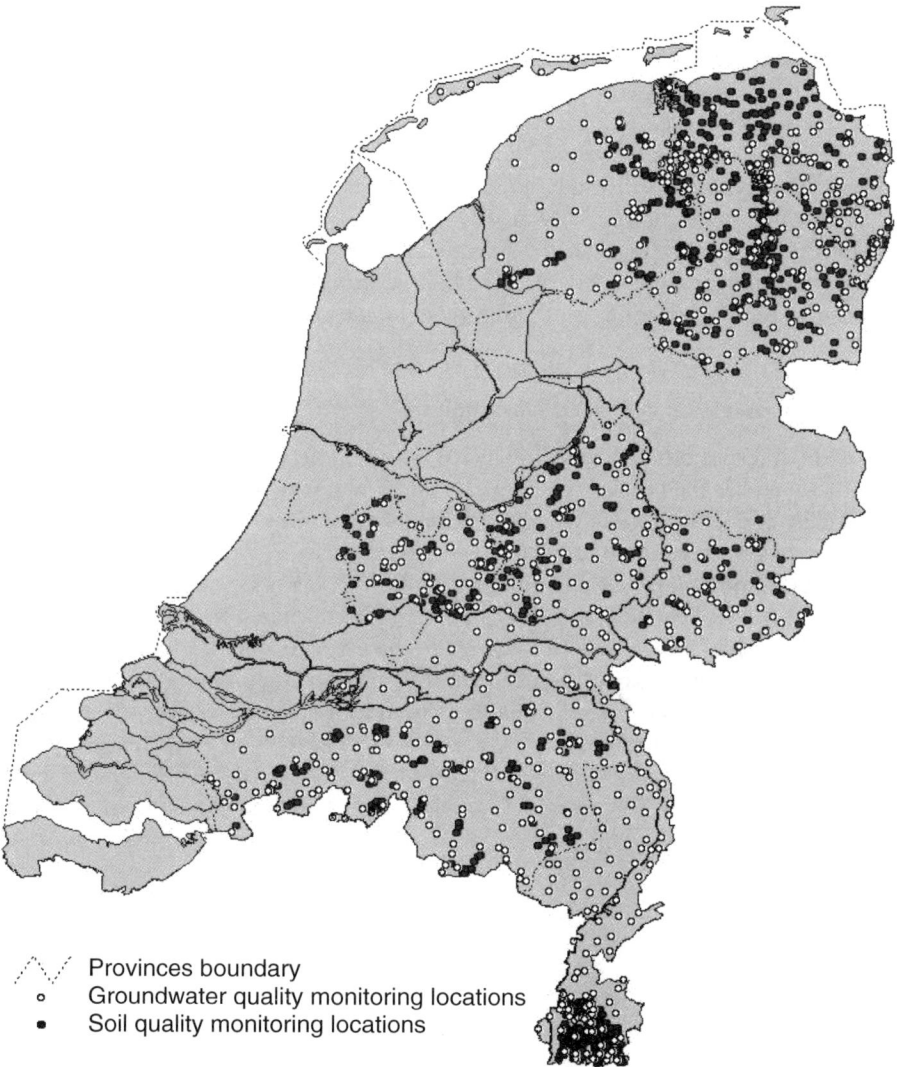

Provinces boundary
o   Groundwater quality monitoring locations
•   Soil quality monitoring locations

Figure 1. Monitoring locations of soil- and groundwater quality monitoring networks for seven out of twelve Dutch provinces: Friesland, Groningen, Drenthe, Gelderland, Utrecht, Noord-Brabant and Limburg.

the sampling sites of both the groundwater quality and soil quality monitoring networks for the seven Dutch provinces. The intention is that the added value of integration will become clear, while maintaining the original monitoring network objectives for the individual provinces. This should result in more uniform monitoring of the entire soil system, and give insight into relationships and (potential) cost reductions in terms of monitoring.

MATERIALS AND METHODS

Data quality control

A quality analysis has been conducted on the soil- and groundwater monitoring data as supplied by the provinces. This revealed a number of bottlenecks regarding the reliability of the data, the units used for reporting and the various formats in which the data were supplied by the provinces. Moreover, missing data or the lack of complete water analyses hampered quality control. A quality control of the groundwater quality data was conducted with a subroutine of the WATEQ4F program (Ball and Nordstrom 1991). With this procedure a quality tag can be assigned to a groundwater analysis. This quality analysis could not be conducted for the shallow groundwater data of the soil quality monitoring networks, since in most cases no full water analyses had been conducted.

Homogeneous areas for integrated data analysis

Based on the present province-specific and monitoring network-specific area types, it is difficult to compare the monitoring network data of the various provinces and of the soil and groundwater quality monitoring networks. At the time, the individual provincial monitoring networks were not designed for this purpose. For example, the land use at parcel level is important for the soil quality monitoring networks, since this focuses on the relationship between leaching and land use. On the other hand, the groundwater quality monitoring networks require information about the average land use in a catchment area over a period of several years. In order to still be able to compare the provinces and the various vertical components (soil − shallow groundwater − deeper groundwater) homogeneous area types have been defined at a supra-provincial level. This creates a frame of reference for both monitoring networks as well as a basis for inter-provincial investigations. These area types have been defined on the basis of land use, soil type and hydrological situation.

RESULTS

Data quality control

Quality control of the groundwater quality data with the WATEQ4F program (Ball and Nordstrom 1991) showed that in 58% of the samples the ion balance appeared to be correct (deviation <10%). An excess of cations appeared to be present in the deeper groundwater. This indicates a systematic error, possibly because the measured bicarbonate concentration in the laboratory was lowered as a result of degassing between the moment of sampling (when Electrical Conductivity was measured) and the moment of laboratory analysis. Consequently, the calculated EC on the basis of the laboratory analysis does not correspond well to the field-measured EC for 72% of the samples.

Homogeneous areas for integrated data analysis

Figure 2 shows the concept of homogeneous areas. In the Netherlands, recharge areas are typically found on the Pleistocene sandy areas in the eastern and southern part of the country. They are characterized by mean groundwater levels between 1 and 2 metres below the surface (groundwater class V–VII) and the absence of a drainage network. Typical groundwater discharge areas are found in the lower parts. They are characterized by mean groundwater levels of less than 1 metre below the surface, mostly peaty soils and a dense drainage network. Intermediate areas are found in the transition zone between recharge and discharge areas. Because of the presence of drains and ditches, local groundwater infiltration systems evolved.

In order to compare the results of the soil- and groundwater quality monitoring networks, homogeneous areas are defined based on the criteria land use, soil type and hydrological situation. Based on similarities in land use, a distinction between agriculture (mainly arable land and grassland) and nature (mainly forest and, to a lesser degree, heath and grassland) was made. Based on similarities in soil type a distinction between

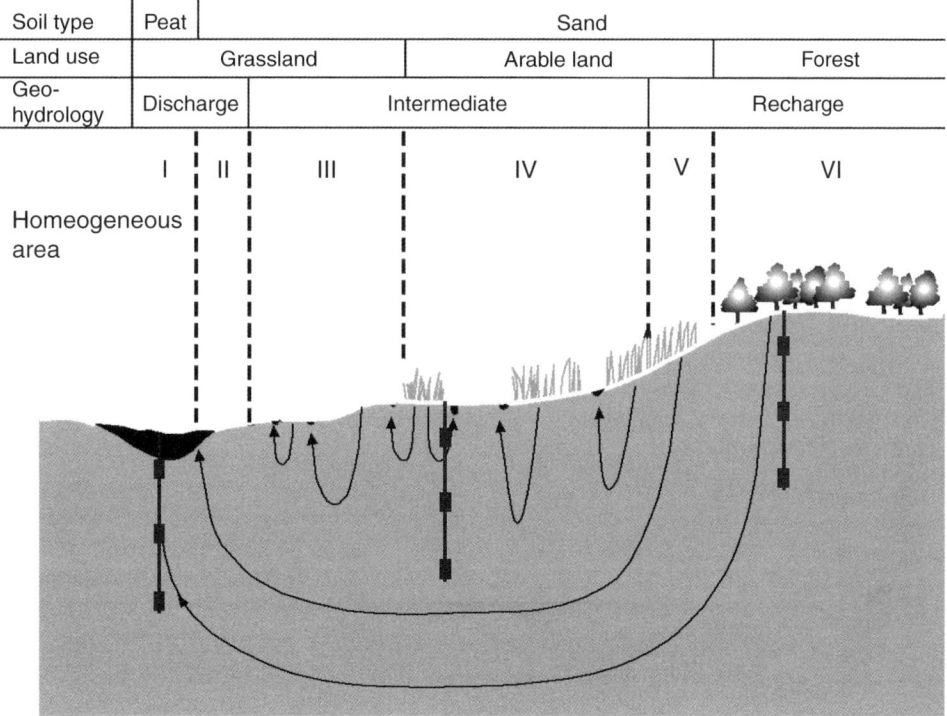

Figure 2. Homogeneous areas for integrated data analyses of soil- and groundwater quality monitoring networks.

anthrosols, podzols and gleysols was made. In order to make a distinction between recharge, intermediate and discharge areas, detailed information about groundwater flow systems is necessary. This information was not available for all seven provinces. Therefore groundwater levels were used as an indicator for the hydrological situation. A distinction was made between 'dry' anthrosols and podzols (groundwater class V–VII) and 'wet' anthrosols, podzols and gleysols (groundwater class I–IV). Based on tritium data, which show that dry soils with groundwater levels V–VII mostly contain groundwater that infiltrated after 1950 at depths between 5 and 30 metres below surface, dry soils could be identified as recharge areas. Based on the tritium data the wet soils were identified as either intermediate or discharge areas.

Concentration depth profiles for homogeneous areas

An integral presentation of the groundwater quality data of the shallow groundwater (soil quality monitoring network) and the deeper groundwater (groundwater monitoring network) shows that the groundwater quality depth profile is related to the combination of soil type, land use and hydrological situation.

Figure 3 shows nitrate concentration depth profiles for homogeneous areas with agricultural land use. The figure shows that nitrate is present in low concentrations in the

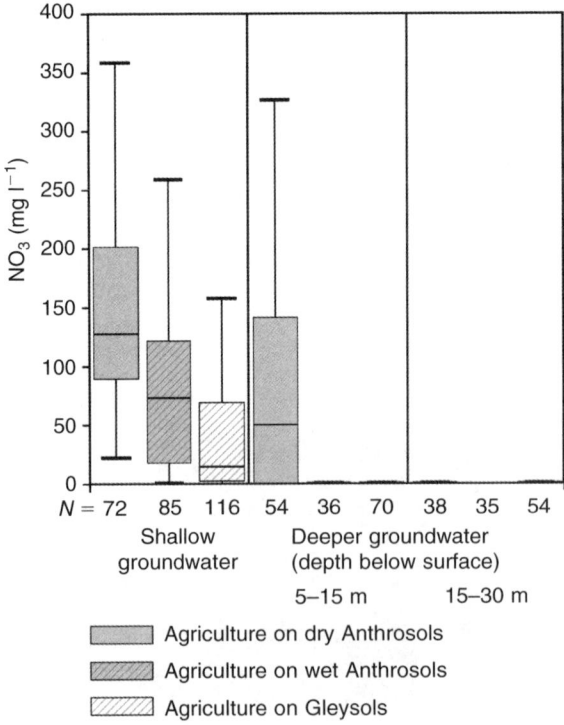

Figure 3. Mean nitrate concentration depth profile for anthrosols and gleysols with agriculture for seven Dutch provinces. Boxplots show median, 25 and 75 percentile, minimum and maximum. The x-axis shows number of measurements per depth class.

shallow groundwater of the clayey and peaty gleysols, while it is absent in the deeper groundwater of these areas. In the shallow groundwater of the sandy anthrosols with agricultural land use, nitrate concentrations are high and often exceed the EU nitrate directive of $50 \, mg \, l^{-1}$. However, clear differences can be observed between dry and wet

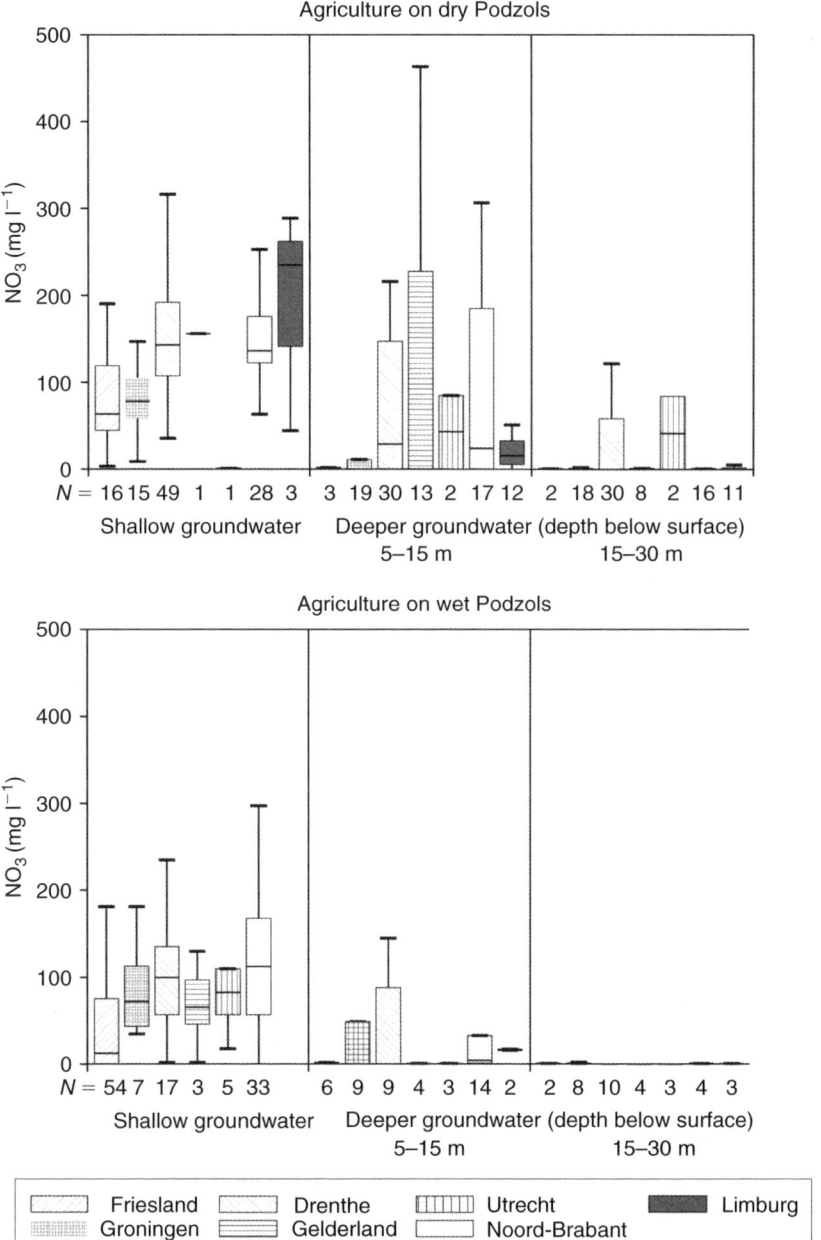

Figure 4. Nitrate concentration depth profile for podzols with land use agriculture, differentiated per province.

anthrosols. In the deeper groundwater these differences are more distinct than in the shallow groundwater. This is related to the hydrological situation: the dry anthrosols occur in recharge areas, whereas the wet anthrosols occur in intermediate/discharge areas. The groundwater in the discharge areas is mostly older than 50 years. Around 1950 manure loads in the recharging groundwater were much lower and therefore this groundwater contains less anthropogenic influences, or more time has passed during which denitrification could occur.

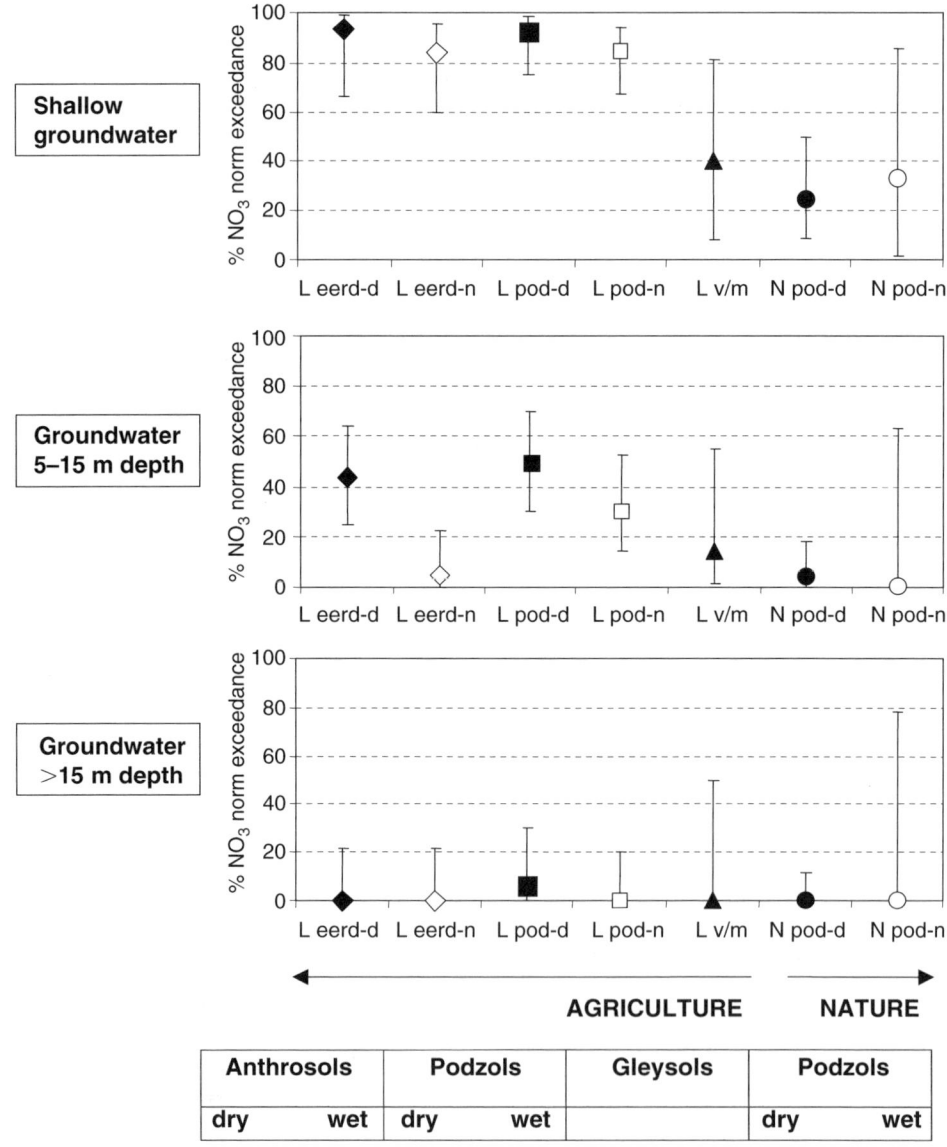

Figure 5. Exceedance of the EU nitrate directive of $50\,mg\,l^{-1}$ in the groundwater of the province of Noord-Brabant, shown with a 95% reliability interval.

Nature areas on dry podzols show a similar nitrate concentration depth pattern, although clearly lower concentrations are involved. Only a few exceed the EU nitrate standard of $50 \, mg \, l^{-1}$. Occurrence of nitrate in groundwater in these forest areas is related to airborne nitrogen deposition.

Differences can also be observed between the individual provinces. Figure 4 shows nitrate concentration depth profiles for podzols with agricultural landuse in the different provinces. The results show that the groundwater quality depth profiles of nitrate in dry podzols in Noord-Brabant and Drenthe show a lot of similarities, whereas manure loads in Noord-Brabant are distinctly higher. This is related to regional differences in geochemical characteristics of the subsoil, which cause nitrate reduction in Noord-Brabant, but not in Drenthe.

Figure 5 shows exceedance of the EU nitrate directive ($50 \, mg \, l^{-1}$) for the province of Noord-Brabant. It shows that in agricultural areas, the directive is often exceeded in the groundwater till 15 metres depth, but not beyond that depth. These data also show a difference between dry and wet area types: in groundwater between 5 and 15 metres depth under dry anthrosols and podzols, the EU nitrate directive of $50 \, mg \, l^{-1}$ is exceeded more often than in groundwater between 5 and 15 metres depth under gleysols and wet anthrosols and podzols.

CONCLUSIONS

It can be concluded that integration of the soil- and groundwater quality monitoring networks is feasible and will offer a number of advantages compared to sector measurements. The data from various compartments can now be compared because a spatial basis for the comparison has been provided in the form of integrated homogeneous units. It provides a better insight into patterns of nitrate leaching to deeper groundwater and regional differences between the different provinces, which are related to differences in manure load as well as differences in the denitrification capacity of the subsoil. The data show that the nitrate depth profile is related to the combination of soil type, land use and hydrological situation. Especially with regard to the groundwater at depths of more than 15 metres, the deeper groundwater of dry sandy soils with agricultural land use in recharge areas (dry anthrosols and podzols) shows higher nitrate concentrations and therefore a greater risk towards nitrate leaching than the deeper groundwater in discharge areas (wet anthrosols and podzols).

REFERENCES

Aa, N.G.F.M. van der, Grift, B. van der, Beusekom, G.W. van, Kremers, A.J.W., Buijs, E.A. and Meima, J.A. 2001: *Integration of soil- and groundwater monitoring networks* (in Dutch). TNO-report NITG 01-190-A.

Ball, J.W. and Nordstrom, D.K. 1991: WATEQ4F – User's manual for WATEQ4F. US Geological Survey Open-File Report 91-183.

Busink, E.R.V. and Postma, S. 2000: Provincial soil-quality monitoring networks in the Netherlands as an instrument for environmental protection. *Netherlands Journal of Geosciences* 79 (4): 429–440.

Drecht, G. van, Reijenders, H.F.R., Boumans, L.J.M. and Duijvenbooden, W. van 1996: *The quality of groundwater at depths between 5 and 30 meter in the Netherlands in 1992 as well as changes during period 1984–1993* (in Dutch). National Institute for Human Health and the Environment.

Dufour, F.C. 1998: *Groundwater in the Netherlands.* Netherlands Institute of Applied Geosciences (NITG-TNO), Leiden.

Mol, G. 2002: *Soil acidification monitoring in the Netherlands.* Thesis Utrecht University – Faculty of Earth Sciences.

Index